Y0-CCH-783

Weather and Climate

CLARENCE E. KOEPPE
San Diego State College

GEORGE C. DE LONG
Eastern Michigan College

Weather and Climate

1958

McGraw-Hill Book Company, Inc.

NEW YORK TORONTO LONDON

Library of Congress Catalog Card Number 58-6687

THE MAPLE PRESS COMPANY, YORK, PA.

Preface

"Weather and Climate" represents what the authors, both of whom are college teachers of geography, believe to be an up-to-date, reasonably complete but nontechnical college-level treatment of modern weather principles and processes, and a clear practical application of those principles and processes to the field of climate. It is essentially new in viewpoint and purpose, and original in organization. It is designed to meet the ever increasing need for a simple, brief, and accurate survey text in the fields of meteorology and climatology. With the expansion of air transportation, both military and civilian, not only in middle and low latitudes but in high latitudes and even over the poles as well, people are becoming increasingly weather conscious and weather wise; and every year greater numbers of people are becoming interested in and concerned with weather conditions both on the ground and aloft. Certainly, a sound knowledge of weather and climate is not alone for the professional meteorologist, aeronaut, or ballistic engineer; it is of real concern to the layman.

This text is intended primarily for three general groups of students: (1) students of geography who need a more detailed and scientific knowledge of weather and climate than that usually offered in the first course in geography, but who do not have the background in mathematics and physics to pursue a standard course in technical meteorology; (2) students who may want to sample meteorology and climatology without the possibility of becoming discouraged because of an inadequate technical background; and (3) students who are interested in a broad and cultural liberal arts training but who are not prospective geographers, meteorologists, or climatologists. And there is no reason why the general reader should not profit well by making a systematic study of the book. Accordingly, all aspects involving a knowledge of higher mathematics and physics are omitted; at the same time, the authors believe they have achieved scientific accuracy. They have tried to present the latest developments in weather science, particularly in the presentation of the newer concepts of the origin, characteristics, and movements of air masses including jet streams, and their use in weather forecasting; also in the use of synoptic charts, both sea level and aloft. Although complete in scope, "Weather and Climate" makes no claim to be other than elementary and introductory; in no sense should it be considered as covering completely any one aspect of meteorology or climatology.

The book departs, in some respects, from the customary order of presentation of topics. Instruments used in observing and recording weather conditions are not taken up together, but are presented in connection with each element studied; this is certainly logical, and it holds student interest—to study some forty weather instruments in succession is tiring and confusing to the beginning student. The authors have striven for strict sequence throughout the book, yet certain topics are presented piecemeal so that the student may readily see the application of principles to weather processes; thus, the thunderstorm is introduced in the chapter dealing with moisture in the atmosphere, and it is again presented in connection with air masses, and later with synoptic weather. In addition, certain facts and principles are repeated occasionally for the sake of emphasis; this seems better than reminding the student to turn back to a particular page or section for review.

In the second part of the book, climate is first presented as a study of general principles and later as descriptive climatology, emphasis always being placed upon genesis or causes and upon geographic position. Since the student of geography is genuinely interested in climate as a factor in the physical environment, concrete application is offered by a brief description of natural vegetation and soils in connection with each type of climate. Some reference is also made to certain cultural aspects, particularly occupations, within the climatic type.

The Köppen system of climatic classification is used as the basis for the description of types of climate, not because the system may seem flawless to some nor because some may consider it the best available, but because it is well known and widely used. It has the further advantage of being available as a large-scale wall map in full color. In consideration of those who might feel that the discussion of climate according to the Köppen classification is too detailed considering the time available, or who have been using the simpler classification developed by the senior author, there has been placed in the Appendix the Koeppe map of "Climatic Regions of the World," together with an explanation of the system and a brief description of each of the sixteen types of climate shown.

The authors necessarily have been influenced by the many books and articles published, dating back to Moore, Davis, and Ward, and continuing down to the present. Almost all of the authors' information has come, directly or indirectly, from their tutors or from reading; their debt to such sources is not only obvious, it is duly acknowledged. They are especially indebted to the United States Weather Bureau for making available certain maps and charts as well as other illustrative material, and to Weather Bureau officials for their helpful advice and information. Space does not permit direct recognition of scores of other sources of assistance in the preparation of the manuscript, but the debt to such sources is gratefully acknowledged.

Clarence E. Koeppe
George C. De Long

Contents

Weather and Climate

1

Weather and Climate

What Is Weather?

Weather is ever changing because air moves and because the chief elements of weather (sunshine, temperature, pressure, winds, and moisture) vary in intensity from moment to moment. A change in wind direction or in wind velocity changes the weather to some degree because a change in one weather element will in turn eventually change the others. A small change in the conditions of the elements may not affect people much, but a larger change frequently necessitates notable human adjustments. Thus a rapid rise in temperature or a steady drop in pressure affects many people sooner or later.

While the nature, characteristics, and actions of each element of weather need to be considered separately, it must be recognized that all the elements are at work at the same time, each one affecting, modifying, or partially controlling the others. While one element may be more noticeable at a given time than the others, they, too, are operating, and the present weather is the result of their interaction. The weather will be determined in five minutes, an hour, or a day by the variations of each element individually and the combination of them all. Therefore, weather may be defined simply as *the state of the atmosphere at any given time and place.* Although this definition is rather restrictive as to duration, it is common to generalize on weather conditions for a day, a week, or even a season by referring to them as "good" or "bad" or "hot" or "unusually rainy." Similarly, we frequently broaden the area having essentially the same weather by stating that the whole state enjoyed fine weather, or that the entire eastern half of the nation was subject to storms with heavy rain.

Weather forecasting relies on the amount and speed of change occurring within a mass or masses of air. Weathermen need to know where a given mass of air will be and how much it is expected to change in detail within a given length of time. Also they need to know when a different mass of air will arrive and what it will be like in terms of sunshine, temperature, pressure, winds, and moisture.

Usually the most difficult element to predict accurately is moisture. A very slight drop in temperature, for example, may cause condensation to occur. If a drop sufficient to form visible moisture has not been expected, the forecast may be in error. On the other hand, an unexpected rise in temperature may prevent the occurrence of precipitation, or a slight drop in the temperature of a mass of air near freezing may change the form of precipitation from rain to sleet or snow. There are many other changes, often slight, that determine the state of moisture in the air.

Who Is Affected by the Weather?

The weather affects everyone to some degree. Unsatisfactory weather conditions frequently hamper the plans of the farmer, the market gardener, the construction engineer, and the lumberjack. Great extremes of weather usually make life difficult for some people; frequently they cause a heavy loss of life. Inclement conditions are likely to cause restlessness among school children, factory workers, and even animals. Either hot or cold weather may necessitate a delay in the shipment of perishable foods. Windy weather may cause severe desiccation of crops or land surfaces, especially if temperatures are high. Lack of rain or too much rain may result in crop failure for a season. Poor visibility leads to the cancellation of air flights and in wartime may even dictate a change in battle maneuvers. Bad weather frequently affects the recreational plans of urban dwellers. Good weather, on the other hand, is likely to bring out the best in everyone.

Who's Who in Weather Forecasting?

Everyone talks about the weather! While few people attempt to change it, many help materially to avert its occasional ill effects.

Private Forecasters

There are, for example, approximately eight hundred private weather forecasters in the United States. Some of them have the very best equipment, maintain staffs of men and women who scrutinize the weather over large areas, and attempt to prevent serious monetary losses to their subscribers. Their forecasts and warnings, by telephone, telegraph, and radio, are sometimes very detailed. Obviously, private forecasters are handicapped when they attempt to make detailed and long-range weather forecasts without the aid of the U.S. Weather Bureau.

The U.S. Weather Bureau

The U.S. Weather Bureau offers most of its services free. The Bureau, under the Department of Commerce, maintains a nationwide network of over 12,000 weather stations operated on a cooperative basis with other governmental agencies, private organizations, and individuals. Over 200 of these are classed as first-order stations. They are usually located in cities or at major airports, are staffed with professional and subprofessional forecasters and observers, and are charged with furnishing complete meteorological data. Most major stations send daily reports and make local forecasts. At a number of our major airports the weather stations work in close conjunction with the Civil Aeronautics Authority, each giving information to the other.

Quite a variety of substations disseminate special information. Thus there are second-order substations that make detailed observations but do not forecast; third-order substations that send out daily observations at stated times for special purposes; river substations; snowfall substations; display substations that warn of possible violent weather; crop substations; airways substations; and the cooperative substations, which are the most numerous of all. The cooperative stations are maintained chiefly for climatological purposes. The unpaid cooperative observer makes daily records of temperature, precipitation, and unusual phenomena. He makes monthly and yearly reports that are incorporated in *Climatological Data*, a monthly summary published by the U.S. Weather Bureau. Because of the many useful functions of the Weather Bureau in providing both weather and climatic data, many references will be made in the following chapters to the Bureau and its work.

The cooperative observer's reports are processed, analyzed, and used in research and prognosis; essential information in them is transmitted to weather stations in the United States and foreign countries and to various civil and military establishments in the form of radio teletypewriter transmissions, facsimile transmissions, and photostat and Bruning copies—the latter especially for governmental use and records. This tremendous expansion and development of our national weather service has occurred in less than 100 years.

In addition to the Daily Weather Map published and distributed by the Weather Bureau from Washington, articles relating to weather are printed on the reverse side of the weather map; to date, there are nearly 100 of these articles. "An Index of Information" giving the general titles of the articles which are published is included in the list of references in the Appendix of this book. Sixteen of these articles deal with the subject of aeronautical climatology, and are so printed they can be cut out, folded, and placed in a loose-leaf binder; they are also printed as separate pamphlets. Unfortunately, the Weather Bureau is unable to furnish many of the back numbers, although Weather Bureau offices and many libraries have bound volumes of the Daily Weather Map. Certainly, the Weather Bureau is one of the best sources for simple and current information on every phase of weather relating directly to the science of weather or to its application to aviation, crops, industry, transportation, or recreation.

Various organizations of the Department of Defense maintain weather observation stations throughout the world, collecting and disseminating weather information and making detailed weather forecasts. Their observation of upper-air conditions is especially significant. All such military installations work in close conjunction with the Weather Bureau.

The scope of the Weather Bureau service is indicated by the fact that every 24 hours the National Weather Analysis Center in Washington receives:

22,000 hourly surface reports
8,000 international six-hourly surface reports
1,000 ships reports
1,600 pilot balloon winds-aloft reports
600 radiosonde reports
400 rawinsonde reports
800 reports from commercial aircraft in flight
300 scheduled military weather-reconnaissance reports
30 encoded transmissions
16 facsimile transmissions

Weather Control

The rain maker's art—and it seems almost as much an art as a science—has received much publicity in recent years. To the thirsty southwest as well as to the dry reservoirs of metropolitan areas, the ability to produce precipitation artificially sometimes seems to be the answer to recurring water deficiencies. Probably rain making will be perfected eventually but now it is still in its infancy. Some proponents hail it with enthusiasm, while others note its limitations with reservations. The fact remains, however, that rain makers are numerous and aggressive, and they have quite a following—to the extent that some of them are employed by cities and farmers with the hope and expectation that they can induce sufficient precipitation. The quest for ways of adjusting the weather to human needs and desires is likely to be intensified rather than curtailed.

Meteorology

Meteorology is the science of the atmosphere and its attendant activities and phenomena. The term is derived from Aristotle's *Meteorologica*, one of the early and many-sided treatises on the natural sciences. The four-volume work delved into astronomy, some phases of geology and physical geography, and even into the physics and chemistry of the various atmospheric phenomena. Today there is a tendency to consider meteorology and the science of weather as synonymous, since much of our present investigation is concerned with the effects of the atmosphere on people and their ways of adjusting to it. Even the term *aerology* frequently means meteorology, inasmuch as the fundamental principles involved are the same. On the other hand, some scientists treat aerology as a subdivision of meteorology: one which embraces a study only of the upper, free air throughout its vertical extent and not of the lower layer of air adjacent to the surface of the earth. Aerological investigations are made by airplanes, various types of balloons, radiosondes, and ceilometers.

There are several subdivisions of meteorology. *Dynamic* or *theoretical meteorology is* the physical science concerned with energy changes and the forces involved in atmospheric motions. It makes use of both laboratory and mathematical physics. *Physical meteorology* is similar to dynamic meteorology, but it seeks to explain all atmospheric phenomena by the principles of physics. It deals with the thermodynamics and mechanics of the atmosphere and also attempts to explain acoustical, electrical, and optical phenomena. Dynamic and physical meteorology are systematic in approach; they explain how and why forces act as they do. *Synoptic meteorology*, on the other hand, is regional in approach and comparative and analytical in nature. It is the study of weather conditions over a large area at a given time and is concerned with the detailed study of air masses and with weather forecasting. The term *synoptic chart* is therefore frequently used for the Daily Weather Map because it shows the conditions of the atmosphere at a given time for a large area. Still another branch is *aeronautical meteorology* which is concerned with the weather as it affects aviation.

While this text strives to be scientific and accurate in discussing and explaining all types of weather phenomena, limitations of space and the desire to be clear, simple, and concise without recourse to technical mathematics and physics leave the authors with definite and rather narrow prescriptions. Hence the simple term *weather*, rather than *meteorology*, seems to describe better the purpose and scope of the first part of this book.

Climatology

Climatology, which is closely related to both meteorology and geography, has a twofold purpose: (1) It seeks to explain the causes of climates, the reasons for their variations, their general and specific locations, their effects on fauna and flora, and the processes producing the various kinds of physical climates, such as marine, continental, mountain, desert, and monsoon. This is *physical climatology*. (2) It seeks to determine and describe the various types of climates that are based on the analyses of climatic statistics. It further details the interaction of the weather and climatic elements upon the life, health, and economics of peoples and areas. This is *descriptive climatology*.

The goal of climatology, then, is to discuss climate, its component elements, the factors which determine and control its distribution, and the actual distribution over the earth. The simple term *climate* is used in this text in preference to the term *climatology*, but the details of climatology are nevertheless given.

Why study climate in addition to weather? Climate is a summary, a composite of weather conditions over a long period of time;[1] truly portrayed, it includes details of variations—extremes, frequencies, sequences—of the weather elements which occur from year to year, particularly in temperature and precipitation. Taking fluctuations into consideration, climate

[1] The geologist is correct when he maintains that the climate of any place is the sum of all the weather that has occurred at that place. Thus not only historical but also geologic time is included in the make-up of climate.

is more than simply an average of weather conditions, although it reflects averages, too. *Climate is the aggregate of the weather.*

In general, meteorologists use *mean* rather than *average* with respect to climatic elements. The normal temperature at a given place is usually the mean temperature for the most recent 30-year period, and the normal precipitation is the aggregate of the precipitation for the 30-year period divided by 30. The fact is recognized, however, that the normal may not be the actual figure for any one of the 30 years because of the fluctuations that may occur annually. This is particularly true in the domain of the westerlies where rapid and marked fluctuations do occur, but it is less true of the tropics or even the subtropics.

There are countless reasons why people want to know what kind of weather they can expect in other parts of the world, not just for one day but throughout the year. They may need to seek a climate where there is little variation in temperature during the year, or where it is constantly dry, or where it is moist enough and hot enough to grow a certain crop, or where they can have a heaven-on-earth holiday. Industrial plants need to know how cold it is likely to be in the winter and thus how much heating will be needed. Cereal-growers cannot succeed where there is insufficient harvesting weather for their crop; they will have to either raise other crops or move to a place where there is a long enough period of dry, sunny weather to permit proper harvesting. Examples are legion to express the desirability, or even necessity, of knowing something about climate or climates. This book, therefore, devotes considerable space to the general and specific details of climate, the various climates of the world, and some of the relationships of those climates to the activities of man.

2

Weather Lore

Weather Proverbs

All countries have amassed folklore, mores, customs, and proverbs that have special meanings locally. Included in the list are likely to be a number of sayings regarding the weather. In the course of the many migrations to this country, numerous weather proverbs have been transplanted and adopted in varying degree as our own. Many of our proverbs are borrowed or assimilated from Germany, Scandinavia, Britain, France, various countries of the Orient, and from other nations. Of special significance is the contribution of seafarers whose lives and occupations are so closely associated with the weather; they observe its changes carefully. It should be borne in mind, however, that weather signs on the ocean and those on large continents are not always the same; therefore, the wholesale adoption of a weather proverb just because it sounds interesting or because it has application in one part of the world may not prove its validity in another place. Undoubtedly much weather lore is spurious and untenable because it is misapplied in one way or another.

The following compilation of weather proverbs is relatively short. Its purpose is to bring together a modest number of proverbs which apply reasonably well in intermediate latitudes, especially in the westerly wind belt of this country. Explanations of them and the specific conditions under which they apply individually are also given.

> What is it moulds the life of man?
> The Weather.
> What makes some black and others tan?
> The Weather.
> What makes the Zulu live in trees,
> And Congo natives dress in leaves
> While others go in furs and freeze?
> The Weather.

This poem is not prophetic; it simply summarizes the truth, namely, that the activities and destinies of human beings are partially determined by the weather and the climate of the region in which they live. It is because of this great influence upon the activities of man that numerous observations, rules, and even omens have been incorporated into people's lives and used in foretelling coming events, some of them long-range in nature but most of them applied correctly only to the near future.

Weather proverbs may be divided into three main categories: (1) those based upon the observing of animals, (2) those based upon the observing of plants, and (3) those based upon the observing of atmospheric phenomena. Those in categories (1) and (2) are, for the most part, unreliable. There are, however, hundreds of proverbs based upon the activities of animals and plants, and every nation has avid adherents to some of them. A few of the proverbs, to be sure, have real physical explanations. The question is, How good are they in weather forecasting?

Animals as Weather Forecasters

According to legend, if the ground hog or woodchuck sees his shadow on February 2, he returns to his lair for six more weeks of winter; if he doesn't see his shadow, presumably spring is just around the corner. Why February 2? Because it is Candlemas Day, and from time immemorial feast days and other religious days have been used one way or another in a prophetic sense. Actually, the forecasting by the woodchuck is simply an interesting bit of fiction. There is, however, this useful, although indirect, connection with his "forecasting": an early spring is usually not greeted wholeheartedly because enough cold weather is usually interspersed with it to cause fruit trees and other vegetation, lured along by unseasonably high temperatures, to freeze and thus be damaged. A late spring, on the other hand, signifies a more gradual march of temperature, and when it finally arrives, there is relatively little danger of temperatures dropping to the freezing point. So, people concerned by the springtime possibility of losing their crops—either sown or indigenous—are really interested in knowing whether there will be a killing frost. But the weather on February 2 is not prophetic of the weather for the following six weeks. Even the weatherman, with all the instruments and data of other stations as well as his own at his command, would not forecast for even a month ahead without reservations; even then such a forecast would be fairly broad in scope, and the weatherman would probably have to alter it later, perhaps even change it drastically.

The power of predicting the general nature of the coming weather has been ascribed to many animals. Burrowing animals, hibernating animals, rabbits, foxes, and a rather lengthy list of other animals are supposed to foretell a hard or closed winter by developing a heavy coat during the fall; if the coat is light, the winter is expected to be mild or open. The biologist explains that animals develop a heavy coat because they are well fed, not because they anticipate hard sledding in the winter. In effect, the behavior of animals, and of plants as well, is largely a reflection of their past and present adjustments to weather conditions, not of their prognostic abilities.

Animal forecasting on a very short-term basis, however, is frequently valid. Their bodies sense changes in pressure, temperature, and moisture—or better still, a combination of all three of them—and they react accordingly. Thus they may take cover, giving a few minutes' warning or, at best, a few hours'. Yet even bees, considered infallible by some, have been observed acting in an unorthodox manner, welcoming an increase of moisture instead of taking heed and flying head-long into the hive. So even their foretelling has been questioned.

Talman[1] asks whether any enterprising journal has ever undertaken to collect scientific data based on animal behavior, whether it has been responsible for equipping biologists with "scales, tape-measures and other appropriate paraphernalia" enabling them to

[1] Charles Fitzhugh Talman, "The Realm of the Air," pp. 286–287, The Bobbs-Merrill Company, Indianapolis, 1931.

foretell accurately the forthcoming weather according to the actions of wild animals. Nor have naturalists attempted such a procedure. He states, however, that Dr. Charles C. Abbott was an exception. Dr. Abbott kept records of the years in which muskrats built winter houses and of the amounts of food that gray squirrels stored up each fall—both supposedly criteria as to the nature of the coming winter—yet he could find no correlation between the animals' behavior and the severity of the winter which followed. Talman concludes that the lower animals have been no more successful in predicting the weather of seasons to come than man has. Many animals, he maintains, are fairly good short-range weather prophets, for some of them scuttle to cover before a shower or when a storm of long duration is sensed.

Plants as Weather Forecasters

The only prognostic abilities latent in plants are in their reactions to certain atmospheric changes such as temperature, sunshine, moisture, and possibly wind.

> When leaves show their undersides,
> Be very sure that rain betides.

This reaction occurs, when viewed laterally or at a distance, as the leaf stalk turns in response to the variations in absorptive power of the different cells of the leaf. On the other hand, the pimpernel flower, called the "poor man's weather glass" and long a favorite forecaster in Europe, has lost its reputation; it is now known to stay open both before a rain and during it, contrary to former belief. Actually, the flower's response is to light; it habitually opens a few hours after sunrise and closes before sundown.

In the last century the *Abrus precatorius*, a species of pea, was hailed in Europe as having power, through the varying position of its branches and leaves, to predict not only storms and weather changes but also such things as the occurrence of firedamp in mines and even earthquakes. Such notions flourished in London until they were completely discredited by the authorities of Kew Gardens. Other signs now discredited are that thick husks on maize foretell a severe winter; abundant blooms on dogwood in the spring indicate a severe winter to come; an oversupply of acorns is nature's way of warning against a hard winter; the number of layers of an onion indicates the nature of the following winter.

Proverbs Based on Atmospheric Observations

While the proverbs based upon the antics of plants and animals have quite limited utility, those based on careful observations and correct applications of many atmospheric phenomena may be very useful. We shall outline some of the best proverbs which have scientific explanations. It is interesting to observe that by far the greatest number of tried and tested proverbs are concerned with atmospheric signs pointing to rain. The signs may be evident in clouds, wind, dew, fog, odors, smoke, or in more tangible items that are in turn affected by conditions of the air, such as the appearance of the moon, the reactions of leaves and hair, and the small changes in barometric readings. These signs are so numerous they will be listed by title.

Clouds

> Rain long foretold, long last;
> Short notice, soon past.

The first line of this quotation refers to a thickening and lowering of clouds which precedes for a day or so a general storm or cyclone or sometimes a well-developed frontal condition. The second line applies to a warm-season situation in which the weather is for the most part fair but in which a cloud, or clouds, develops rapidly, bringing a shower of relatively short duration followed soon by clearing conditions.

When the carry goes west,
Gude weather is past;
When the carry goes east,
Gude weather comes neist.

The "carry," or the direction in which the clouds move, is often good evidence of the kind of weather that is on its way. East winds (that is, winds from northeast, east, or southeast) commonly herald the approach of a *Low* or cyclone; west winds indicate the retreat of the cyclone, or good weather in general.

Mackerel scales and mares' tails
Make lofty ships carry low sails.

The first signs of a prolonged storm include high, wispy, cirro-cumulus clouds which look like the scales of a mackerel, or cirrus clouds which look like horses' tails in the distance. When the wind is easterly and the barometer is going down, and cirrus clouds are succeeded by lower and thickening clouds, any experienced weatherman knows we are in for a prolonged storm, perhaps of two or three days' duration.

When Cheviot ye see put on his cap,
Of rain ye'll have a wee bit drap.

Since mountain peaks and ridges are higher than the base of rain clouds, one can frequently see mountains enshrouded in clouds a few hours before rain begins at the site of the observer. There are many citations of specific mountains or peaks to illustrate this fact.

When the clouds appear like rocks and towers,
The earth's refreshed by frequent showers.

This reference is clearly to towering cumulo-nimbus clouds or thunderheads. These clouds are caused by rising air currents and are evidence that air is being raised and cooled sufficiently to cause thunderstorms. Thus on a warm summer day there may be numerous local storms in the vicinity, and the weatherman predicts "local showers and thunderstorms."

In the morning mountains,
In the evening fountains.

This reference simply verifies the fact that if the clouds attain great vertical depth early in the day—that is, if it is hot enough to develop cumulo-nimbus clouds in the morning—they will surely cause enough cooling of the air in their upper elevations to give rise to rain in the evening or even in the afternoon.

The higher the clouds, the finer the weather.

Except for the cumulo-nimbus cloud, which is correctly classed as a cloud of great depth rather than a high cloud, this is true. In general, high clouds are not precipitation makers; low clouds frequently are. Consequently, as long as high clouds prevail, precipitation is not expected except where a mountain may intercept the clouds, but the probability of precipitation increases as clouds become lower and as more and more of the sky becomes clouded.

Fogs

When fog goes up the rain is o'er;
When fog comes down 'twill rain some more.

As a general rain approaches, hills and mountains become enshrouded in clouds, while the lee portions of these clouds are drawn out as a scud cloud which is eventually dragged to lower levels. On the other hand, the ending of the storm is accompanied by cool and drying winds from a different direction; these drive up the mountainsides, clearing away both the remaining low-flying clouds and the fog.

Dew

When the grass is dry at morning light,
Look for rain before the night.

Also:

When the dew is on the grass,
Rain will never come to pass.

In a sense these two proverbs are paradoxes: when the morning grass is dry you expect rain, but when there is dew in the morning you go merrily on your way expecting no precipitation that day, or at least not until a new set of weather conditions is ushered in. The reason: when there is no dew in the morning it means that the air is relatively moist up to a sufficiently high altitude to prevent surface cooling—since moisture retards radiation and thus a drop in temperature—thereby preventing the formation of dew and aiding in the formation of precipitation. On the contrary, dew forms only under dry conditions (low relative humidity) when the sky is clear; thus rapid radiation from the ground permits the temperature of the grass to drop to the dew point fairly rapidly, especially when there is little or no wind to raise the temperature above the dew point. Therefore, *when it is dry enough for the formation of dew*, it is dry enough to preclude the formation of precipitation until there is a change of weather conditions.

Rainbows

A rainbow in the morning
Is the shepherd's warning;
A rainbow at night
Is the shepherd's delight.

This quotation originated in the westerly wind belt and applies there where the prevailing wind and therefore most local showers, necessary for rainbows, move from west to east. The rainbow is caused by the refraction and reflection of sunlight by raindrops; thus if the sun is able to shine through intervening clouds, the colors of the spectrum appear in bow form. The sun must be fairly low in the sky in order to produce a rainbow, and the viewer is always approximately between the sun and the bow. In the morning a rainbow is always to the west (windward) of the observer, and clouds will normally be proceeding eastward toward him, bringing showers. In the evening he views a rainbow to the east (leeward) and knows that at least that particular shower has already passed. Technically, a morning rainbow indicates a fairly high relative humidity because a shower has occurred with only a limited ascent of air and therefore with only slight cooling. An evening bow, on the contrary, means the air is comparatively dry because the existing cumulonimbus cloud has required considerable cooling in its build-up. Consequently, the air temperature has been reduced considerably, and cooler air cannot contain so much moisture. As a result, an evening rainbow usually signifies fair weather.

Halos

The moon with a circle brings water in her beak.

The refraction of light from the sun or moon by the ice crystals of high cirrus clouds causes the halo (the ring) around the moon or sun. The angle from which one sees the luminary, either solar or lunar, produces either a 22-degree or a 46-degree halo if the thin cloud covering is complete enough; if the covering is incomplete, only portions of the halo may be visible. The significant fact is that halos indicate cirrus clouds which may be forerunners of a storm. If the halo, and hence the cirrus clouds, is followed by lower and thickening clouds, it is the beginning of the cloud succession on the leading edge of a cyclone or a warm frontal disturbance. Occasionally, however, cirrus clouds appear, with an accompanying halo or portion of a halo, only soon to disappear. In the latter case they are merely ephemeral clouds with no portent.

Moon

Pale moon doth rain,
Red moon doth blow,
White moon doth neither rain nor snow.

The pale moon signifies a thin, high cloud covering such as precedes a storm condition, again being the forerunner of lower, rain-bearing clouds that are located farther west. The red moon also indicates high humidity attendant upon the approach of a storm center. The white moon means clear and dry atmospheric conditions.

Red and Gray Skies

There are many significant quotations concerning sky coloring. Often they are confusing because they fail to distinguish between the various hues of colors, notably those of red. Pink, for example, indicates little moisture in the air and hence good weather, but deep, lowering red, morning or night, means high moisture or dust or smoke content and possible rain, the clouds being so thick as to block the sun's rays.

> A red sun has water in his eye.

If the western horizon in the evening is greenish or yellow, then the chance of clear weather is exceptionally good because these colors indicate little moisture available for condensation. If, however, the evening sky is a somber sheet of gray overcast, indicating a thick cloud covering, it signifies that the great number of particles of dust in the air have become saturated with water in the form of countless droplets; hence,

> If the sunset is gray
> The next will be a rainy day.

Or

> If the sun goes pale to bed
> 'Twill rain tomorrow, it is said.

A gray morning is a good sign. At that time the gray is close to the surface and is caused in part by the refraction of light on dew and on dust particles in the lower atmosphere only. It should be noted that when dew forms, the atmosphere in general is dry; no precipitation, therefore, is expected when the morning dawns gray, as indicated by:

> Evening red and morning gray
> Help the traveler on his way;
> Evening gray and morning red
> Bring down rain upon his head.

The grayness is due also to a low and relatively thin layer of stratus or sheetlike clouds which result from the cooling of the lower air by nighttime radiation. Radiation occurs because the air is actually quite dry. As soon as the sun rises, the warmth dissipates the clouds, and fair weather follows. On the other hand, "evening gray" indicates sufficient moisture in the air to permit cloud formation in spite of warming by the sun during the daytime. Since this cloudy condition lies to the west, it is quite likely that the eastward movement of the air will bring the clouds, and possibly rain, to the observer's region.

Odors

> When the ditch and pond offend the nose,
> Then look for rain and stormy blows.

Humphreys,[1] explaining this phenomenon, states that the decaying vegetable matter of stagnant bodies of water inevitably produces gas which is evidenced by bubbles forming on the bottom of the pond, lake, swamp, or stream. With the approach of a storm and its relatively low pressure, there is a decrease in the weight and pressure that imprison the gases on the bottom of the water mass. This allows the bubbles to migrate to the surface of the water, expand, and break, thus releasing their putrid odors. Therefore, a particularly odorous pond is frequently indicative of lowering pressure as a storm moves into the area.

Smoke

> The smoke from chimneys straight ascends,
> Then spreading back to earth it bends.

This occurs under conditions providing much moisture, a supply of smoke and soot, and little wind. Under a relatively dry condition of the atmosphere, smoke particles, acting as condensation nuclei, have very little moisture available to condense on them; consequently, they are light and easily buoyed up into the air where they scatter and soon disappear. But when there is a large amount of moisture in the air with little wind, the smoke nuclei soon become laden with condensed moisture which weights them down, and they trail along close to the ground in a long train. Here again is evidence of high moisture content preceding a storm.

[1] W. J. Humphreys, "Weather Proverbs and Paradoxes," p. 72, The Williams and Wilkins Company, Baltimore, 1934.

Hair

But I know ladies by the score
Whose hair, like seaweed, scents the storm;
Long, long before it starts to pour
Their locks assume a baneful form.
—HERBERT

Women are aware that, unless the curl is natural, a rain may undo a coiffeur's handiwork. The reason is that hair is made up of many small cells which absorb moisture when the humidity is high. Dry hair shortens; wet hair elongates, putting in curls or losing the curl when they are not wanted, or making it unruly in some other way. Thus, to many women, unruly hair means bad weather is on the way.

St. Elmo's Fire

Last night I saw Saint Elmo's stars,
With their glittering lanterns all at play,
On the tops of the masts and the tips of the spars,
And I knew we should have foul weather today.
—LONGFELLOW

This phenomenon, called *corposant* and recorded frequently in both prose and poetry, is likely to occur on the darkest nights at the tops of ships' masts and on high mountains, such as Pikes Peak; Krakatau had unusually fine displays as it erupted in 1883. It may occur before, after, or during a thunderstorm, or it may even appear during snowstorms. It is an electrical discharge, akin to, but weaker than, lightning, which proceeds from pointed, high objects which have acquired an electrical charge because of the proximity of strongly electrified clouds or masses of air overhead. The superstitious belief of sailors that it was a manifestation of the souls of dead seamen returning to warn of a coming storm is a bit discredited nowadays. As indicated above, however, it is a harbinger of stormy weather.

Barometer

When the glass falls low,
Prepare for a blow;
When it rises high,
Let all your kites fly.

The "glass" refers to an old type of liquid barometer, such as the Cape Cod barometer, that is still used to some extent. The reference clearly indicates that a barometer which is decreasing, or lowering, steadily indicates worse weather is approaching, while a rising barometer denotes fairer weather is coming.

Winds

The most inclusive truth about winds is, as Bacon stated,

Every wind has its weather.

No more significant fact can be kept in mind by the weather student than this statement, for it sums up the truth that whenever the wind changes (except in the case of gusty conditions), so does the weather. Each wind has its own significance, especially when connected with the passage of a storm center.

The wind from the northeast,
Good for neither man nor beast.

This quotation indicates clearly that when the wind is from the east or northeast (or, usually, from the southeast) it indicates the approach of a storm because the wind is blowing toward the storm center which lies somewhere to the west and is an area of low pressure. It should be mentioned that not *all* easterly winds, however, indicate a storm, inasmuch as easterly winds can come from a polar continental air mass and be relatively dry. But when easterly winds coordinate with a steadily falling barometer and with thickening and lowering clouds, bad weather usually follows. On the vernal equinox in 1958, the "northeaster" along the North Atlantic Coast brought the heaviest snows in 40 years to parts of Pennsylvania, New Jersey, and New England.

The whispering groves tell of a storm to come.

When the forest murmurs, and mountain roars,
Then close your windows and shut your doors.

All these sounds indicate considerable wind,

and thus, in turn, the conditions connected with a general storm.

The wind in the west
Suits everyone best.

Do business with men when the wind is in the northwest.

The statements attest the fact that westerly winds (except along west coasts) are often relatively dry winds, seldom accompanied by precipitation except under quite warm conditions that may lead to convection. Northwest winds are the coolest winds in the belt of westerlies; they are usually dry and exhilarating, in both summer and winter. On occasion they bring the lowest temperatures of the winter, but they are most welcome respites from the hot southerly winds of summer.

A high wind prevents frost.

"High" in this case means high in velocity, or strong. Both dew and frost fail to materialize if there is much wind, even though other requisite conditions are present: (1) because wind continually brings a new supply of air into the area affected and, since moving air does not cool so rapidly as quiet air, aids in preventing the grass's cooling enough to reach the dew point; (2) because wind causes friction. The latter condition generates heat, and this in turn commonly prevents a temperature drop sufficient to form visible moisture.

Miscellaneous

All signs fail in dry weather.

Little rain, less rain;
Much rain, more rain.

In wet weather it rains without half trying.

These merely indicate that during wet weather, moisture is readily available for continuing rain, but once the ground becomes dry, there is less moisture available, and therefore the drought is likely to continue.

Moonlight nights have the heaviest frosts.

The reason for this is that on clear nights (moon or no moon) there is little moisture in the air; consequently, rapid radiation of the lower air causes the temperature to drop to the dew point early in the evening, thus producing either dew or frost (depending upon the temperature) for a relatively long period.

The following proverbs indicate that an early spring is likely to be injurious to the growth of certain plant life.

January warm, the Lord have mercy.

January blossoms fill no man's cellar.

All the months of the year
Curse a fair Februeer.

March damp and warm
Does farmer much harm.

A cold April,
The barn will fill.

A late spring never deceives.

Weather Paradoxes

According to Webster, a paradox is "a tenet contrary to received opinion; also, an assertion or sentiment seemingly contradictory, or opposed to common sense, but that yet may be true in fact." While there are many meteorological paradoxes, the following are some of the more interesting ones with their explanations.

Air Pushed North Blows Eastward

This applies only in the northern hemisphere and embodies, in effect, Ferrel's law, which states that all moving masses of air move to the right (as they proceed) in the northern hemisphere and to the left in the southern hemisphere. This is because air, in moving from higher to lower pressure areas, is subject to the effect of the earth's rotation. Therefore, all moving air (wind) eventually deviates from its original course.

To Cool Air, Heat it! To Warm Air, Cool It!

These two paradoxes go hand in hand. They are part and parcel of local convection. Thus, during a summer day, contiguous areas may heat at different rates of speed. A warm mass of air is lighter than an adjoining cool one, and therefore it is forced by the cooler to rise. In doing so it attains heights where the pressure is less; hence it expands and cools. Paradoxically, air that has become cooler than surrounding or contiguous air settles because cool air is heavier than warm air. But, in the process of descending, air is compressed because pressure increases downward toward the earth's surface. In turn, the compression of air is a warming process.

The Closer the Sun, the Colder the Season

This applies to the northern hemisphere only. It is a fact that at *perihelion* (January 3) we are closer to the sun than at *aphelion* (July 3). More important than our distance from the sun (which varies by 3.3 per cent) is the fact that during our winter the rays are less direct than during the summer; this means that considerably less radiation is received in the winter than in the summer.

The Sun Rises before It Is Up; the Sun Sets after It Is Down

When light from the sun reaches the earth's atmosphere, it is bent or refracted according to well-known laws. Refraction has the net result of raising the rays and making the sun appear to be on the horizon when it is geometrically about one-half of a degree below the horizon. Similarly, in the evening the rays are bent upward, allowing one to see the sun as though it were above the horizon when geometrically it is already below it.

More Air Goes Up than Ever Comes Down

Rising air contains water vapor, a constituent of air just as nitrogen and oxygen are. Because condensation of part of the water vapor and the resultant precipitation usually occur in rising air, it follows that the air which eventually comes back to the earth's surface will be less than that which originally ascended.

The Almanac

A consideration of proverbs invites mention of the almanac, for the almanac thrived a long time on the application of proverbs as well as on nebulous weather forecasts. The better almanacs have now dropped forecasting, but others still have clientele that follows their forecasts and advice religiously. Some journals also give long-range forecasts similar to those of almanacs. The secret of the pseudo success of the almanac lay in its dealing in broad generalities. Thus a particular publication may have been distributed throughout the area east of the Mississippi. Reference to rain on June 21 was certain to be correct *somewhere* in such a large area, so the prediction was infallible! The avoidance of too definite a location plus vagueness as to the exact time of occurrence also heightened chances of success in a given region.

In arriving at the various daily forecasts which, after all, are printed months ahead, most almanacs predicting weather employ figures for the various weather elements. Thus, if the data for Columbus, Ohio, are considered an average for the area of distribution, the next problem is to determine what the average weather conditions have been for, say, the past twenty years. So, if it has rained in Columbus on April 28 for 4 of the 20 years, the weather prediction for that date may well be, "Possible rains in the Ohio Valley." If it has rained 12 out of the 20 years, the prediction may be, "Probable rains in the Ohio Valley."

Such "forecasting" cannot possibly take into consideration the departures from the normal which occur regularly and which cannot be foreseen very far in advance by even the best equipped weathermen. Little by little the almanac has sunk from its lofty peak as best seller and is being supplanted by more accurate scientific forecasting via newspaper, radio, and television. The quackery of guesswork is on its way out.

3

The Atmosphere: Extent, Composition, and Pressure

Height of the Atmosphere

The atmosphere is a mixture of various gases completely surrounding the land and water of the earth, penetrating the smallest crevices in the earth's crust, and even dissolving in the ocean waters to considerable depths. While the vertical extent of the atmosphere is unknown, its upper limit is being calculated at higher and higher elevations. Not so many years ago its height was given as 200 miles, then 600 miles; but now, in highly attenuated form, it is certain to extend as high as 800 miles. In fact, it is considered probable that traces of it extend to cosmic heights of 20,000 to 250,000 miles.

Knowledge of the upper air is still mostly indirect. Information comes, however, from such phenomena as meteors, the aurora, sound, light and radio reflections, semicosmic clouds, and from other astronomical observations. More direct knowledge of the lower atmosphere comes not only from actual flights by man in airplanes and balloons but also from several types of unmanned, helium-filled balloons that are observed and tracked. The simplest of these is the *pilot balloon* tracked by means of a theodolite (similar to a surveyor's transit) which furnishes data on the speed and direction of winds at various elevations. If self-recording meteorological instruments are attached, they are called *sounding balloons;* the value of these for observations other than those made directly by tracking, depends upon the recovery of the instruments when the balloon bursts and they descend to earth by parachute. Another type of balloon which can transmit data on pressure, temperature, and relative humidity by means of radio pulses while in flight is the *radiosonde;* two types of radiosonde equipment and a radiosonde recorder are shown in Figures 3–1 and 3–2. By tracking these radiosonde balloons, the speed and direction of the wind can also be determined. The temperature data shown in Table 5–1 were taken from radiosonde sound-

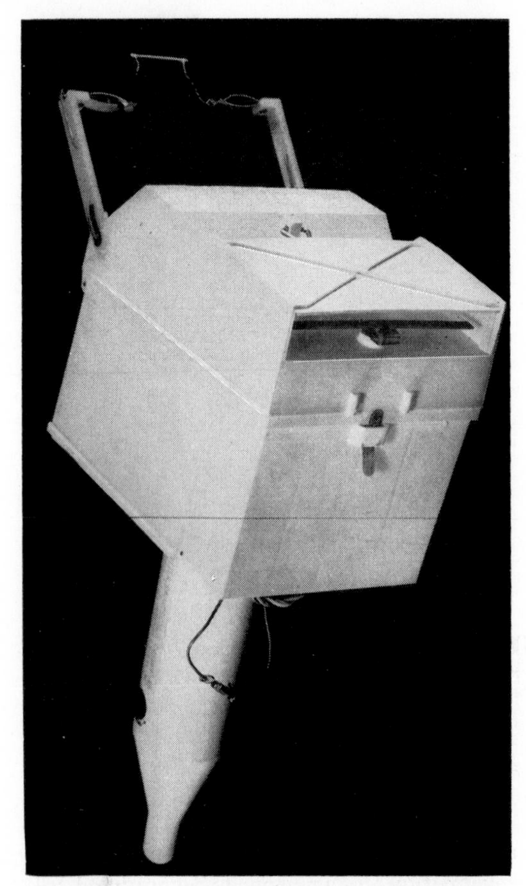

Fig. 3–1. Two types of radiosonde equipment used in determining upper-air conditions. When balloon bursts, such equipment descends to the earth by parachute, and much of it is eventually recovered. (*Courtesy Bendix-Friez.*)

ings. Some sounding balloons have been known to ascend to heights of about 30 miles, indicating there is appreciable density of the air at that high altitude.

Meteorites, or shooting stars, become luminous because of the heat resulting from compression of the air in their paths; they are known to become visible at 135 miles above the earth's surface. The aurora (Figure 3–3), believed to be the result of electrical bombardment of the upper layers of the atmosphere by charged particles or corpuscles shot out from the sun, is evidence there is still some atmosphere at altitudes of 400 to 500 miles. Whatever the upper limit of the atmosphere, its density at such high altitudes must be very inappreciable and certainly unable to sustain life or to support air flight under present means of ascent; this does not, however, exclude the unlimited heights which various types of rockets, guided missiles, and earth satellites may attain. Whatever the depth of the air, it is probably small if we consider it as having measurable density, structure, and composition comparable with conditions at the earth's surface.

Man attained his first really high altitude—at the time considered phenomenal—in the summer of 1935 when the balloon gondola, *Explorer II*, manned by Air Corpsmen and financed by the National Geographic Society, reached an altitude of 13.71 miles; in August,

1957, an occupied plastic balloon reached an altitude of more than 19 miles. In 1953, an airplane gained an altitude of 15.7 miles; and in 1955, the manned rocket plane X_2 reached an altitude of 22.3 miles with a flying speed of 1,900 miles per hour, the high speed indicating very light air resistance. Photographs have been taken automatically at even higher elevations by apparatus in unmanned rockets or guided missiles; such photographs show, among other things, the distinct curvature of the earth's surface. With jet streams being explored and even used commonly in certain flights, a vast new era is undoubtedly being initiated in which new data will be added to our present knowledge. Also, the successful orbiting of the "Sputniks," "Explorers," and "Vanguards" is unquestionably opening up an era of enormously expanded research and knowledge concerning the upper atmosphere.

Layers of the Atmosphere

Troposphere

The atmosphere is usually considered as being divided horizontally into strata or layers of which there are four or five chief ones; a graphic pictorial representation of some of those layers is shown in Figure 3–4. In the lower atmosphere called the *troposphere*, indicated by layer *A* in the diagram, temperature normally decreases upward from the earth's surface. This lower layer varies in height from place to place and from season to season but averages perhaps 40,000 feet. At the poles it is from 23,000 to 33,000 feet, but at the equator it extends to approximately 60,000 feet. It is higher in summer than in winter, attaining on the average a height of 37,000 to 40,000 feet at the 45th parallel.

The upper limit of the troposphere is called the *tropopause*. It is a rather thin and uneven transitional zone. The tropopause is not only the upper limit of ordinary clouds but also the upper limit of moisture and dust particles, except those originating above the troposphere. On rare occasions, however, *mother-of-pearl clouds* have been reported above the troposphere, especially during the occurrence of chinooks in the Rocky Mountain area, although the exact relationship has not been determined. Although the troposphere occupies in depth a very small portion of the total atmosphere, it is the most important part because it is the realm to which most of man's activities are confined and to which weather changes are limited—all of the various weather phenomena are believed to occur in this layer or zone. In Table 5–1, the elevation of the tropopause is indicated for both January and July at the foot of each column.

Increasing exploration of the nature and cause of *jet streams*, those narrow though clear-cut bands of high-speed winds near the upper

Fig. 3–2. A type of radiosonde recorder. Devices of this sort are both complicated and expensive, although accurate in recording upper-air data. (Courtesy Bendix-Friez.)

Fig. 3–3. The aurora borealis. (From a painting; courtesy American Museum of Natural History.)

limits of the troposphere, have added much to our knowledge of that portion of the atmosphere in which they occur. At times, jet streams appear to go beyond the limits of the troposphere, causing an upward bulge in the tropopause. This bulge, a purely hypothetical one, is indicated in the diagram, Figure 3–4; it may break, causing the tropopause to reform at lower altitudes, thereby indicating that there may be definite lack of continuity in the tropopause. A more detailed account of jet streams will be taken up in connection with the general circulation of the atmosphere in Chapter 8.

Stratosphere

Ascending above the troposphere, one would encounter a layer in which temperatures at first would be essentially static (isothermal) but might increase slightly. This stable, stratified layer, called the *stratosphere*, extends upward to about 20 miles. As in the troposphere, the temperatures of the stratosphere are lower above the equator than above the poles—at times perhaps as much as 80 degrees lower.[1] The temperatures vary seasonally as well as from one locality to another. The stratosphere is indicated by layer *B* in Figure 3–4.

Ozonosphere

Just above layer *B* is a layer of ozone concentration extending upward to about 30 miles; this zone is commonly called the *ozone layer*, although *ozonosphere* is better terminology; it is indicated by layer *C* in the diagram. Ozone absorbs more solar radiation than do other permanent gases of the atmosphere, filtering particularly the ultraviolet rays, thereby adding heat to the ozone layer. It is said that if the thin and rarefied ozone stratum were absent, we would be subject to severe sunburn and perhaps to blindness. It appears that temperatures increase in the ozonosphere at the rate of about 16 degrees per mile.

[1] Unless otherwise indicated, all references to temperature will be given in degrees Fahrenheit.

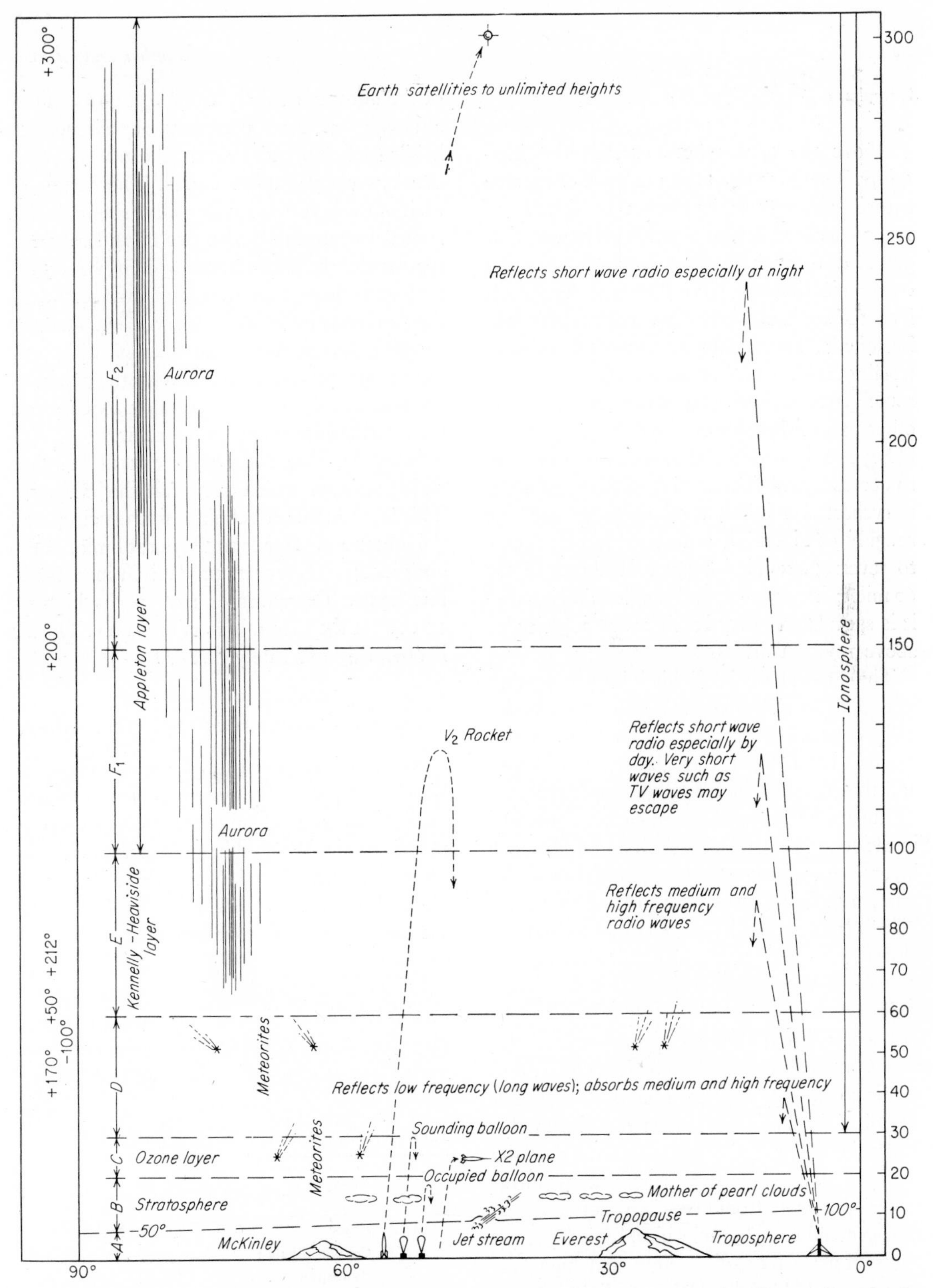

Fig. 3-4. A pictorial representation of the various layers of the atmosphere. Elevations in miles are shown by the scale on the right margin of the diagram; some observed and theoretical temperatures are indicated on the left margin without scale, and degrees of latitude are shown by the numbers across the bottom, the equator being on the right and the north pole on the left. It should be noted that some authorities consider layers C and D as parts of the stratosphere, thereby limiting the ionosphere to the E and F layers. (Original diagram drawn by Carl Gewalt, La Jolla, Calif.)

Ionosphere

Above the ozonosphere extends the *ionosphere*. There is very little evidence of its true nature, although it is known to consist of several ionized layers. Information was first obtained through the development of radio communication and, since that time, by soundings and rockets which have reached far into its domain. The various layers seem to be electrically conductive, but selectively so, because some parts reflect long radio waves while other parts reflect short waves.

Beyond the ozonosphere, temperatures seem to decrease with increased height, dropping to a suspected minimum of about $-100°$ at roughly 50 miles. Then the temperature seems to reverse, attaining a level of about 50° at 60 miles; it continues to rise indefinitely and—it is speculated—may reach 1300° at altitudes above 1,900 miles.

The ionosphere is divided into a number of regions or strata based largely upon electrically conducting properties. They are designated as the *D* layer, the *E* layer, the F_1 layer, and the F_2 layer, including the upper portions of the aurora borealis region (aurora australis in the southern hemisphere). The *D*, *E*, and *F* layers are each apparently separated from the other by a layer having no distinct property. The *D* layer, around 50 to 60 miles altitude, is closely associated with solar radiation since it disappears at sunset. It reflects low-frequency radio waves but absorbs medium- and high-frequency ones. During pronounced solar flares, all medium- and high-frequency radio traffic stops. The *E* layer, roughly centered at 80 miles, is sometimes called the "Kennelly-Heaviside layer"; it is better defined than the *D* layer and is probably produced by ultraviolet photons from the sun acting upon nitrogen and oxygen molecules. This layer reflects radio waves of medium and high frequency; and, like the *D* layer, disappears at night. The F_1 and F_2 layers, collectively referred to as the "Appleton layer," extend upward to 300 miles or more; they are important in long-distance radio transmission. The realm of the aurora is chiefly from 50 to 165 miles, although on rare occasions it extends to heights of 600 miles; thus the aurora layer overlaps several other layers. The height of each of the layers has wide daily, seasonal, and annual variations; it also has considerable irregularities in temperature and in the reflection and absorption of radio waves. Noctilucent clouds have been observed between the *D* and *E* layers. Meteorites seem to disappear most frequently at approximately the height of the noctilucent clouds, or in the ozonosphere; the relationship, if any, is not clear.

Some authorities recognize still another major stratum which they call the *exosphere*. This layer in rather nebulous form encircles all the others and extends to the outer limits of the atmosphere. Hydrogen and helium increase in this layer. Temperatures may be as high as 10,000° at the extremities of the layer. The air is obviously extremely rarefied.

Composition of the Atmosphere

Pure, dry air at sea level is a mixture of at least nine gases. By volume, two of these, nitrogen and oxygen, make up 99.02 per cent of the lower air. While this combination changes somewhat with altitude, the change is negligible, at least in the troposphere. Table 3-1 lists the known gases in per cents at the surface of the earth.

Of these, oxygen is the most obviously essential to human life, carbon dioxide is necessary for plant growth, and nitrogen serves mainly as a diluent and as the source of nitroge-

Table 3-1. Composition of the Atmosphere in Per Cents

Gas	Per cent
Nitrogen	78.03
Oxygen	20.99
Argon	0.9323
Carbon dioxide	0.03
Hydrogen	0.01
Neon	0.0018
Helium	0.0005
Krypton	0.0001
Xenon	0.000009
Ozone	0.000001

nous compounds in organisms. Far aloft in the atmosphere, ozone becomes important, while helium and hydrogen increase at the expense of nitrogen, oxygen, argon, and carbon dioxide.

Air near the earth is never completely dry, not even in the driest of deserts; there is always some water vapor in it. The amount of water vapor varies from scarcely a trace when temperatures are very low and little moisture is available to about 5 per cent by volume when temperatures are high and maximum moisture is available. Water vapor decreases as altitude increases: at approximately 6,500 feet, on the average, one-half of the water vapor is below, while at an altitude of five miles the water vapor content is extremely small. The total amount of water vapor in the atmosphere, even though confined largely to surface air, must be prodigious; it is said that 16 million tons of water vapor are evaporated from the earth's surface each second. On the other hand, it is believed that if all the water vapor in the atmosphere at any given time could be condensed and precipitated, it would amount to only about 1 inch of rain over the earth's 200 million square miles of surface.

The amount of water vapor required for saturation depends upon the temperature of the air. At a temperature of 80°, air can include about four times as much water vapor as air at 40°, or, for about every 20-degree rise in temperature, a mass of saturated air can include approximately twice as much water vapor. The actual amount of water vapor in saturated air at various temperatures is shown in Table 3–2. The amount of water vapor in the air under the same conditions of temperature may determine whether the region will be a steaming jungle or a parched desert. The moisture content of the air is as essential to human existence as any of the other gases.

Table 3–2. Water Vapor Content of Saturated Air

Air temperature, degrees F	Water vapor content of saturated air, grains per cu ft
−30	0.095
−25	0.126
−20	0.167
−15	0.220
−10	0.286
− 5	0.371
0	0.479
5	0.613
10	0.780
15	0.988
20	1.244
25	1.588
30	1.942
35	2.375
40	2.863
45	3.436
50	4.108
55	4.891
60	5.800
65	6.852
70	8.066
75	9.460
80	11.056
85	12.878
90	14.951
95	17.305
100	19.966
105	22.966
110	26.343
115	30.130

The other variable constituent of the air is dust consisting of tiny particles of clay, smoke, sea salt, pollen and spores of plants, dry and pulverized bits of organic matter, volcanic dust of various sizes, and other materials. Some of these particles, particularly smoke, sea salt, and volcanic ash, are more or less hygroscopic and are thus very important since they act as condensation nuclei for water droplets in the air. The variegated colors of sunrises and sunsets as well as the phenomenon of twilight also result largely from the presence of dust particles. Absence of large quantities of foreign particles, including moisture, is responsible for the clearness of the skies of dry regions.

Characteristics of Air

Weight

If you were to compute the number of square inches in the surface of a 30- by 30-inch

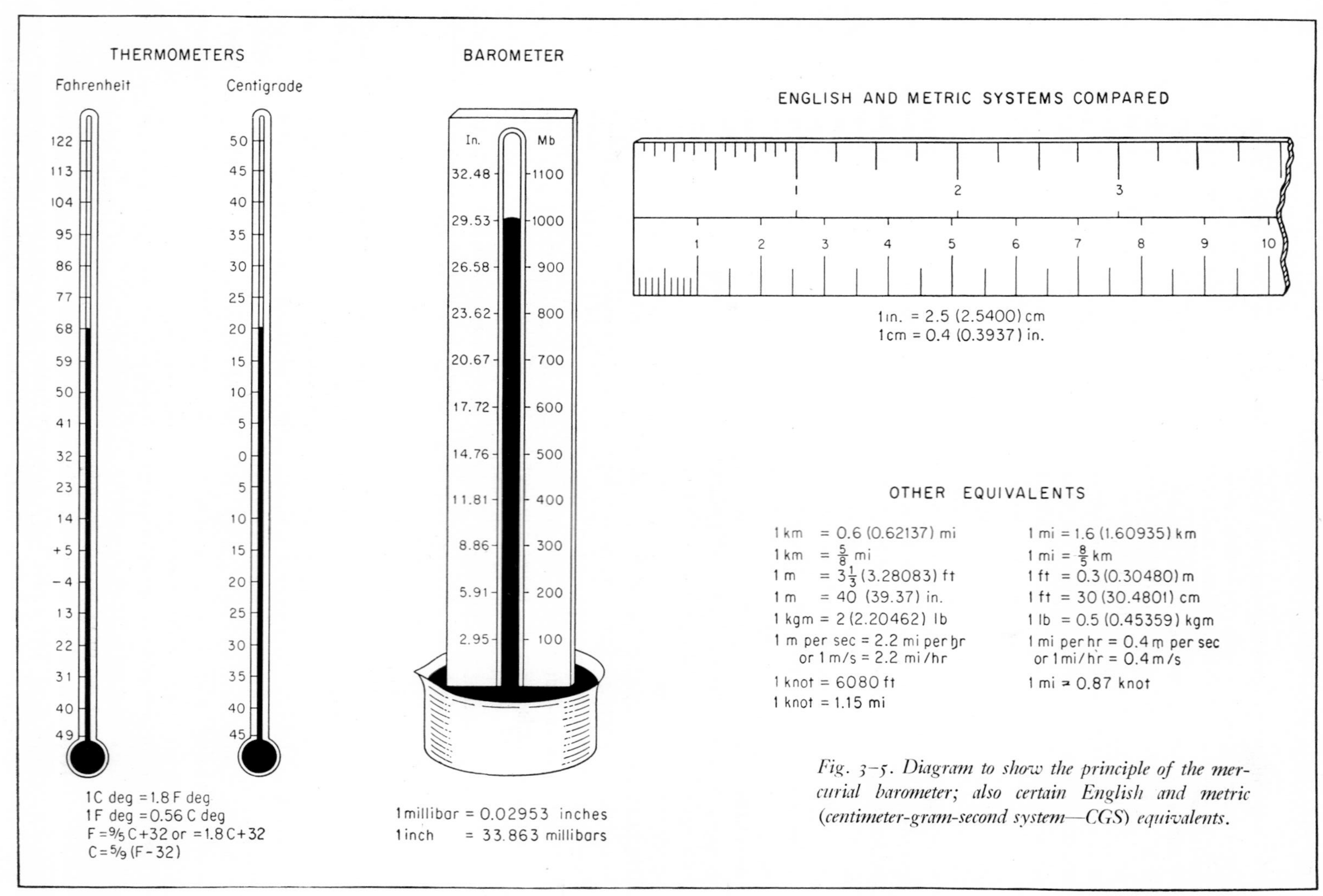

Fig. 3–5. Diagram to show the principle of the mercurial barometer; also certain English and metric (centimeter-gram-second system—CGS) equivalents.

card table and multiply that figure by 15, you would obtain the approximate weight in pounds of the air that would press upon the table at sea level; it might surprise you to know the figure would be about 6½ tons. It becomes obvious, then, that the earth's surface, including all natural and cultural forms, is able to withstand the weight of thousands of trillions of tons; bodies or structures unable to support weights of great magnitude would be crushed. It has been calculated that the total weight of the atmosphere is 5½ quintillion tons. Some authorities claim that a large decrease in pressure over a relatively large area, as with the passage of a typhoon, can cause a readjustment of stresses in the earth's surface, resulting in earthquakes and increased volcanic activity.

Since a cubic foot of water *vapor* weighs approximately three-fifths as much as a cubic foot of dry air at the same temperature and pressure, it follows that moist air is appreciably lighter than dry air; the presence of any amount of water vapor in the air causes a corresponding decrease in the percentage of the other constituents, as shown in Table 3-3. This fact is of great significance in accounting for certain weather phenomena which will be discussed later.

Table 3-3. Composition of Moist Air by Volume in Per Cents

Water vapor	Nitrogen	Oxygen	Argon
0	78.03	20.99	0.93
1	77.25	20.77	0.92
2	76.47	20.51	0.91
3	75.69	20.36	0.90
4	74.81	20.15	0.89
5	74.13	19.94	0.88

As one rises above sea level, he moves from under part of the weight of the air and thus has fewer pounds per square inch pressing upon him. If he were to ascend 3½ miles, he would find that there would be only about 7½ pounds per square inch pressing upon him and that one-half of the weight of the atmosphere would be below him.

Since air at 3½ miles has only a little more than one-half the density of air at sea level, each breath a person takes brings in only about one-half as much oxygen as he normally requires, unless he takes a deeper breath. Consequently, he becomes weak and faint at high altitudes, largely because of the decreased supply of oxygen. Furthermore, his body is adjusted to the normal weight of 15 pounds per square inch, and when he ascends to high altitudes his internal pressure is so much greater than the external pressure on his body that he may suffer from nosebleed, earache, nausea, headache, and other troubles. Gradually, however, he may become adjusted to the decreased pressure, and he may also breathe more deeply, so that considerable activity at altitudes of 3 or 4 miles is possible. To ascend much higher than that, one needs to carry an additional supply of oxygen. Furthermore, if one descends to the surface of the Dead Sea (almost 1,300 feet below sea level), so much more weight is added to him that he may find some difficulty in supporting it, particularly if he is ill. Again he may react unfavorably to the weight or pressure of the air.

Pressure

It is more common and more useful to speak of the *pressure* exerted by the air than of its weight. The weight of one's hand is one thing; the pressure it can exert is another. When the weatherman says the pressure is 29 inches, he refers to the pressure exerted by the weight of the atmosphere at the time and place of reference. Pressure of the air is one of the most important factors in weather forecasting, although it is the least noticeable of the weather elements to the average person. It is true that either great or rapid changes may cause marked physiological disturbances, but most of us are not conscious of the ordinary changes from day to day. Pressure is very closely intertwined with the other weather elements; it is a cause-and-effect phenomenon, as will be shown later.

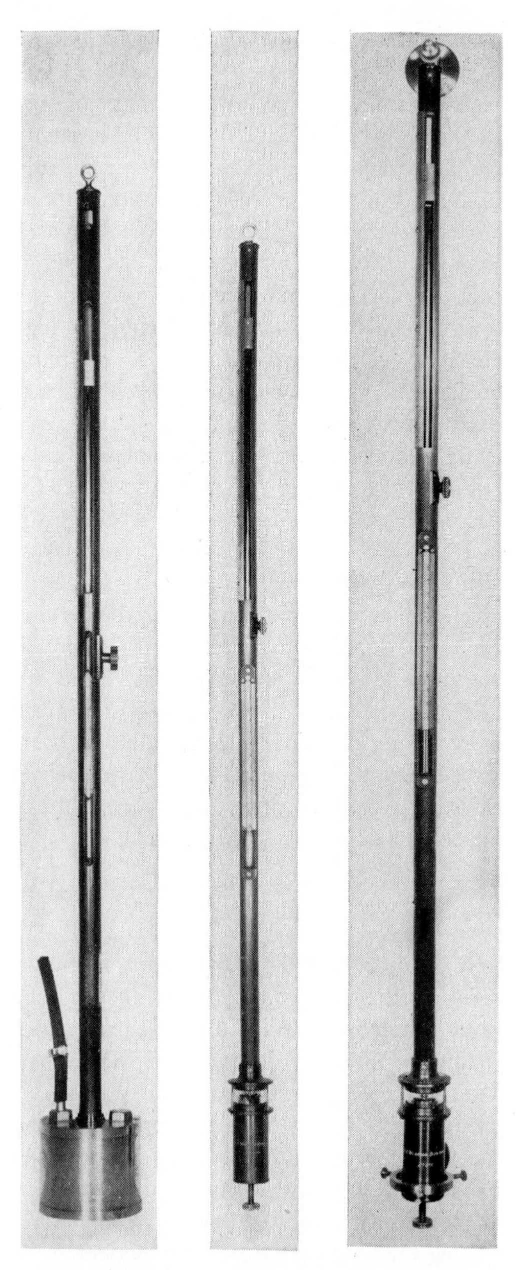

Fig. 3–6. The Fortin type of mercurial barometer housed in a metallic case for protection. Note the vernier operated by the milled thumbscrew on the right side, the milled thumbscrew at the bottom for adjusting the level of the mercury in the cistern, and the attached thermometer, since a temperature correction must be made for every reading. (*Courtesy U.S. Weather Bureau.*)

Mercurial Barometer. If we were to use a tube, 1 square inch in cross section and closed at one end, and pour in 15 pounds of mercury, the length of the mercury column at sea level would be approximately 30 inches, since a cubic inch of mercury weighs just about ½ pound. That is essentially what Torricelli did in 1643 when he showed his teacher, Galileo, that a short column of mercury would do what a 35-foot column of water would do. Therefore, we may say that a column of air, having a cross-sectional area of 1 square inch—or any other area—and extending to the limits of the atmosphere, will weigh the same as a column of mercury having the same cross-sectional area and extending vertically only 30 inches. This may be demonstrated by taking a glass tube, 32 to 34 inches in length and closed at one end, filling the tube with mercury, placing a finger over the open end, and inverting the tube in a dish of mercury (Figure 3–5). A part of the mercury will flow back into the well or dish of mercury; at sea level the length of the mercury column in the tube will measure, on the average, 29.92 inches. If we were to try the experiment at an altitude of 18,500 feet, the length of the column would be only about 15 inches. An instrument of this sort is called a *barometer*, a weight measurer.

Fortin-type Barometer. The Fortin type of barometer, as used by the Weather Bureau, is housed in a brass case (Figures 3–6 and 3–7). At the top of the column of mercury are scales indicating inches and millibars of pressure. While measurement of pressure in inches seems more practical to the layman, in scientific work millibars are frequently preferable to their equivalent in inches of mercury; furthermore, millibars are now commonly used by the Weather Bureau. A millibar is one one-thousandth of a bar; a bar is defined as the pressure exerted by a force of 1 million dynes on a square centimeter of surface, a dyne being the force which, acting on a gram for one second, imparts to it an increase in speed of one centimeter per second.[1] The bar is, therefore, really a unit of *force* rather than a mere linear measurement. A bar is equivalent to 29.53 inches of mercury, or the normal atmospheric pressure at a height of 350 feet above sea level.

Although it varies from time to time and from place to place, average sea level pressure is considered to be approximately 29.92 inches or 1013.2 millibars (commonly referred to as *one atmosphere*). Since pressure at sea level normally varies only about 3 inches, the scale actually appearing on the barometer is usually only from 26 or 27 inches to 32 inches. If sea-level pressure drops quickly below 28.5 inches, it is good evidence that a rather violent storm is imminent. Any good mercurial barometer indicates pressure to one one-hundredth of an inch and will give a fair approximation to one one-thousandth of an inch. Except during a very rapid drop in pressure, such as that effected by the sudden visitation of a tornado or possibly a hurricane or tropical cyclone, the pressure inside and outside a building is virtually the same, although temperature differences inside and outside would cause a slight difference in pressure readings on the barometer.

Aneroid Barometer. Less accurate than the mercury barometer but easier to handle is the *aneroid* (meaning without liquid) which has been in use for more than a century (Figure 3–8). It consists of a corrugated vacuum box

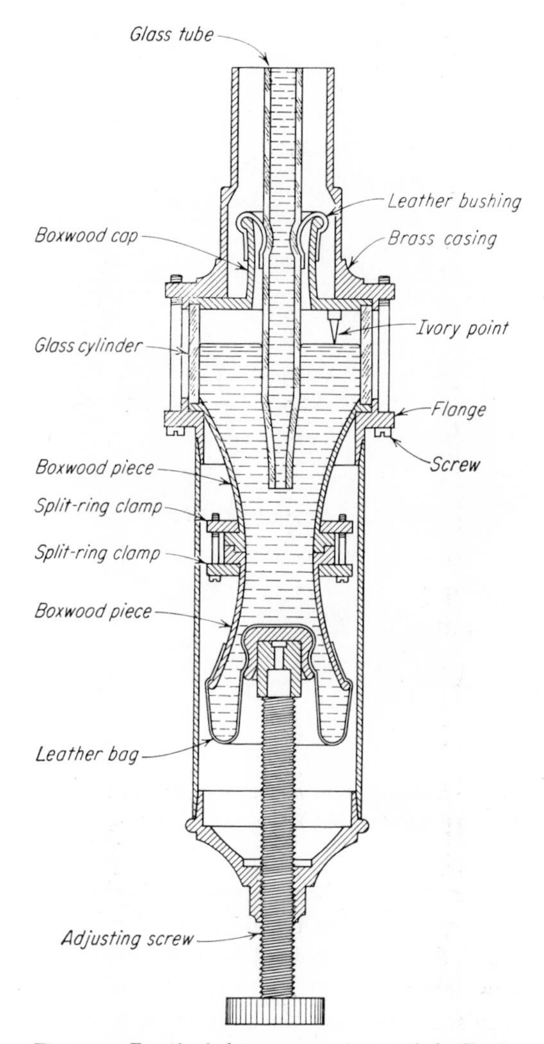

Fig. 3–7. Detail of the mercury cistern of the Fortin-type barometer. The adjusting screw is turned right or left until the ivory point just touches the level of the mercury in the cistern. (Courtesy Prentice-Hall, Inc. and Colonel George F. Taylor.)

or chamber made of silver or some similar thin metal or alloy that reacts sensitively to changes of pressure. As pressure increases, the sides of the box or chamber are depressed, and this in turn moves a system of levers connected to a pointer which moves to a higher reading toward the right on the face of the instrument. With decreasing pressure, the sides

[1] The value of a bar is readily computed by use of the simple formula

$$F = m \times a$$

in which F equals the force in dynes, m equals the mass of the mercury column expressed in grams on a square centimeter of surface, and a equals the acceleration of gravity. Thus, if we multiply 76 (which is standard pressure in centimeters, equivalent to 760 millimeters or 29.921 inches) by 13.5951 (the density of mercury in grams per cubic centimeter) and then multiply that product by 980.665 (the acceleration of gravity in centimeters per second per second), we obtain the result of 1,013,250 dynes or 1.01325 bars (1013.25 millibars).

Fig. 3–8. Aneroid barometer with cover removed; pressure is indicated on the face by millibars. (Courtesy Bendix-Friez.)

of the vacuum box expand, being forced by a spring within the box, and the pointer moves downward toward the left, indicating a lowering barometer. An *altimeter*, which is standard equipment on aircraft, is a special aneroid barometer on which altitudes commensurate with decreasing pressures are indicated. A pressure of 29 inches would indicate an elevation of about 900 feet above sea level; 26 inches would mean approximately a 3,500-foot elevation.

Descriptive terms, such as "stormy," "rain," "fair," "windy," or "dry," found on the face of some aneroids are only relative and should not be taken too literally. Those conditions may or may not occur with the actual pressure indicated. What *is* true is that, in general, a continually rising barometer indicates the approach of better weather conditions, while a steadily falling barometer frequently heralds less propitious weather—perhaps cloudiness, rain or snow, or bad weather of some kind. The student of weather should recognize that many meteorological readings of specific pressure, actual temperature, amount of moisture in the air, and definite wind velocity and direction, if taken alone or without regard to preceding conditions, may have little meaning with respect to the immediately ensuing weather. On the other hand, knowledge of all of them together, plus the changes they may have been undergoing, can be very significant. Pressure *changes* have much more meaning than simply a spot reading such as 28.9 or 30.1 inches; these alone do not indicate any trend.

Corrections Applied to Barometer Readings. To all observations of pressure, certain corrections need to be applied in order to make them comparable with one another: (1) A correction to bring the temperatures of both the mercury and the scale to a standard (32°); since mercury expands with heat and contracts with cold, the correction will be subtracted if the temperature is above freezing, because the mercury column will be too long, and added if below freezing. An aneroid reading requires no temperature correction because no expanding or contracting mercury is involved; (2) a correction for altitude to reduce all barometer readings to sea level; this is done by adding to the station pressure the difference of pressure between that particular station and sea level. If weather station High Tower is located 2,700 feet above sea level, approximately 3 inches will have to be added to the actual station pressure recorded on the barometer because the normal decrease in pressure is approximately 1 inch per 900 feet of altitude; (3) a correction to reduce readings to standard gravity; since an object weighs slightly more in high than in low latitudes (owing to the flattening of the earth near the poles) and since an object weighs very slightly less at high altitudes than at sea level, a gravity correction for both latitude and altitude must be made, the standards being the 45th parallel and sea level. These corrections, however, are very small, e.g., about −0.044 inch for a station on the 30th parallel and at an

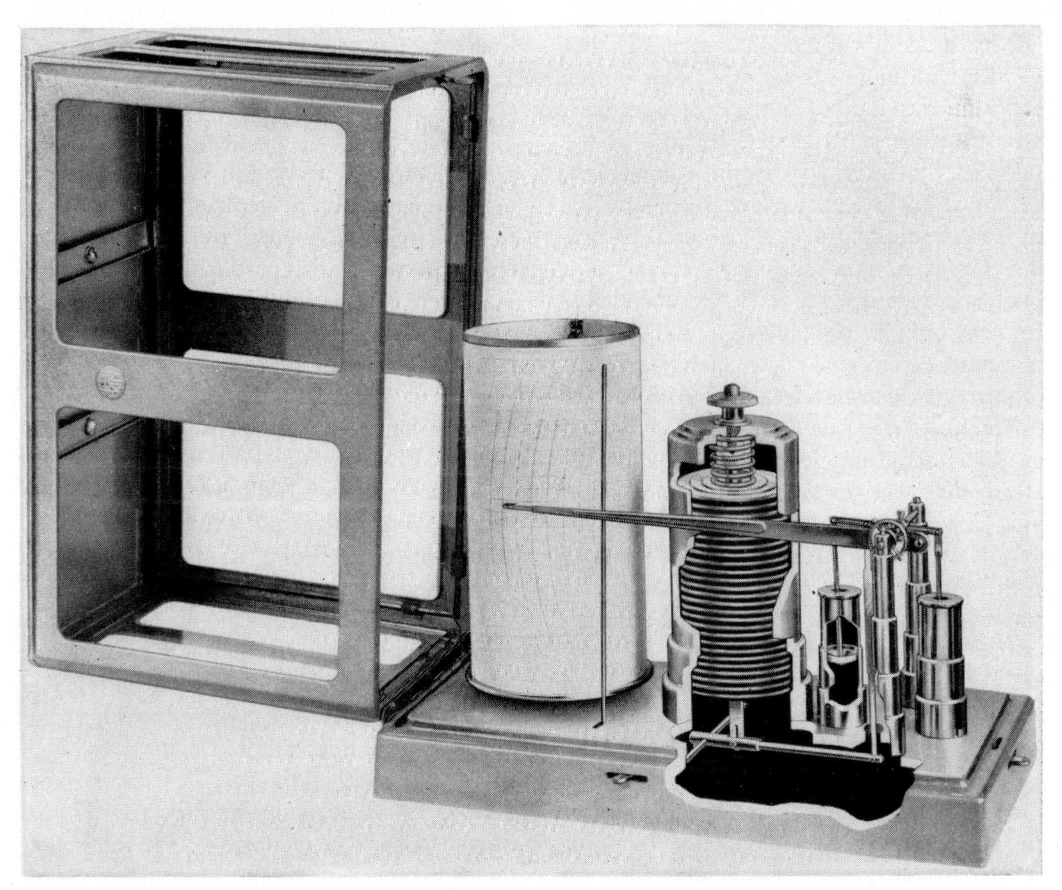

Fig. 3–9. Microbarograph with the protective metal cover cut away in order to show the battery of hollow metal discs. This instrument will record quite minute changes in pressure. (Courtesy Bendix-Friez.)

elevation of 7,000 feet. This *gravity* correction for altitude should not be confused with the standard correction for altitude as mentioned in (2) above.

Barograph. For a continuous record of pressure, consult a *barograph*—an aneroid with levers that move an inked pen which, in turn, moves upward with increasing pressure and downward with decreasing pressure. The pen records on a seven-day barogram—a sheet of paper marked with the days of the week and the hours of the day and fastened around a clock-mechanized brass cylinder. As the cylinder makes a complete rotation in a week, a line graph results which represents the pressure for seven days. A *microbarograph* (Figure 3–9) has a battery of metal boxes and a taller cylinder with more minute gradations on the barogram; thus it gives more detailed readings than an ordinary barograph. Because of the multiplicity of moving parts in a barograph, the instrument is somewhat less accurate for any specific reading than a mercurial barometer, but it has the advantage of giving a continuous record for a week. Also available are numerous cheap, colorful, and usually inaccurate barometers which may work reasonably well if under constant temperature; it should be remembered, however, that good scientific instruments are not cheap.

Vertical Changes in Pressure. Pressure is not static either vertically or horizontally. Normally it decreases upward because the weight of the air decreases upward. In the lower atmosphere, the amount of decrease at ordinary but uniform temperatures is at the *geometric* rate of about one-thirtieth of the pressure reading for every 900-foot increase in elevation. Beginning with a sea-level pressure reading of 30 inches, at 900 feet it would be 1 inch less, or 29.0 inches; at 1,800 feet it would be less by $\frac{1}{30}$ of 29.0 inches (0.97), or 28.03 inches; at 2,700 feet it would be less by $\frac{1}{30}$ of 28.03 (0.93), or 27.10 inches; at 9,000 feet it would be 20.67 inches; and at 18,500 feet it would be 15.01 inches. This indicates, therefore, what was stated earlier, that one-half of the atmosphere is below 18,500 feet; above that altitude, pressure becomes increasingly small until it is practically nothing at the upper edge of the ionosphere. At 90,000 feet—the altitude allegedly attained by the rocket plane X_2 in 1955—it is estimated that the pressure would be less than 0.3 inch. In general, however, at each successive 900-foot interval the pressure is $\frac{29}{30}$ of what it was at the beginning of the interval. There are many occasions and conditions when the normal decrease in pressure upward is not too evident; some of these will be noted later.

Another important consequence of vertical decrease in pressure is the effect upon the boiling point; in the case of water this is lowered approximately 1.8 degrees for each 1,000 feet of ascent.

Mobility

Air—"wind," as we call it—is ever on the move horizontally. Sometimes it is in a hurry to get elsewhere, sometimes it slows down, sometimes it stalls, producing a calm. But whatever its speed, ease of mobility is a characteristic of air just as it is of any other gas. It obeys definite laws of physics. The science of weather, then, is based upon the characteristics—the ever-changing characteristics—of moving air. With a change of air, i.e., a new air mass, a different set of weather conditions arrives. Ability to analyze air correctly—its temperature, its moisture content, its pressure, its velocity, its probable or possible changes in detail, and its effects upon given areas when it arrives—is the paramount goal of weathermen; they in turn can transmit accurately what is to be expected weatherwise. If air moves faster or more slowly than anticipated, the weather forecast may be partially wrong, or at least incorrectly timed; that often happens when unforeseen variables creep in. Air moves vertically, too. For several reasons it may rise or descend, along a mountainside or on the broadest plain; at other times it may remain in a horizontal position with no tendency to ascend or descend. Because of the great importance of wind as a transportation factor, as a weather element, and as a weather control, much more will be added to this subject in a later chapter and throughout the book.

Expansibility

Air is capable of expansion. When more heat is applied, or whenever pressure is decreased, air expands. When it does, it occupies more space and is cooled. As we shall see later, if air did not have the ability to expand and cool, probably there would be no precipitation, not even clouds.

Compressibility

Air can be compressed, or squeezed, into smaller space. Under such conditions it is warmed because the molecules making up the air are concentrated in a smaller area—they bombard each other with greater velocity and create more heat. When air is compressed, its pressure increases. Compressed air in an air hose may be warm; allow it to escape, and the nozzle of the hose is heated as the air is released, but once outside the confining walls of the compressor—hose and nozzle—the air expands and is rapidly cooled. In nature, there are masses of compressed air, or air at higher pressure, as well as areas of expanded air, or air at lower pressure. It is the ability of air to be compressed and to expand that causes it to flow, and air normally flows (or the wind *blows*) at or near the earth's surface from areas of higher pressure to areas of lower pressure. How and where areas of differing pressure develop will be an important facet of the weather story.

4

Heating and Cooling of the Atmosphere

Source of Heat

The earth is a radiator; if it were not, life as we know it could not be supported for lack of heat. When we are close to a radiator, we receive more heat than when we are far away from it. Similarly, as we ascend above the earth, we find the air becomes cooler as we move farther away from the "radiator." Thus we consider the earth as the direct source of our heat, while the sun—the ultimate source of practically all our heat—is considered the indirect source. The amount of heat given out by the stars, planets, and the moon is so slight it is negligible. The only other source of heat is the interior of the earth, but this is so small it is sufficient to raise the average temperature of the atmosphere only about three-fourths of one degree. It has been said that our outdoor temperature would be close to absolute zero (459.6° below zero) if it were not for the sun. The heating of the atmosphere is explained both by the nature of solar and terrestrial radiation and by the absorptive power of the earth and its atmosphere.

Solar Radiation

Insolation

The heat, or energy, which is given out by the sun is known as "solar radiation." The solar radiation received on a horizontal surface is known as *insolation*, in[coming]-sol[ar]-[radi]ation. The insolation received at the earth's surface is readily measurable. It is found to vary (1) with the length of the path of the sun's rays through the atmosphere and (2) with the area of horizontal surface which a "bundle" of rays of a certain cross section spreads over, both of which are determined by the changing altitude of the sun above the horizon in the course of a day; (3) with the distance of the earth from the sun (we are approximately 3 million miles closer to the sun early in January than we are six months later); (4) with the amount of dust, water vapor, and clouds in the atmosphere; and, if we consider total insolation during a 24-hour period, (5) with the duration of sunlight.

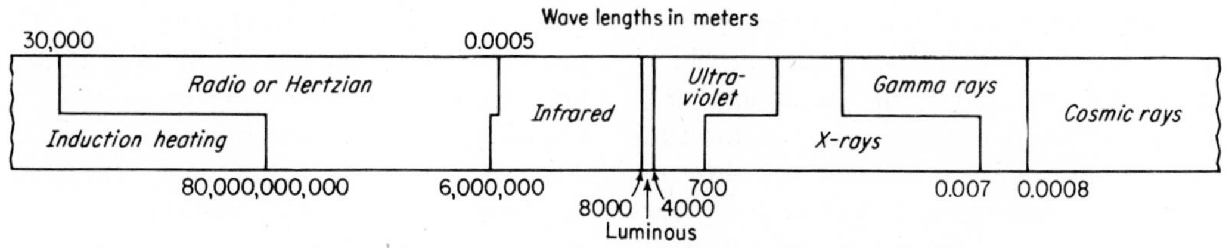

Wave lengths in Angstroms (one hundred millionth centimeter)

Wave-Lengths of Various Parts of the Spectrum

Cosmic Rays	less than 0.0008	Angstroms
Gamma Rays	0.0008 to 2	"
X-Rays	0.007 to 700	"
Ultraviolet Rays	40 to 4000	"
Luminous Rays	4000 to 8000	"
Infrared Rays	8000 to 6,000,000	"
Radio or Hertzian Waves	0.0005 to 50,000 meters	
Induction Heating	more than 8	"

Fig. 4–1. The electromagnetic spectrum. (Adapted from a wall chart furnished by the Westinghouse Research Laboratories, Pittsburgh.)

Nature of Solar Radiation

Solar radiation is made up of rays which have waves of different lengths, varying from a few billionths of an inch to over 1 million inches (about 17 miles). Only a part of the rays are visible; these are called *luminous rays.* In ordinary sunlight, all the luminous rays together make up *white light.* When a beam of sunlight passes through a triangular glass prism, it is broken up into colored bands ranging from red through orange, yellow, green, and blue to violet, each color merging very gradually into the next color. The distinction between these colors is merely one of wavelength, the red rays having the longest waves and the violet rays having the shortest waves. In the entire electromagnetic spectrum, luminous rays occupy an extremely narrow band, as shown in Figure 4–1. Rays which have somewhat longer waves than the red rays are called *infrared rays;* these we cannot see, but we can readily feel them. The rays just beyond the violet are known as *ultraviolet rays;* these we can neither see nor feel, although they cause such chemical effects as a tan or sunburn.

A nonluminous body gives out only infrared rays; and the hotter a body becomes, the shorter are the wavelengths of its principal rays. The sun's rays vary only in the length of their waves.

When luminous or other rays fall on an object, they may be partially *absorbed,* and this absorbed energy becomes heat. Also, the long infrared rays, if concentrated, may be transformed into shorter ones which we can see; that is, they become luminous. The shorter ultraviolet rays are neither luminous nor heat rays; we can neither see them nor feel them. Yet they may be absorbed and converted into heat in the same manner as luminous rays. Thus the difference between these three classes of rays is physiological rather than physical.

Effect of the Atmosphere on Solar Radiation

General Effect

As we have said, the longer the path of the sun's rays through the atmosphere, the smaller are the values of solar radiation received at the earth's surface in a given length of time. This is because of the *absorbing, scattering,* and *reflecting* of solar radiation as it passes through the atmosphere. Of the solar radiation which enters the outer layers of the atmosphere, perhaps one-third is completely lost through reflection from dust and clouds and through scattering. Another one-third is probably absorbed by the atmosphere as the radiation passes through it. The remaining one-third reaches the surface; almost all of this radiation is absorbed. These are averages for the earth as a whole. The ultraviolet rays and the shorter portions of the luminous rays suffer the greatest loss in passing through the atmosphere; the longer luminous rays and the infrared rays pass through with comparative ease, especially if the air is dry. Scattering is most pronounced with the blue luminous rays; hence the blue of the sky.

It has been found that the atmosphere weakens and scatters sunlight in inverse ratio to the fourth power of the wavelength of the rays, the scattering and weakening of violet light being about 10 times that of red light. Thus, in general, the atmosphere tends to scatter more of the short waves and to transmit more of the long waves so that, with increase in altitude, the short ultraviolet rays become more intense than at the earth's surface. Also, as the sun mounts higher in the sky, the intensity of the ultraviolet radiation increases much more rapidly than the general intensity of the sunshine.

Selective Effect

The various constituents of the air have a *selective* tendency in their action on solar radiation. Dust and water droplets reflect and scatter; carbon dioxide, ozone, and water vapor are great absorbers. Water vapor is particularly effective in absorbing and radiating the infrared rays. Consequently, water vapor acts as a blanket or screen to long wavelength radiation, whether coming from the sun or from the earth. It is very much like glass in this respect. The glass of a greenhouse or sun parlor readily transmits the luminous rays of sunshine. These are absorbed by the walls, floor, or ground and changed into heat. The glass does not readily transmit the infrared rays, and so the house or room becomes warm. Ordinary glass and water vapor are unlike, however, in their effect upon ultraviolet radiation: ordinary glass absorbs almost all that it does not reflect; water vapor transmits it. Ozone, which increases with increasing altitude up to certain limits, is particularly absorptive of ultraviolet rays.

Dust particles, including smoke, and water droplets not only scatter sunlight but also reflect and absorb it in varying degrees; and, as already pointed out, the shorter the wavelength of the rays, the greater their diminution, or weakening, by these constituents. As a consequence, when the sun is low in the sky, as at sunrise or sunset, most of the short wavelengths are completely lost, so that sunrise and sunset colors are predominantly red or yellow, occasionally (though only momentarily) green, and never blue or violet.

The reflection and *diffraction* of sunlight, especially the rays of long wavelength, by all particles of the atmosphere up to several miles in height cause *twilight*. It is estimated that astronomical twilight continues until the sun has gone 18 degrees below the horizon, while civil twilight, the period after sunset when there is sufficient light for outdoor work, continues until the sun is 6 degrees below the horizon. Since the sun rises and sets almost perpendicularly near the equator and very obliquely in high latitudes, it is evident that twilight, both civil and astronomical, is of much longer duration in high latitudes than in low.

In humid climates, the moist air during the day absorbs the long wavelength radiation so that the earth's surface has less to absorb; at night, the same moist air prevents the heat of the earth's surface from being transmitted into space and lost. The *diurnal*, or daily, ranges of temperature, therefore, are small. In arid climates, the dry air transmits solar radiation with comparative ease, and the earth's surface becomes greatly heated during the day.

At night, the long wavelength, *terrestrial radiation*, is readily transmitted by the dry air, so that temperatures fall relatively low. Such climates, therefore, are characterized by large diurnal ranges of temperature. Clouds, of course, have much the same effect as water vapor but to a more marked degree; they also reflect radiation. Strangely enough, the diurnal variation of heat received decreases rapidly with an increase in altitude, so that, at an elevation of 1 mile above the earth's surface, the difference between the highest temperature in the daytime and the lowest temperature during the night is, on the average, only 1 degree.

Disposal of Insolation on the Earth's Surface

Land Surface

The nature of the earth's surface has a marked effect on the actual disposal of solar radiation incident upon it. A snow or ice surface, for example, will absorb much less heat than will bare ground. For this reason, when there is a snow cover that remains for a number of days, temperatures will be lower than when there is no snow. Snow not only reflects insolation but also acts as a blanket on the earth's surface, keeping in whatever heat the ground already has and preventing further addition of heat when the sun is shining. Thus a heavy, early snow blanket, falling upon unfrozen or only slightly frozen ground, will frequently prevent the ground from freezing deeply and will also cause air temperatures to be relatively low, especially at night. Winters during which very little snow falls are likely to be warmer but with deeper frozen ground than winters in which there is an early and persistent snow cover.

Black soil will absorb well, while lighter-colored soils will reflect much heat and absorb little. People in hot, dry lands characteristically wear white clothing because it reflects heat; black clothing absorbs it.

Reflected heat is not greatly absorbed but rather is simply sent back into the air, or to clouds, and adds little, if any, heat to the lower atmosphere. But heat that is once absorbed may be radiated and will in turn add heat to the lower air. Under average conditions, it has been found that from 10 to 15 per cent of the radiation reaching a land surface is reflected; the remaining 85 to 90 per cent is absorbed. Of this absorbed part, only a little is conducted to deeper layers or is used in evaporating water from the surface; most of it is sent back into the air as terrestrial radiation.

Water Surface

A water surface acts somewhat differently, although it is like land in that it reflects a relatively small part of insolation, the rest being absorbed. But of the part which is absorbed, about one-half is used in evaporating the surface water and about one-half penetrates to deeper layers, though not to unlimited depths. Since more heat is required to raise a given quantity of water a given number of degrees, the *specific heat* of water being two to five times that of land, it is evident that a water surface, and consequently the air over it, will not be warmed so quickly as a land surface and the air over it, other conditions being the same. Also, because of the great mobility of water, the process of mixing prevents the surface water from reaching as high a temperature as a land surface under the same conditions of sunshine. For the same reasons, it takes longer for a water body to cool; *marine climates*, therefore, have smaller ranges of temperature, both diurnally and annually, and are less extreme than *continental climates*, that is, climates in which the land is the dominant control.

Conduction and Convection

Conduction, particularly by day and in summer, is also of some significance in the heating of the atmosphere whenever the earth's surface is warmer than the air in contact with it. The layer of air next to the ground becomes heated by contact—that is, by conduction. Since air itself is a relatively poor conductor, the heat is not conducted more than a few feet above the ground. Air is seldom motionless, however, so that the heated layer is constantly being replaced by, or mixed with, cooler air from above. Likewise, at night and in winter, conduction from a warm layer of air in contact with the ground to the cooler earth is of some importance in cooling the air.

In addition, *convection*—the transference of heat by moving air—becomes a factor of considerable consequence in heating and cooling the atmosphere: as soon as the surface layer of air becomes heated through radiation and conduction from the ground, as well as by the absorption of direct and reflected solar radiation, it expands, becomes lighter in weight (therefore less dense), and is pushed up by the cooler and denser air which settles from above. Rising air cools by virtue of its expansion, thereby being an important factor in accounting for lower temperatures aloft. Conversely, descending air warms by virtue of its compression, and this in part accounts for higher temperatures at low altitudes. Furthermore, if it were not for convection on a sunny day, particularly in summer, the surface air would become unbearably hot. Because convection is so important a factor in our daily weather—temperature and pressure changes, winds, and precipitation with all its attendant phenomena—a more complete discussion will be taken up in Chapters 5 and 6.

5

Temperature

Thermometry

Fahrenheit and Centigrade Scales

Since heat is a form of energy, it is measurable in terms of temperature by thermometers. *Thermometer* means literally "a measure of heat." While a number of thermometer scales are used, the two most common ones are the *Fahrenheit* and the *centigrade* or *Celsius*. As indicated earlier, the Fahrenheit scale is used in this book unless otherwise noted. For very special but limited purposes, the *absolute scale*, in which 0°A is equivalent to −273°C or −459.4°F, is employed. For recording air temperatures, mercury and alcohol are the ordinary liquids used in standard thermometers. There are also nonliquid thermometers; these are usually metal springs or sleeves that react sensitively to changing amounts of heat.

A comparison of the Fahrenheit and centigrade thermometer scales (Figure 3–5) shows that the freezing points on the two scales are 32° and 0° respectively, and the boiling points are 212° and 100°. Thus the difference between the boiling and freezing points on the Fahrenheit scale is 180 degrees; on the centigrade scale, the difference is 100 degrees. Therefore, 1 centigrade degree, is always equivalent to 1.8 Fahrenheit degrees; or, 5 centigrade degrees are equivalent to 9 Fahrenheit degrees. Since the freezing point on the Fahrenheit scale is 32 degrees above the value for the freezing point on the centigrade scale, it is evident that a reading of 5°C will not be equivalent to 9°F but rather 9 plus 32, or 41°F. Thus to change centigrade readings to Fahrenheit, multiply by 1.8 (or by $9/5$) and add 32. To change Fahrenheit readings to centigrade, use the reverse process: subtract 32 and divide by 1.8 or multiply by $5/9$. Tables for converting Fahrenheit to centigrade and centigrade to Fahrenheit are given in the Appendix.

Thermometers

In making liquid thermometers, uniform calibrations called *degrees* are first indicated on the glass between the established boiling and freezing points and then are added proportionally above and below these fixed points. Thermometers with the degrees indicated only on the metal or wooden mountings to which they are attached may not always be accurate because the glass tube may shift upward or downward if the brackets holding it are not absolutely tight.

Accurate thermometers must meet certain specifications by having (1) accurately determined fixed points, usually the boiling and freezing points; (2) gradations correctly indicated on the glass or stem; (3) a uniform bore from the bulb to the top of the stem (including the vacuum area at the top); (4) a fluid that will not freeze, vaporize, or decompose; and (5) a good grade of glass that will not add or subtract heat from the real heat of the air. Thermometers should also be easy to read.

Fig. 5–1. Standard thermometer shelter. (Courtesy Bendix-Friez.)

Instrument Shelters

Another factor of prime importance in gaining correct temperatures is the location and exposure of the instrument. For comparative purposes, a thermometer should never be exposed to direct sunlight. To prevent such exposure, all official temperature data, whether from Borneo, Timbuktu, or the U.S. Weather Bureau, are taken from weather instruments which are properly housed in an *instrument shelter*. While shelters vary in design, in size, and in the instruments they contain, they have certain details in common. The Weather Bureau shelter is a cubical box about a yard square and two feet deep, with a double sloping roof, a closed bottom, louvered sides, and with the whole shelter raised at least 4 feet above the ground (Figure 5–1). This type of construction prevents undue heating from above or below, and it also allows for the free circulation of air. The shelter should not be close to buildings, although in large cities the top of a high building may provide the best exposure available. A shelter usually houses a standard thermometer, a thermograph, frequently wet- and dry-bulb thermometers, and occasionally other instruments. One of its chief purposes is to obtain correct temperature data.

Maximum and Minimum Thermometers

The purpose of maximum and minimum thermometers is to record the highest and lowest temperatures occurring during any given period of time. The maximum thermometer (Figure 5–2, lower) has a constriction in the lower part of the bore just above the bulb; it is very similar in construction to a clinical thermometer. As the heat increases, so does the molecular activity of the mercury in the bulb, causing the mercury to expand; this expansion forces the mercury to rise in the bore past the constriction. When the temperature goes down, the mercury column does not: the mercury thread breaks at the

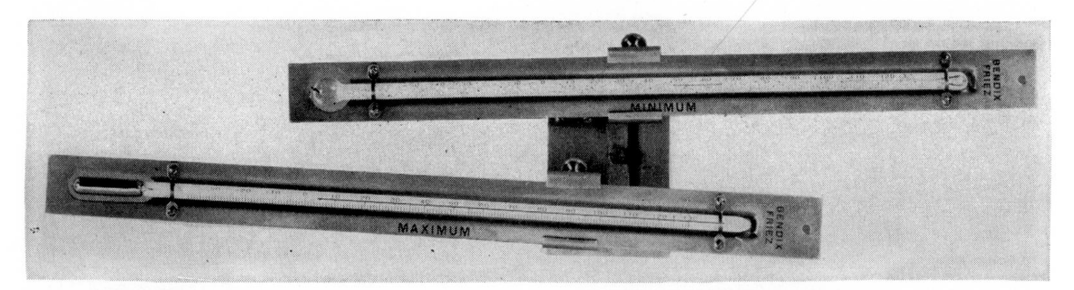

Fig. 5–2. Maximum and minimum thermometers properly set. (Courtesy Bendix-Friez.)

constriction, leaving the mercury in the stem. The maximum thermometer is reset centrifugally—whether daily, monthly, yearly, or otherwise—by whirling it rapidly, forcing the mercury through the constriction into the bulb area.

The minimum thermometer (Figure 5–2, upper) uses alcohol which does not freeze under ordinary air conditions (mercury freezes at $-39.4°$). The thermometer contains a small, dumbbell-shaped index of glass or some magnetic material. When the temperature falls, the alcohol contracts, dragging the index down with it by surface tension. Thus the upper end of the index gives the lowest reading for any given period of time. When the temperature rises, the liquid flows past the dumbbell, leaving it at its lowest position. The thermometer is reset by turning its bulb end farther upward; this returns the index to the top of the liquid.

Thermographs

For many purposes a continuous record of temperature is desirable. This is obtained from a *thermograph* (Figure 5–3), an instrument which uses a broad but thin curved metallic, or bimetallic, sleeve that is sensitive to changes in heat. The sleeve is located outside the housing and is connected to a system of levers controlling an inked pen. As the heat increases, the sleeve straightens (one metal expanding more than the other) and slowly pushes the pen upward; as the heat diminishes, the pen drops. The pen is placed so that it records on a thermogram which surrounds a revolving cylinder. The cylinder makes one complete revolution in a week, giving the temperature record for a seven-day period. For more detailed readings, a taller cylinder with a more minutely divided thermogram may be used. Because of the multiplicity of moving parts, the thermograph does not produce a record so accurate as that of liquid thermometers, but it has the advantage of providing a continuous record and of giving a graphic representation of temperature changes for the period covered.

Other Means of Measuring Temperature

One frequently sees a nonliquid thermometer in homes, schools, and offices. Although not quite so accurate as a good liquid thermometer, it is portable and readable, has an open face with the temperature figures increasing from left to right, and a hand which moves over its face, indicating the temperature. The thermometer, like the thermograph, has a bimetallic sleeve, or even a spring, which reacts to increases and decreases in heat and in turn controls an indicator on the face.

Suppose you have no thermometer on a summer evening but crickets are chirping distinctly. Then, it is said, if you count the number of chirps of the black field cricket for 14 seconds and add 40, you will have a fairly accurate temperature reading.

Variability of Temperature

Surface Variability

While the increase in temperature during the day and the decrease during the night are by no means uniform at all places and at all

Fig. 5–3. Thermograph. (Courtesy Bendix-Friez.)

times, there are, nevertheless, an increase and a decrease almost daily. This rhythm is called the *march*, or *course*, *of temperature*. If the sun were the only direct control of the heat we receive, the march or course of temperature would be very even; the temperature would increase evenly and steadily every day until the sun was on the meridian at noon and then decrease steadily until sunrise, at which time the minimum daily temperature would occur. But as we have shown, other factors help to determine our temperature: the angle of incidence of the sun's rays, our location with respect to land or water, the time required to heat the land or water, vegetation, soils, elevation, winds, and even population distribution with such attendant, cultural features as pavements and buildings, air pollutants, plowed ground, and timber removal. Many of these factors will be considered later in connection with world distribution of temperature, winds and precipitation, mountain and plateau climates, and continental and marine climates. Because of the control of these additional factors, the diurnal (daily) rise or fall of temperature is not quite so steady as one might expect; it is pronounced, nevertheless. Maximum temperature usually occurs several hours after the time of the highest sun, and the minimum usually just before sunrise.

In the summertime, the highest temperature of the day occurs from two to five o'clock in the afternoon and, at times, even later. In winter, the maximum temperature is somewhat earlier, from one to three o'clock. Minimum temperatures usually occur near sunrise, although in winter a cold air invasion may cause the lowest temperature to occur at any time of day, even in midafternoon. The retardation of highest and lowest temperature—after high noon and midnight respectively—is called the *retardation*, or *lag*, *of temperature*.

In almost all parts of the country, there is also a seasonal lag of temperature. The highest temperatures of the year lag behind the summer solstice, June 21, and usually occur in July; that is, July is the warmest month in most places, although under certain marine conditions the warmest month may be delayed until August or even September. Similarly, the coldest month is likely to be January—somewhat after the winter solstice, December

21[1]—when the sun is lowest in the southern sky, even though the sun is some 3 million miles closer to us at that season.

Adiabatic Changes

If a mass of dry air is greatly heated at the earth's surface, it expands and becomes so much lighter than the surrounding colder and denser air above it that it is forced to ascend. As it does so, the weight of the air over it becomes less, and so it expands still further. The expansion of any substance requires heat, and if this heat does not come from an outside source, the substance cools. Consequently, the mass of ascending air becomes cooler with continued expansion. Under such conditions, it has been found that the rate at which ascending air cools is very nearly 5.5 degrees per 1,000 feet of ascent (1 degree C per 100 meters); this rate is said to be *adiabatic* because it is assumed that the mass of air neither receives heat from nor gives heat to the surrounding air.

Similarly, descending air *warms* adiabatically by compression and at the same rate of 5.5 degrees per 1,000 feet of descent. It should be emphasized that rising or descending air becomes cooler or warmer without any radiation or conduction to or from the surrounding air and that the changes in temperature are the result only of the changes in pressure. Line *A* in Figure 5–4 is a graphic representation of adiabatic cooling (or warming, if the air is descending). It is assumed in this graph that the mass of air starts to ascend (or, more accurately, is forced to rise) at a temperature of 90°—it could, of course, start at any other temperature. When the rising air reaches, say, an elevation of 6,000 feet, it will have cooled to 57°; if conditions are such that it continues to rise, its temperature will drop to −75° at the 30,000-foot level.

A mass of rising air does not continue to ascend indefinitely because eventually it reaches a point where it has the same temperature, and therefore the same density, as the surrounding air, provided both masses of air have the same water-vapor content and provided also there has been no condensation of water vapor in the rising air. This is possible because the adiabatic rate is nearly twice the rate of the average decrease in temperature with increase in altitude.

If for any reason a mass of moist air is forced to rise, it will expand and cool at the rate of 5.5 degrees per 1,000 feet until a temperature is reached at which some of the moisture is condensed. Because heat is required for the evaporation of water (known as the *latent heat of vaporization*), it is evident that heat is given up in the condensation of water vapor (known as the *latent heat of condensation*). It has been found that the latent heat of condensation retards the adiabatic rate so that the latter becomes on the average about one-half the regular rate, or roughly 3 degrees per 1,000 feet. This is often called the *retarded* or *condensation adiabatic rate* or, more frequently, the *wet adiabatic rate*. The degree of retardation depends upon several factors, such as pressure, the amount of moisture condensed—a function of temperature—whether the moisture condensed is in the form of rain or snow, and whether the precipitation is carried aloft in the rising column of air.

Referring again to Figure 5–4, let us suppose that condensation begins when the temperature of the rising air (starting at a temperature of 90°) drops to 57°; this will occur at the 6,000-foot level, where a cloud, indicated by *H*, will develop. From this point on, the rate of decrease of temperature will be so retarded by the receipt of heat from condensation that, if the air continues to ascend to the 30,000-foot

[1] Owing to the varying lengths of our months and to the introduction of an extra day every four years, as well as to other factors, the equinoxes and solstices do not necessarily occur on the twenty-first day of the month concerned, nor do they occur at the same hour each year. For convenience, we arbitrarily use March 21 and September 21 as the dates of the equinoxes and June 21 and December 21 as the dates of the solstices. Actually, the autumnal equinox occurs most often on September 23, and the winter solstice occurs most often on December 22.

level, it will have a temperature of only −15° instead of −75°; this condition is represented by the line *B*. It should be noted that this retardation rate of 3 degrees per 1,000 feet is not a constant rate but only an average; it could be as low as 2 degrees or as high as 4 degrees per 1,000 feet. It should also be observed that, since descending air is warmed, there could be no condensation in a mass of descending air because condensation occurs as a result of cooling, not warming. On the other hand, there could be evaporation if water were present in the descending air, as in the case of falling rain, particularly in connection with a thunderstorm; this topic is discussed in the next chapter. Since evaporation is a heat-consuming, and therefore a cooling, process, the descending air's rate of warming could also be retarded—retarded so much, in fact, that it could be as cold upon reaching the ground as it was at the start of its descent, e.g., as when cold air sweeps out in front of a thunderstorm. Because of the dynamic aspects of heating and cooling and because of condensation and evaporation, the ramifications of the subject are so great that further discussion at this point would serve to complicate rather than to simplify.

Vertical Temperature Gradients

The fact that Mount Kenya, situated almost on the equator in Africa, is permanently snow-capped is an indication that temperature decreases with increase in height, and it certainly cannot be assumed that the below-freezing temperatures at this elevation are due primarily to the cooling of rising air currents. Furthermore, observations based on the use of kites, sounding balloons, and radiosondes indicate there is also a fairly uniform decrease of temperature with increase in elevation in the *free air*—air not in contact with the ground; here again this cooling cannot be attributed entirely to rising currents because there must be descent of air to compensate. It has been found that the rate of temperature decrease up to a certain limit is, on the average—considering day and night, summer and winter, high latitude and low latitude, over land and over water—approximately 3.5 degrees per 1,000 feet (0.65 degree C per 100 meters).[1] This rate is called the *normal lapse rate*, or *vertical temperature gradient*. Strictly, the term "lapse rate" applies to any vertical decrease of temperature; we may refer to dry adiabatic lapse rates, wet adiabatic lapse rates, or simply average or normal lapse rates. When temperatures decrease with increase in altitude, they are considered as *positive lapse rates;* if temperatures *increase* with increase in altitude, the lapse rates are said to be *negative*, or *inverted*.

Line *C* in Figure 5–4 represents the average decrease of temperature with increase in altitude; in this instance, the temperature at the ground is taken as 94°. The temperature at the 20,000-foot level would be about 24°, and at the 30,000-foot level about −11°. It is important to note the difference between adiabatic lapse rates and normal or average lapse rates. Adiabatic changes in temperature are the dynamic result of changes of pressure on the air which is actually ascending or descending. Normal or average lapse rates do not necessarily involve actual ascent or descent of air. For example, suppose you were to make a series of balloon ascensions at various places, at various times of day, and at various seasons and that you were to make temperature recordings every 1,000 feet of ascent; you would probably find, upon averaging all your observations at each level, that the temperature drop would be close to 3.5 degrees per 1,000 feet. If we consider only the 12 columns of figures in Table 5–1, we find that the average decrease, including both summer and winter,

[1] With few exceptions, one being in connection with line *C* of Fig. 5–4, the approximate rate of 3 degrees per 1,000 feet is used in this book because estimating temperatures at various altitudes is much easier when the temperature at any given altitude is known or assumed. For example, if we say the average annual temperature in Denver at approximately 5,000 feet elevation is 50°, then we can reasonably assume that the average annual temperature on Longs Peak or Mt. Evans, each at approximately 14,000 feet elevation, would be fairly close to 23° (3 × 9 thousands, or 27 degrees lower). The computation is simple and easily made, and it is reasonably accurate.

is nearly 3.2 degrees per 1,000 feet, the decrease being three- or four-tenths of a degree greater in summer than in winter.

Fig. 5–4. Graph to represent various temperature conditions with increase in elevation.

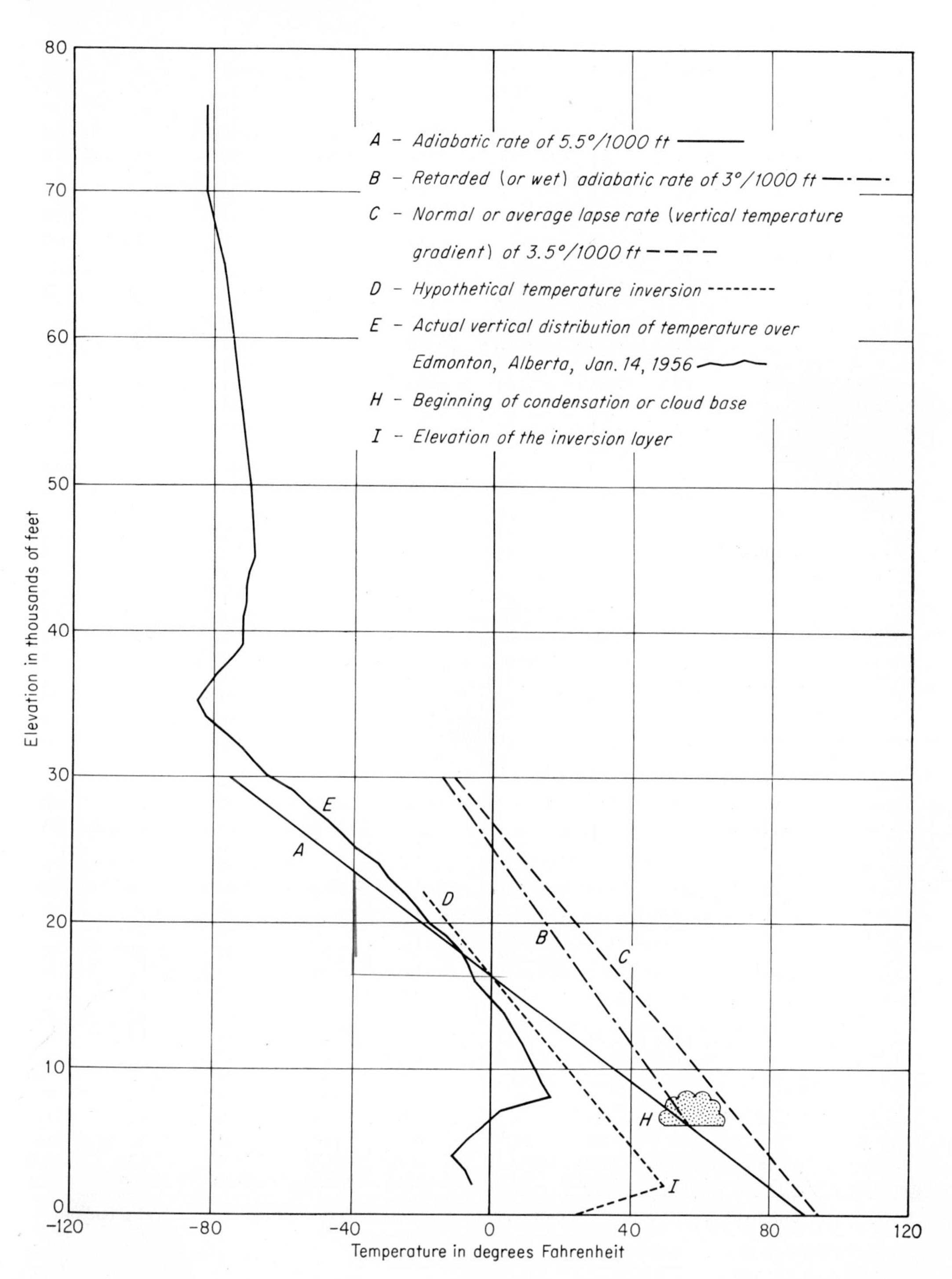

*Table 5-1. Radiosonde Observations of Temperature for Selected Stations**

Elevation, ft.	Balboa C.Z.	San Juan P.R.	Washington D.C.	Edmonton Alta.	Goose Bay Newf.	Cold Bay Alas.	Balboa C.Z.	San Juan P.R.	Washington D.C.
85,000	−70					−65			
80,000	−74					−60			
75,000	−80	−88	−74	−82		−59			
70,000	−86	−88	−74	−82		−59	−80	−63	−67
65,000	−94	−94	−78	−77	−85	−61	−85	−77	−72
60,000	−103	−116	−73	−75	−83	−64	−97	−88	−77
55,000	−109	−111	−72	−72	−81	−64	−102	−97	−76
50,000	−98	−102	−69	−70	−82	−60	−105	−92	−74
45,000	−87	−88	−65	−69	−85	−58	−92	−81	−75
44,000	−82	−85	−64	−70	−85	−58	−89	−79	−75
43,000	−79	−79	−63	−71	−85	−59	−84	−78	−76
42,000	−75	−76	−62	−71	−85	−60	−81	−76	−79
41,000	−72	−75	−61	−72	−91	−61	−76	−71	−83
40,000	−68	−71	−60	−72	−97	−62	−72	−69	−84
39,000	−63	−69	−60	−72	−96	−63	−68	−67	−77
38,000	−58	−63	−60	−75	−92	−66	−64	−65	−69
37,000	−56	−58	−60	−79	−88	−80	−58	−62	−67
36,000	−53	−53	−60	−82	−81	−78	−54	−58	−60
35,000	−49	−47	−60	−85	−78	−75	−49	−54	−54
34,000	−45	−41	−60	−82	−73	−71	−46	−50	−52
33,000	−40	−32	−61	−77	−68	−68	−41	−47	−49
32,000	−34	−31	−63	−73	−65	−65	−35	−43	−43
31,000	−27	−26	−64	−69	−61	−62	−31	−38	−39
30,000	−22	−22	−62	−65	−56	−58	−28	−33	−35
29,000	−19	−20	−56	−58	−51	−55	−23	−29	−30
28,000	−15	−18	−52	−53	−48	−51	−20	−24	−23
27,000	−13	−17	−48	−48	−43	−49	−15	−20	−18
26,000	−8	−15	−43	−43	−37	−45	−9	−17	−15
25,000	−6	−12	−38	−39	−31	−40	−6	−13	−11
24,000	−3	−8	−34	−33	−28	−36	−2	−7	−6
23,000	−1	−3	−29	−30	−23	−32	2	−3	−5
22,000	5	3	−25	−25	−19	−27	6	5	−2
21,000	10	8	−22	−21	−15	−24	10	8	2
20,000	12	14	−19	−18	−11	−19	14	11	5
19,000	19	18	−14	−13	−8	−15	16	15	9
18,000	24	19	−11	−9	−4	−12	20	18	13
17,000	28	21	−6	−7	−1	−7	23	22	18
16,000	31	24	3	−5	3	−3	26	25	22
15,000	31	27	2	−1	6	2	30	28	25
14,000	32	28	6	3	9	5	33	31	29
13,000	35	30	10	5	13	10	38	34	32
12,000	38	34	15	8	15	14	43	38	35
11,000	42	39	18	10	18	16	47	44	35
10,000	47	43	21	12	20	18	49	50	37
9,000	48	47	25	14	22	19	52	53	40
8,000	48	50	28	16	24	20	55	57	45
7,000	49	53	31	2	26	24	57	61	50
6,000	53	57	34	−2	28	27	60	64	55
5,000	57	62	37	−7	29	29	63	63	60
4,000	60	66	32	−11	30	32	66	63	66
3,000	64	69	31	−7	23	36	69	70	71
2,000	68	71	33		23	39	72	71	74
1,000	70	74	38		25	31	74	76	79
Surface	78	70	40	−5	27	18	76	82	74
Alt. of Sta.	21	82	65	2219	144	99	21	82	65
Date	1-13-56	1-13-56	1-13-56	1-14-56	1-14-56	1-14-56	7-13-56	7-13-56	7-13-56
Tropopause	57,000	59,300	63,400	35,000	39,300	37,000	49,000	53,000	40,900

* Most of these figures are not actual soundings but are taken from plotted curves of the actual soundings.

Edmon- ton Alta.	Churchill Man.	Thule Grnld.
		−37
		−38
		−40
−58	−49	−41
−58	−52	−42
−58	−53	−44
−58	−55	−46
−57	−56	−47
−53	−57	−48
−52	−57	−49
−55	−58	−50
−58	−58	−51
−61	−58	−52
−63	−59	−53
−66	−60	−54
−69	−64	−54
−66	−68	−56
−63	−72	−63
−62	−68	−68
−58	−65	−64
−51	−62	−61
−48	−57	−59
−43	−53	−56
−40	−48	−51
−36	−43	−49
−31	−40	−44
−24	−36	−40
−21	−33	−37
−16	−27	−34
−11	−24	−28
−5	−20	−23
−2	−16	−18
2	−11	−15
5	−8	−11
9	−3	−8
10	1	−4
14	6	0
18	9	4
20	13	8
24	15	10
29	17	14
32	18	18
36	20	20
41	24	22
44	27	20
47	30	24
50	33	27
53	38	30
56	42	33
57	47	36
58	49	40
	53	43
	59	47
57	64	51
2219	115	36
7-13-56	7-13-56	7-13-56
38,000	37,000	35,300

At the equator, vertical decreases of temperature continue up to about 12 miles; at or near the poles, the limit is only 4 to 6 miles. High over the equator, the temperature averages probably 50 degrees lower than over the poles. In this regard, compare the data in Table 5–1 for San Juan and Goose Bay in January and for Balboa and Thule in July. Beyond these altitudinal limits of 12 miles over the equator and 6 miles over the poles, the temperature decreases only slightly, if at all, or it may actually increase a little (note line *E* in Figure 5–4). It is only in the lower stratosphere that such fairly regular and uniform temperature conditions seem to exist, so the lower stratosphere is said to be essentially *isothermal* (having equal heat or temperature).

Even though cold air is normally denser or heavier than warm air, the cold air at high altitudes does not continually settle down to the earth's surface and force the warmer air aloft because, as has already been shown, the decreased pressure at high altitudes more than compensates for the decreased temperature in its effect upon the density of the air.

Inversions of Temperature

If for any reason there is an increase in temperature with increase in altitude—a negative lapse rate—an *inversion of temperature* exists. In Figure 5–4, the line *D* represents an inversion of temperature in which the temperature at the ground is taken as 25°; the temperature is shown as increasing to 50° at 2,000 feet and then as dropping again at the normal lapse rate to 22° at the 10,000-foot level and to −13° at the 20,000-foot level. Inversions are quite common in some localities, particularly in valleys in fall and winter. In Table 5–1, note the indicated inversions for January: Washington from 3,000 to 7,000 feet, Edmonton from 4,000 to 8,000 feet, Goose Bay from 3,000 to 4,000 feet, Cold Bay from the surface to 2,000 feet, and San Juan from the surface to 1,000 feet. In July there was also a distinct inversion in Washington from the surface to 2,000 feet. The winter inversion at

Edmonton is graphically indicated by line *E* in Figure 5–4; note also the position of the tropopause at approximately 35,000 feet, or 7 miles.

Local inversions are likely to occur on any clear, calm, and cool night in hilly or mountainous terrain. Since the hillsides usually cool faster after sundown than the air in the open valley, the air lying close to the valley walls also cools, by both conduction and radiation, faster than the air above the valley floor. Inasmuch as cool air is denser than warm air, the cooler air along the hillsides slowly drains or slides down into the valley bottom. By three or four o'clock in the morning, the difference in temperature between the air on the valley bottom and at the tops of the hills may be considerable, the bottom lands thus being subject, on occasion, to frosty or even freezing conditions when the uplands may be well above freezing. For this reason, agriculturists frequently plant hardier crops in valleys and more exacting ones on the gentle slopes above the valley floor; orchardists in particular are well aware of this weather phenomenon, and so they set their trees on the slopes rather than in valley lowlands. If one were to ascend straight up from the valley bottom when an inversion occurs, he would find the temperature increasing up to a certain height and then eventually decreasing under normal lapse-rate conditions. Since the cooler air underlies the warmer air, there is no disposition for the air on the bottom to rise; thus inversions signify stable air as well as negative lapse rates.

Or suppose that a thick layer of warm air at an elevation of 1,000 feet drifts in. The lower air would have no tendency to rise because it is cooler, yet an inversion would exist which would not be encountered until one had risen to the height of the warmer layer of air. The same stable inversion occurs when a cool mass of surface air is transported along the ground while a warmer mass or layer is present above it. Because inversions of temperature are more frequent in the cool season, it is obvious that lapse rates are less in winter than in summer; they are also less by night than by day. *Air drainage* down a slope takes place in spite of its theoretical warming by compression because the air on the upper slopes is so much colder at the start and continues to lose so much heat by radiation on the way down that it may reach the valley floor still colder than the air already there.

Convection

It has been shown that air near the earth's surface is heated mainly by absorption of the long wavelength radiation from the ground and to some extent by conduction and that the amount of such heating depends in part upon the character of the radiating surface. On a summer day with bright sunshine, the air over an open sandy plot becomes greatly heated and expands more than the surface air over some nearby cornfield, meadow, or forest. The result is that the air over this sandy plot is pushed up by the cooler, denser air of the surroundings, or by cooler air which is descending from aloft. Such a transfer of heat by an upward movement of the heated substance itself is termed *convection* and always occurs when one area becomes more heated than an adjoining area. If the heating is intense enough, the air will continue to be pushed up until it attains the temperature of the neighboring air at the same level. Since there is always some moisture in the air, there would eventually be a point where condensation would occur and clouds would form. From that point on, the decrease in temperature would be the retarded, or wet, adiabatic, and the air would continue to rise—if it had time before late afternoon cooling began and if it was not mixed with other air—until it reached the tropopause, the boundary zone between the troposphere and the stratosphere, the troposphere being determined largely by the limit of convection. Some mixing, however, is likely to take place, thus a convective column usually extends only a few thousand feet above the ground. Occasionally heating at the surface may be so intense that the convective column, aided by the latent heat of condensation, will extend upward a mile or more, thereby causing a cloud of considerable extent to develop. This cloud, with all its attendant phenomena, is called a *thunderhead*, a subject discussed in the next chapter.

There is another type of convection in addition to that resulting directly from surface heating. Air which is forced, by whatever cause, to move horizontally may pass over ground which is quite rough. The result is that the air is forced to rise in some places and to descend in others. Where the ascent is high enough, cloudlets form. This is characteristic in country where the wind happens to be blowing perpendicularly to a series of ridges and valleys. This is a type of *turbulence* and is especially marked on a windy and sunny day because of convective irregularities and the buoyancy of the heated air.

If the topography is very rough—if it is marked by some high hills or mountains—the air, in passing over, is forced to rise to such great heights that clouds form on a large scale and rain or snow results. Such precipitation is said to be *orographic*. There are other types of convection associated with large mass movements of air in all of which adiabatic changes of temperature are present. All these types of convection, and the various weather conditions resulting from them, will be explained later. They are simply mentioned here because they play such an important part in accounting for the vertical decrease of temperature.

Reasons for Vertical Temperature Gradients

More than 50 per cent of solar radiation is in luminous rays or in rays of even shorter wavelengths, and these pass through the atmosphere with little loss; in fact, fully one-third of the radiation which enters the outer layers of the atmosphere passes through it and reaches the earth's surface where it is absorbed. As pointed out earlier, approximately one-third of the total radiation is completely lost through reflection and scattering, leaving the remaining one-third, principally long wavelength radiation, to be absorbed directly by the atmosphere. Almost all the shorter wavelengths that reach the earth's surface and are absorbed are immediately given back to the atmosphere as long wavelength terrestrial radiation. It is evident, therefore, that the atmosphere nearest the earth's surface receives the greatest amount of heat; thus there is necessarily a decrease of temperature with an increase in altitude.

Furthermore, the amount of heat which a body may receive is strictly limited by the heat available, while the amount of radiation which a body can give out is practically unlimited since absolute zero, the temperature at which heat supposedly does not exist, has never been reached. It has been found that insolation increases about 20 per cent for every 2-mile increase in altitude, while radiation increases about 40 per cent in the same distance. Thus radiation increases at a much more rapid rate than insolation does with an increase in altitude. The upper atmosphere, therefore, gives out more heat than it receives, so that the higher one goes, the lower the temperature.

As indicated in the preceding section, convection in the atmosphere is also an equally important part of the explanation for vertical decreases in temperature: heating at the ground causes the expansion and ascent of the air in contact with it and the consequent cooling of the ascending air.

Stability and Instability

Stability

When air is in a state of equilibrium—when it tends to resist vertical movements—it is said to be *stable*. A stable condition of the atmosphere is often compared to the stability of a cone resting on its base: if the cone is tilted slightly, it returns to its original position when the tilting force is removed. When air near the ground is cold, when its water vapor content is low, and when its density is likely to be greater than that of the overlying air, considerable force would be required to push up that mass of cold air. Under such conditions, the air is said to be in a state of *stability*. Obviously, this is the condition when there is an inversion of temperature. Now suppose that some force were able to push this mass of air aloft for some distance. The air would cool adiabatically at the rate of 5.5 degrees per 1,000 feet, and would become still colder than it was at the ground; consequently, it would tend to sink back to the ground as soon as the force causing its upward movement was withdrawn. Air, therefore, is said to be stable whenever the vertical decrease in temperature—the lapse rate—is less than 5.5 degrees per 1,000 feet. Under such conditions, convection does not take place, and precipitation is practically nil. On the other hand, radiation could occur readily through the cold dry air, resulting in considerable condensation in the form of dew, frost, or fog.

Instability

When air can be pushed up easily—when it does not tend to resist vertical movement—and when, if moved, it will continue its vertical movement for an indefinite period, it is said to be *unstable*, just as a cone would be unstable if it were resting on its apex instead of its base. *Instability* results if air over heated ground is warmed enough to cause its density to become less than that of the surrounding air; the latter would push under the warm air and force it upward. The air over the heated ground would continue to rise as long as cooler air was moving in to push it up, or until the rising air reached a level where its own temperature and that of the free air were the same. This process might require some hours because condensation would probably occur, thereby retarding the adiabatic rate to a point which might be less than the lapse rate of the surrounding air. Instability is expected on a sunny day when there is considerable heating of the ground locally, indicating a rapid decrease of temperature with an increase in altitude—high lapse-rate conditions—so that strong local convection will result. Whenever a large quantity of humid marine air is carried over a warm continental area, strong heating of the surface air takes place, resulting in strong convection and also in an unstable condition. Furthermore, whenever humid air is forced to ascend mountain slopes, the rapid condensation of water vapor adds so much latent heat to the ascending air that instability results. In any case, instability occurs whenever lapse rates in the free air are greater than 5.5 degrees per 1,000 feet. And if condensation occurs, then the air becomes still more unstable because the adiabatic rate is retarded and there is a still greater disparity between the lapse rate in the free air and that in the ascending air. Instability usually gives rise to precipitation, and the greater the instability, the greater will be the precipitation with the probability of hail, lightning, thunder, and strong surface winds locally.

At what altitude will the vertical movement cease, or to put it another way, at what altitude will instability give way to stability? The answer is this: at the point where the temperature of the surrounding air—the free air—is the same as, or greater than, the temperature of the vertically moving air.

The heating effect of condensation on the column of upward-moving air gives rise to what is frequently referred to as *conditional instability* because the stability is conditioned by the degree of humidity. Hence the rule: rising air in which moisture is being condensed is stable if the lapse rate of the free air is less than the retarded, or wet, adiabatic rate, and unstable if it is greater.

Summary

Stable conditions are usually found in quiet air where there is little or no precipitation but often radiation fogs, dew, or frost. In some instances, warm, moist air may drift in over cold, residual stable air on the ground, causing condensation to occur in the warmer air aloft and resulting in clouds, light rains, drizzles, or even in ice or glaze storms if ground temperatures are below freezing. Such conditions, however, do not violate the principle of stability as applied to the cold stationary air on the ground. Instability usually results in winds of varying strength, in much cloudiness and precipitation, and frequently in other and, at times, violent weather conditions.

6

Moisture

Variability and Measurement

We have already seen that the amount of water vapor in the air is a highly variable quantity and that the amount which the air can include depends upon the temperature, the possible water-vapor content being approximately doubled with every increase of 20 degrees. Table 3-2 indicates the amount of moisture a cubic foot of air is able to hold at various temperatures. The amount of moisture in the air depends also upon the character of its source—ocean, desert, grassland, forest, tundra, or ice and snow. If the air is over ice, water, or extremely wet land, so much heat is used in evaporating the water that the air temperature does not reach so high a level as over dry land where the amount of water available for evaporation is less. Thus the air in arctic and mid-latitude marine climates usually has a higher degree of saturation than air in continental climates, although the actual water-vapor content of the air in the former may not be so great as in the latter. The moisture in the air also changes with altitude, the average amount at 6,500 feet altitude in the free air being only one-half that at sea level.

Humidity

Absolute Humidity

It is sometimes desirable to measure the actual water-vapor content of a given sample of air. This is commonly done by determining the weight of the water vapor in a given volume of air. It may be expressed in grains per cubic foot, in grams per cubic meter, or in any other unit of weight per unit of volume. This expresses the *absolute humidity*. Since the mass (or, as we will say, *weight*) of a given volume of gas (or air) is proportional to the pressure of the gas (or air) at a constant temperature, it follows that *vapor pressure*—usually expressed in barometric units, such as inches, centimeters, or millibars—is an indirect measure of the absolute humidity. Vapor pressure may be defined as that portion of the total pressure of a volume of air that is due to the water vapor in the air. If the air is *saturated*—that is, if it contains all the moisture possible—the pressure of the water vapor is termed *saturation vapor pressure*. It is evident, therefore, that the vapor pressure for saturated air is always the same for the same temperature. In the table of dew points in the Appendix, the actual saturation vapor pressures at various temperatures are shown in inches. The terms "vapor pressure" and "saturation vapor pressure" are commonly used by the U.S. Weather Bureau and by professional meteorologists in general.

Absolute humidity has a physiological significance for it determines the efficiency of the water-evaporating phase of our body-cooling mechanism. It is also necessarily an important factor in the amount of precipitation any locality may receive: where there is little moisture in the air, the precipitation will be light unless larger quantities are imported.

Specific Humidity

For upper-air observations, *specific humidity* has come into common use. It is defined as the weight of water vapor per unit *weight* of air (including the water vapor) and is usually expressed in grams of water vapor per kilogram of air; it is sometimes valuable because when a parcel of air is expanded or compressed, the total pressure and the vapor pressure change in the same ratio. Specific humidity, like absolute humidity, is constant unless there is an addition or subtraction of water vapor.

Relative Humidity

If the amount of water vapor in the air at a given temperature is divided by the amount of water vapor which the air would have at that temperature if saturated, the result is a decimal or percentage to which is given the name *relative humidity*. It is always expressed in per cent. For example, suppose a quantity of air at a temperature of 70° contains 4.033 grains of water vapor per cubic foot. According to Table 3–2, it could contain 8.066 grains of water vapor per cubic foot if it were saturated at the same temperature. The relative humidity would be the ratio of those two numbers: 4.033 divided by 8.066 which would equal one-half or .50 or 50 per cent. The absolute humidity would be 4.033 grains per cubic foot.

Relative humidity is rarely measured in this manner; tables are used instead, as will be shown later. In giving weather forecasts, the weatherman usually refers to relative humidity. If he does not give it as an actual percentage, he uses a descriptive term such as "high," "low," or "very high," based on percentage.

Condensation

The Dew Point

Air at a given temperature is said to be *saturated* when it includes all the water vapor possible. Even a slight cooling of saturated air will cause *condensation* of a part of the moisture. If unsaturated air is cooled sufficiently, there will be a point at which condensation will begin. The temperature at which water vapor begins to condense is known as the *dew point*. On almost any clear night in autumn, radiation from grass and leaves—especially those near the ground—is relatively rapid and the temperature of their surfaces soon reaches the dew point of the air in contact with them, resulting in the formation of *dew*. If the dew point is at the freezing point or lower, then condensation[1] is in the form of *frost*. Every mass of air has its own dew point—the temperature to which it must be reduced in order to start the formation of visible moisture. Under certain quiet conditions, and when the air may contain only small dust particles (condensation nuclei), condensation may *not* occur when the dew point is reached or until the air is cooled several degrees below the dew point; the air is then said to be "super-saturated."

[1] Some meteorologists use the term *sublimation* rather than "condensation" when water vapor is transformed directly into frost, ice, or snow.

This may account for the very sudden appearance of clouds, particularly high clouds, where temperatures are below freezing.

Measuring the Dew Point

The temperature of the dew point may be measured readily and directly by means of a shiny tin or aluminum cup of water progressively cooled by ice. The dew point is, however, obtained most often from tables similar to the abbreviated one shown in the Appendix. To use such tables we need to know the temperature of the air—termed the *dry-bulb temperature*—and the temperature to which evaporation cools the well-ventilated moist surface of a thermometer bulb—termed the *wet-bulb temperature*. The latter temperature is found by whirling a thermometer which has its bulb covered with a close-fitting piece of wet muslin. When two thermometers, one with a wet bulb and one with a dry bulb, are attached to a frame, the instrument is called a *psychrometer* (cold measurer); and when this frame has a handle by which it can be whirled, it is called a *sling psychrometer* (Figure 6–1). Some Weather Bureau stations have *telepsychrometers* that automatically record, by electrical means, both wet- and dry-bulb temperatures. Still another type of psychrometer (Figure 6–2) circulates the air around the wet- and dry-bulb thermometers by means of an electrically driven motor; like the telepsychrometer, this device can be installed in a thermometer shelter. With any type of psychrometer, the greater the difference between the dry- and wet-bulb readings, the more rapid the evaporation and, for a particular air temperature, the drier the air, relatively. Wet-bulb temperatures are frequently referred to as *sensible temperatures* because they reflect more accurately temperatures as they are felt by the body.

The Weather Bureau is in the process of establishing "automatic teletypewriter weather observer" stations which have a device known as a *dewcell* mounted on a thermometer shelter;

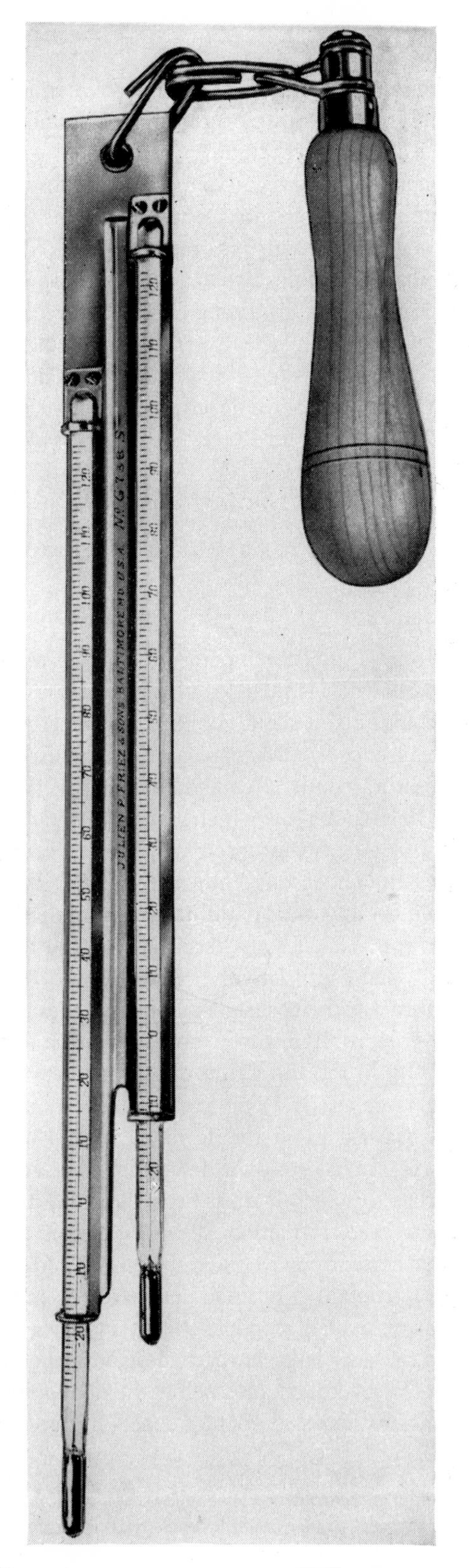

Fig. 6–1. Sling psychrometer. (Courtesy Bendix-Friez.)

this instrument registers the dew point directly. It is interesting to note that these automatic observer stations require no personnel other than that required to service the instruments. The stations record not only the dew point but also precipitation, temperature, wind direction, wind speed, visibility, cloud heights, and other data. The information is fed into a coder and ultimately into a teletypewriter system. The observations are transmitted whenever the control station in the circuit uses certain signals, and the station makes a punched tape of each observation transmitted, thus making a permanent record available. The great advantage of automatic stations is that they can be used in remote areas where it is difficult for people to live.

Measurement and Distribution of Relative Humidity

Dry-bulb readings are always the same for the same air temperature; wet-bulb readings vary with the amount of moisture in the air. Thus the difference between the two readings gives some indication of the humidity of the air. The difference alone, however, is insufficient; but when used with air temperatures, it is very significant. These two values are not as easy to keep in mind as one value—that of relative humidity—which takes both factors into account. Tables from which relative humidity can be determined have been worked out in great detail; a simple one is shown in the Appendix. Dry-bulb temperatures are shown in the left-hand column, differences between the dry- and wet-bulb readings (depression of the wet bulb) across the top, and relative humidities in the other columns. Thus, if the dry-bulb reading is 60° and the wet-bulb is 50°, the depression of the wet bulb will be 10 degrees. Proceeding horizontally along the 60-degree line to the column of 10 degrees depression, one finds that the relative humidity is 48 per cent.

Fig. 6–2. Electric motor-driven psychrometer. (Courtesy Bendix-Friez.)

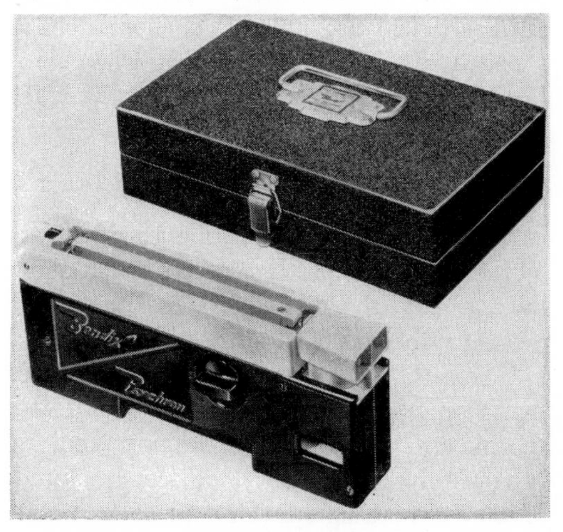

Relative humidity is frequently, though somewhat inaccurately, measured directly by an instrument known as a *hygrometer* (moisture measurer). A common form of hygrometer consists of one or more strands of a child's hair, freed from oil and so arranged that their contraction and expansion with changes in humidity can be shown by an indicator. The hair lengthens and shortens in direct proportion to the relative humidity as long as the hair is kept clean. A hygrometer which makes a continuous record of relative humidity is a *hygrograph* (Figure 6–3).

Absolute, specific, and relative humidity are important in the science of meteorology. Relative humidity determines whether or not there can be *any* precipitation, while absolute or specific humidity determines *how much* there will be.

Maps of average noonday relative humidity for the United States for January and July are shown in Figure 6–4,[1] In spite of the fact that,

[1] It should be noted here that the average distribution of weather elements indicates climate rather than weather, but maps such as these illustrate weather principles so well it seems desirable to present them as the elements are discussed. The same applies to other maps of this type shown in Figs. 6–18, 6–21, 6–30, and 6–32.

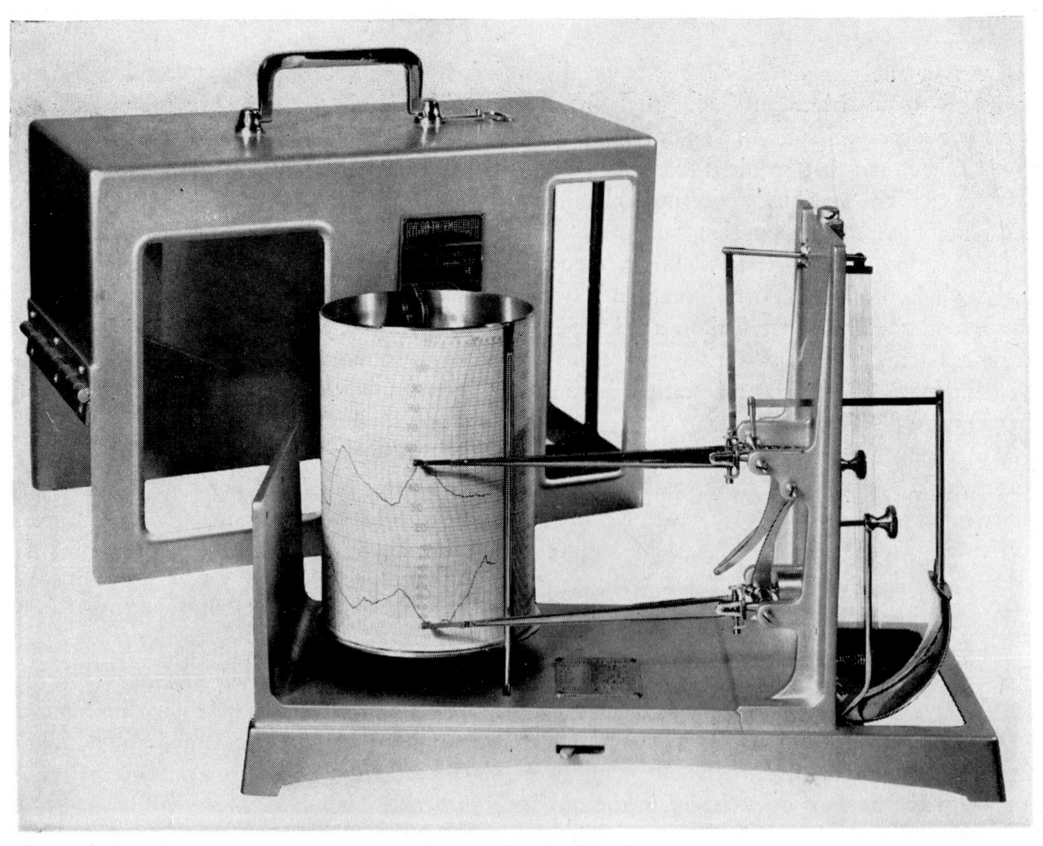

Fig. 6–3. Combination hygrothermograph. (Courtesy Bendix-Friez.)

on the average, absolute humidity is lower in winter than in summer, relative humidity generally takes the opposite course and is higher in winter than in summer, with the low temperatures of the cold season more than compensating for the reduced water-vapor content of the air. Probably the most notable exception to this rule is found along the coast of California, although winter is the rainiest season there; the higher incidence of fogs in the warm season accounts for this seeming anomaly. In both January and July, relative humidity is highest where there are large bodies of water, as along the Atlantic, Gulf, and northern Pacific coasts and around the Great Lakes. In the northern Rocky Mountain states, the high relative humidity in January is a reflection of the low sun, low temperatures, and high frequency of cyclonic storms in that area.

Forms of Condensation

Clouds

If the cooling of ordinary air is carried far enough and if condensation nuclei (usually dust particles) are present, condensation always results. We have seen that when air is forced to rise, it expands and is cooled at the regular adiabatic rate until the dew point is reached. Some of the water vapor condenses and *clouds* are formed. There are three general, or basic, types of clouds, distinguishable partly by their form and partly by their height above the ground: *cirrus*, *cumulus*, and *stratus*. To these may be added a fourth, *nimbus*.

Cirrus (Ci), meaning "curl," is recognized by its veil-like, fibrous, or feathery form. It is the highest type of cloud, ranging from approximately 20,000 to 35,000 feet in altitude (Figure 6–5). Cumulus (Cu), meaning "heap,"

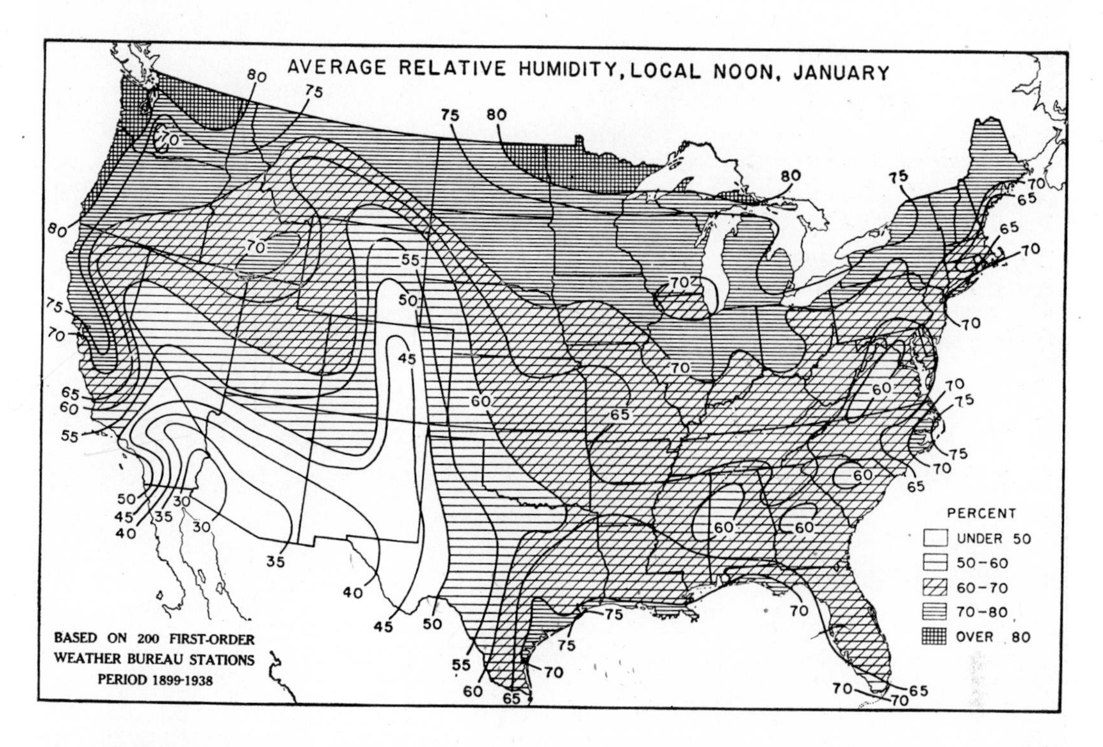

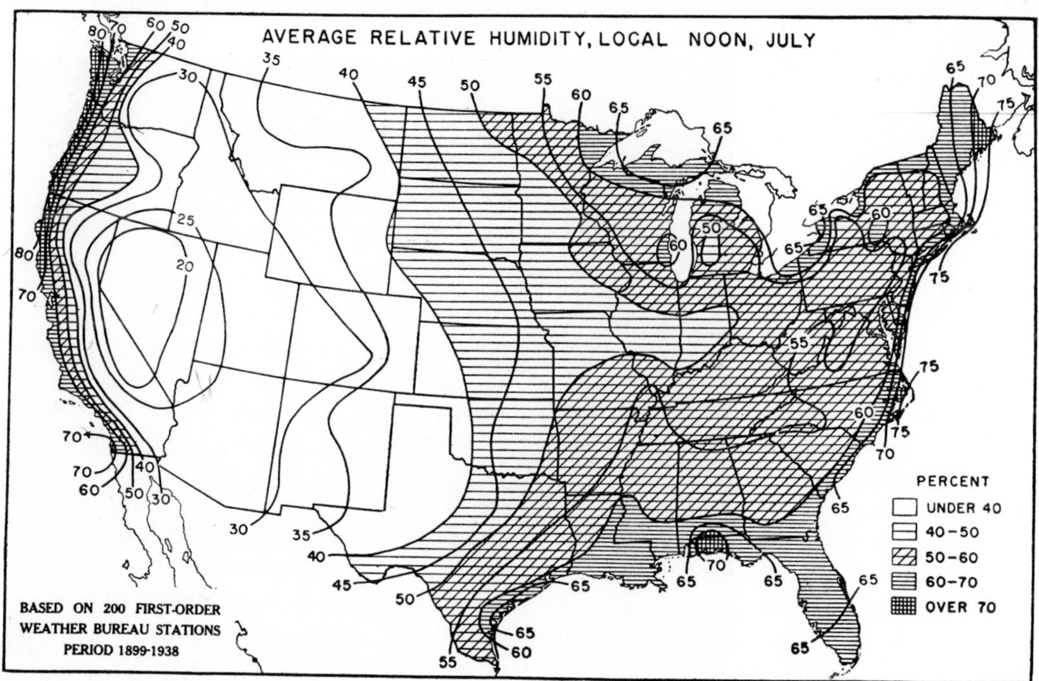

Fig. 6–4. Average relative humidity at local noon for January and July. (Courtesy U.S. Weather Bureau.)

Fig. 6–5. Cirrus clouds. (Photograph by F. Ellerman, U.S. Weather Bureau.)

Fig. 6–6. Cumulus cloud. (Courtesy U.S. Weather Bureau.)

Fig. 6–7. Stratus clouds. (Courtesy U.S. Weather Bureau.)

is the woolly, bunchy cloud with rounded top and flat base (Figure 6–6); it is most common in the summer season and in latitudes where high temperatures prevail, and it always results from convection. While its height is variable, depending upon the relative humidity of the air, its base is usually relatively low. Stratus (St) is a sheet cloud without any other form to distinguish it (Figure 6–7). It is usually lower than cumulus. Nimbus (Nb) is any dark and ragged cloud from which precipitation is, or has been recently, falling, or is threatening to fall. Although there are notable exceptions to the use of the term "nimbus," most persons interested in weather do not refer to it as a single cloud; instead they use it in combinations with other clouds, as is done in the international classification of clouds that is in general use throughout the world. Table 6–1 summarizes the 10 clouds as set forth by the International Meteorological Committee, adding some useful details.

The three types of cirrus clouds represent the high-cloud family. While the pure cirrus is common, the two combination types appear just as often. As their combined names suggest, the *cirro-cumulus* (Cc) (Figure 6–8) and the *cirro-stratus* (Cs) (Figure 6–9) also look like cumulus clouds and stratus clouds respectively but are at higher elevations and are much less dense. All cirrus-type clouds are composed of minute ice crystals and are not considered precipitation makers; that is, precipitation does not fall from them. It is the refraction and reflection of light by the relatively thin layer of ice crystals that cause the common solar and lunar *halos*. Cirrus clouds may appear simply as fleeting clouds which, under relatively clear conditions, are not connected with other forms. In this case they merely indicate sufficient local cooling at high elevations to form ice crystals for a short period. On the other hand, pure cirrus clouds, if coupled with increasing amounts of cirro-stratus and/or cirro-cumulus (mackerel sky) forms, are commonly in the cloud succession on the leading edge (beginning) of middle-latitude storms. In this case they have very definite meaning.

Clouds that occur at middle elevations (approximately 6,500 to 20,000 feet) have the prefix *alto* attached. While they appear to be similar to the cirrus forms, a comparison indicates they are in larger masses. The *alto-cumulus* (Ac) (Figure 6–10), for example, has more distinct edges and is larger than Cc. *Alto-stratus* (As) is thicker and lower than Cs (Figure 6–11); it also shuts out more light and usually appears darker. Instead of forming a distinct ring at a distance from the sun or moon, the alto-stratus forms around the luminary a fuzzy and indistinct edge—called a *corona*—that is caused by diffraction of light by water droplets. Neither the alto-cumulus nor the alto-stratus cloud is an important precipitation maker, although both may produce slight amounts on occasion. Alto-stratus clouds are more likely to produce precipitation than the alto-cumulus, especially as they thicken and become lower.

There are three members of the low-cloud family: the basic stratus cloud (St), *strato-cumulus* (Sc) (Figure 6–12), and *nimbo-stratus* (Ns). Stratus clouds, frequently covering the entire sky in humid middle latitudes, consist of a low uniform layer of cloud which resembles a fog but which occurs at a somewhat higher elevation and usually with an even and distinct lower level of condensation. At times, stratus clouds actually descend to the earth and are then usually classed as *fog*. Similarly, the fog which settles over San Francisco may be seen as a low cloud rolling in through the Golden Gate. Stratus clouds that are scalloped on the bottom, thus having the cumulus in addition to the stratus appearance, are called strato-cumulus (Sc). These also often cover

*Table 6–1. Cloud Families and Types**

Cloud family and height, ft.	Cloud type and abbr.	Phenomena connected with types
High clouds 20,000 to 35,000	1. Cirrus (Ci)	High; icy; fine; "mares' tails"; wispy; sun shines through without making a shadow; ice crystals; do not produce precipitation
	2. Cirro-cumulus (Cc)	Mackerel sky; often forerunners of cyclonic storms; often look like rippled sand; good examples rather rare
	3. Cirro-stratus (Cs)	Thin, whitish veil or tangled web; produce halos
Middle clouds 6,500 to 20,000	4. Alto-cumulus (Ac)	Often separate little woolpacks at a moderate altitude; "sheepback" clouds
	5. Alto-stratus (As)	Fibrous veil or sheet, gray or bluish; look like dark and thick cirro-stratus; produce coronas; may be thin or thick; usually cast a shadow
Low clouds ground to 6,500	6. Strato-cumulus (Sc)	Long parallel rolls pushed together, or broken masses which look soft and gray, but with darker parts; air is smooth above, but strong updrafts occur beneath
	7. Stratus (St)	A low uniform layer, resembling fog but not resting on the ground; can yield a drizzle; chief winter cloud
	8. Nimbo-stratus (Ns)	Low stratus commonly producing continuous precipitation; one of the chief precipitation makers
Clouds with vertical development 1,600 to 35,000	9. Cumulus (Cu)	All cumulus clouds are evidence of convection; the fair-weather cumulus may develop into cumulo-nimbus. Cumulus is billowy, looks like a woolpack; is dark underneath because of shadow
	10. Cumulo-nimbus (Cb)	This is the thunderhead, the second chief precipitation maker. Towering, developing from a simple cumulus; evidence of strong convection, with violent up-and-down drafts; aviators avoid them

* Based on the International Meteorological Cloud Classification.

Fig. 6–8. Cirro-cumulus clouds. (Photograph by E. E. Barnard, U.S. Weather Bureau.)

Fig. 6–9. Cirro-stratus clouds. (Courtesy U.S. Weather Bureau.)

Fig. 6–12. Strato-cumulus clouds. (Courtesy U.S. Weather Bureau.)

large parts of the sky, and the cumulus effect is evidence of some convection. The nimbo-stratus (Ns) cloud is one of the chief precipitation makers. It occurs under fairly stable conditions during cool spells or in the cooler season. It is responsible for the prolonged drizzly rains that soak into the ground slowly, refreshing the vegetation and adding to the level of the underground water (water table).

Two types of clouds develop thickness because of convection. They are always evidence that in them and underneath them are air currents that may be strong, and even disastrous at times, for airplanes. Interestingly enough, the basic type, cumulus (Cu), is frequently referred to as the "fair-weather cloud" (Figure 6–13). While this appellation sometimes appears to be a misnomer, since the cloud *may* develop into a cumulo-nimbus cloud, it is so called because it frequently appears when convection is not increasing noticeably and so is not likely to build up to a great height; furthermore, it is not very common as a cloud type in the cyclonic succession of clouds. Therefore, it is not usually a forerunner of a storm, except sometimes a local thunderstorm. Cumulus clouds are most likely to appear on a moderately warm and moderately dry day, covering sometimes about one-half the sky, only to disappear toward evening. Any typical cumulus cloud will have a flat base because it is formed at the dew-point level in a convective column.

Fig. 6–10. (Above, left.) Alto-cumulus clouds. (Photograph by J. W. Johnson, U.S. Weather Bureau.)

Fig. 6–11. (Below, left.) Alto-stratus clouds. (Courtesy U.S. Weather Bureau.)

Fig. 6–13. Fair-weather cumulus clouds. (Photograph by H. T. Floreen, U.S. Weather Bureau.)

Fig. 6–14. Cumulo-nimbus cloud. (Photograph by H. T. Floreen, U.S. Weather Bureau.)

Fig. 6–15. Alto-stratus mammatus clouds. (*Courtesy U.S. Weather Bureau.*)

Fig. 6–16. Alto-cumulus lenticularis clouds. (*Photograph by Maxwell Parshall, U.S. Weather Bureau.*)

The *cumulo-nimbus* (Cb) (Figure 6–14) cloud is the thunderhead. It develops under unstable conditions and exemplifies well-developed convection. Responsible for hail, lightning, thunder, and torrential downpours at times and given to lesser displays at other times, it is nevertheless frequently viewed with concern and is certainly one of the two chief precipitation makers. Thicknesses of a mile and, in extreme cases, 4 miles or more have been reported. They may cover several square miles in extent.

While the single term "nimbus" (Nb) is still sometimes used by weathermen and airmen to denote a cloud from which precipitation is falling, or is likely to fall, it does not occur alone in the international classification. Its significance is noted, however, since it does appear as part of the two main precipitators (Ns and Cb). The term *fracto*, meaning "broken," is frequently applied to combinations, although it is usually applied only to lower clouds, such as fracto-cumulus (Fc), fracto-strato-cumulus (Fsc), or fracto-stratus (Fs). Other types appearing in limited areas or under special conditions are: mammato, rotor, lenticular, banner, laminated, billow, crest, scarf, and turreted. Two of these are shown in Figures 6–15 and 6–16.

A knowledge of clouds may be readily obtained by outdoor study of the clouds themselves. Such a study could be greatly facilitated by reference to the Weather Bureau publications on clouds; these pamphlets contain complete descriptions and very excellent photographs of the more common cloud forms. They may be obtained for nominal fees from the Superintendent of Documents, Washington 25, D.C.

Measurement and Distribution of Cloudiness

Cloudiness is usually measured by the number of tenths of the sky that the clouds cover. After a little practice, this can be estimated accurately with the eye. The state of the sky is usually designated as follows: if the tenths of the sky covered range from 0 to 3, it is said to be *clear;* from 4 to 7, *partly cloudy;* and from 8 to 10, *cloudy*. Average or mean cloudiness is sometimes and in some countries expressed in per cents of the sky covered.

Cloud heights (or ceilings), so significant in connection with air transport, particularly in landing areas, can be determined accurately by a very sensitive photoelectric device called the *rotating beam ceilometer*. As indicated in Figure 6–17, a beam of light from the *projector* is reflected from the cloud and picked up by the *detector* connected electrically to the *indicator*. Both the projector and detector are rather large and complicated instruments that need to be protected from the weather elements by metal housings or covers; the indicator is connected to the projector by a simple-wire telephone line which permits the indicator in the observatory to be any distance—even several miles—from the projector and detector.

The average annual number of cloudy and clear days for the United States is shown by the two maps, Figure 6–18. Note that in many respects these two maps are quite similar, although it does not follow that if the days are not clear, they must be cloudy; for the United States as a whole probably about one-third of the days should be classed as partly cloudy. The maps do show, however, that the fewest number of cloudy days and the greatest number of clear days are found in the southwest, and that the Pacific Northwest, the Great Lakes area, and northern New England have the greatest number of cloudy days and the lowest number of clear days. An interesting point is that the Rocky Mountain region has only 100, or fewer, clear days while the number of cloudy days are relatively few, indicating this is an area of intermittent or scattered clouds—that is, an area predominantly of partly cloudy days.

Fogs

In these days of high-speed transportation by airplane, train, automobile, and ship, fogs become of such great significance they merit

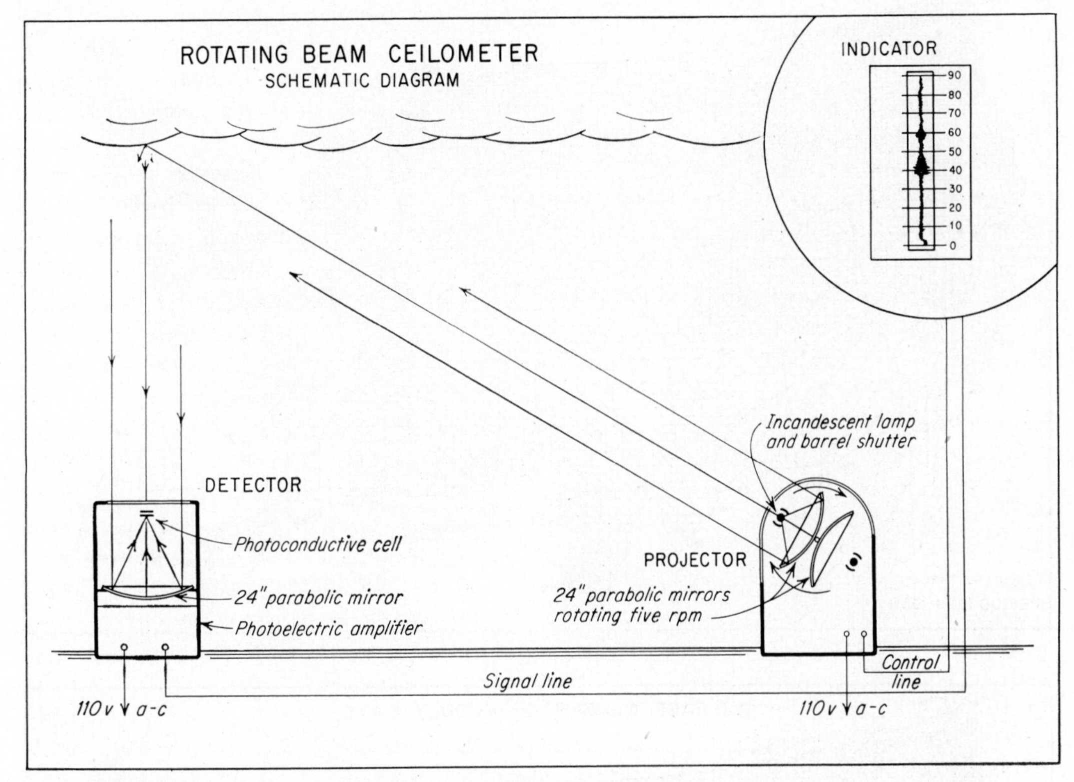

Fig. 6–17. Rotating beam ceilometer. (Courtesy U.S. Weather Bureau.)

more than casual mention as a form of condensation. Fog is frequently defined as "a cloud in contact with the ground (or water surface)," whether over lowlands or on mountains. There are fogs, however, which are not in contact with the ground; in such cases they are referred to as *high fog* or as *high inversion fog*. But it must be admitted that any fog is a type of cloud. Fogs generally have no particular form, structure, or shape. They necessarily develop where air is stable; if air is unstable—that is, if the vertical temperature gradient exceeds the dry adiabatic rate of 5.5 degrees per thousand feet—any condensation occurring will give rise to clouds rather than to fog.

For convenience we may classify fogs under two main headings: *radiation fogs* and *advection fogs*. Radiation fogs result from rapid nighttime radiation either from the ground or from the lower air. They are especially common in autumn in lowlands or valleys, particularly those occupied by streams, ponds, or swamps; they usually occur in quiet weather; and they are relatively shallow—it is not uncommon to see church steeples or the tops of tall buildings rising above the fog level if the observer is on a hill overlooking a valley. Another type of radiation fog is the *high inversion fog* resulting from condensation of water vapor in a mass of warm moist air lying over a layer of cold air near the ground; again, this condition is usually found at night with calm weather. In either of the above types, there is no other cloud layer, and the morning sun soon dissipates the fog.

Advection fogs ("advection" meaning "carried in from elsewhere") occur when warm, moist air rides in over a cold surface, land or water (wind to some extent is therefore implied), and, it should be noted, they may occur at any time of day. This is the type

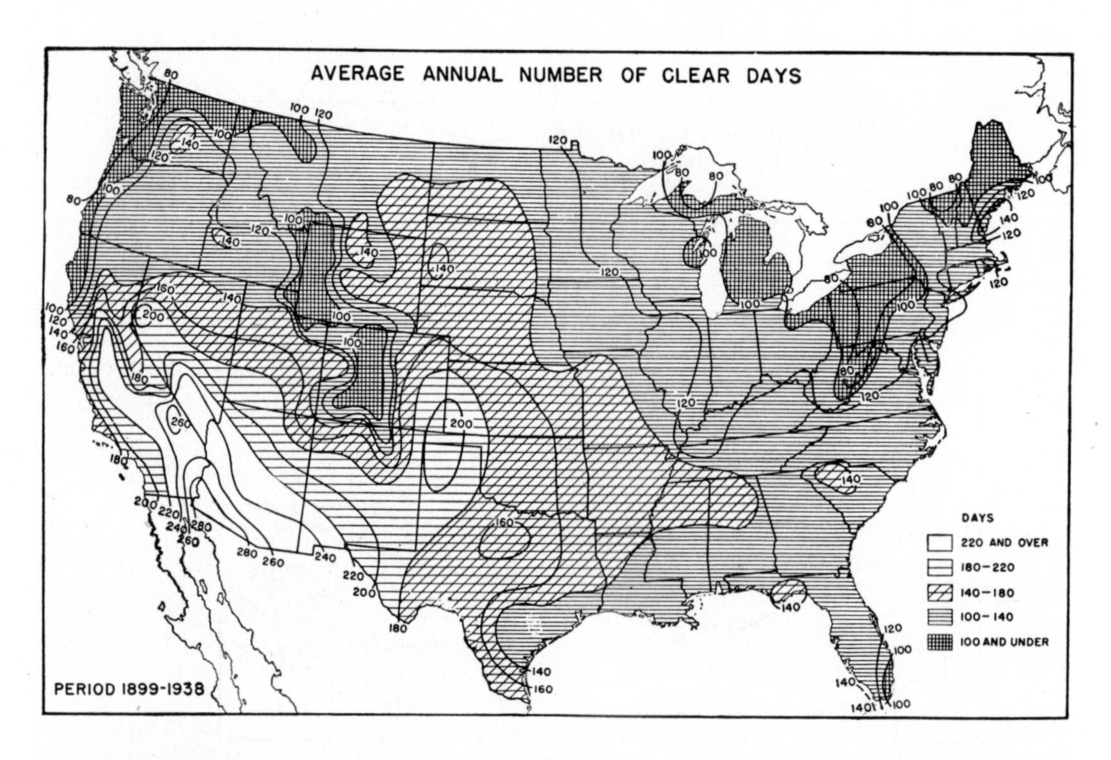

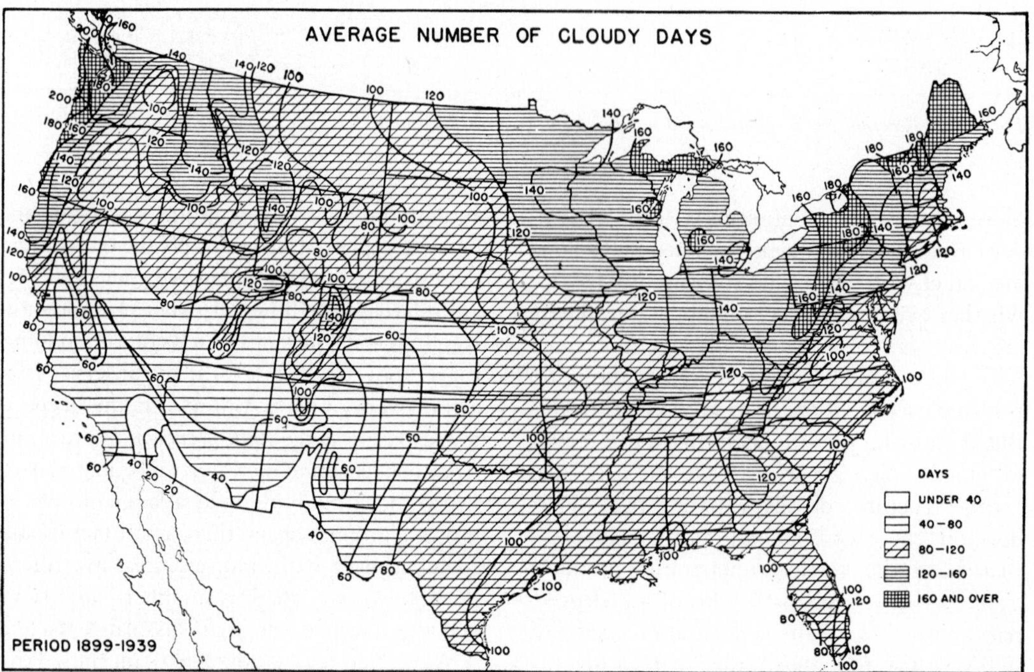

Fig. 6–18. Average annual number of clear days and of cloudy days. (Courtesy U.S. Weather Bureau.)

Fig. 6–19. Fog bank rolling in over Lake Huron. (*Courtesy U.S. Weather Bureau.*)

found over the Grand Banks off southeast Newfoundland where southerly winds bring in warm, moist air that has passed over the part of the Atlantic warmed by the Gulf Stream and then moves on over waters chilled by the icy Labrador Current. The Great Lakes region and many coastal areas are also frequently shrouded in advection fogs (Figure 6–19). Persistent fogs may cover large inland, or even coastal, areas when a slowly moving mass of moist air moves northward over a relatively cool surface. Occasionally the conditions are reversed: a slowly moving cool mass of air moves southward over moist and warmer ground, thereby causing widespread fog with low visibility. There is still another condition: if cold air moves over a warm-water surface, the moisture evaporated from the water may produce saturation in the cold air and cause fog. Since it appears as steam rising from the water, it is sometimes referred to as *steam fog*.

One other type of advection fog merits attention, the *upslope fog*, especially noteworthy on plains abruptly bordering mountains, as along the eastern slopes of the Rockies. Here stable air may drift in from the plains, causing a rapid formation of dense fog even if there is a strong wind.

Wherever there are cold ocean currents, accompanied by an upwelling of colder water immediately offshore—as along west coasts in low or lower middle latitudes—advection fogs are frequent. Moist air from the relatively warm ocean waters blows in over the cold current and the colder shore waters onto the warm land, causing fog to develop over the cold water surface. Along the Pacific Coast, especially in southern California (Figure 6–20), this type of fog is frequently conveyed inland. Upon reaching the coast where the air is unsaturated, the ground portion of the drifting fog is dissipated while the upper portion continues, drifting inland for perhaps several miles. This upper portion is referred to as *high fog*. On the map showing the average number of days with dense fogs in the United States, Figure 6–21, it will be seen that fogs are most frequent along the Pacific Coast, off the coast of the North Atlantic states, and over the northern Appalachians. Fogs are rare in the Rocky Mountain and Great Basin states.

Closely associated with both the radiation and advection fog types are the *smogs* that hover over large cities, particularly industrial cities or those lying in valleys or partially hemmed in by hills or mountains. Smogs may continue for days, causing great inconvenience and discomfort to the populace, and may occur with either the radiation or advection type of fog, although they are generally associated with the latter during long and widespread inversions of temperature. In fact, the smoke near the ground prolongs the inversion, acting as a blanket and preventing the sun from warming the ground sufficiently to cause convectional overturning. The London smog of December, 1952, is said to have caused 4,000 deaths. A recently coined word to indicate another fog type is *smaze*, a combination of smoke and haze. Such words may not be too scientific, but they are highly descriptive.

In certain parts of the arctic—as around the borders of Hudson Bay when the temperature is well below freezing, although the water of the bay itself will not be frozen—fog in the form of supercooled water droplets or tiny ice particles will condense over the water and drift inland; this is referred to as *frost smoke*. In extremely cold weather, the condensation may be in the form of *ice spicules*, a clearly glistening but unpleasant and allegedly harmful phenomenon to those who are exposed to it and forced to breathe the air. A similar type occurs in our western mountain areas and is known locally as *pogonip*.

Still another unusual form of fog is called *arctic sea smoke* because the fog is best observed in arctic seas, especially in the Bering Sea area. It occurs only under conditions of extreme temperature contrasts between land and water surfaces. When rather strong gusts of wind proceed from the cold land onto the much warmer ocean surface, the fog which results is whipped upward into columns that resemble rising smoke. These columns sometimes attain

Fig. 6–20. Advection fog as seen from Mount Wilson, Calif. (Photograph by F. Ellerman, U.S. Weather Bureau.)

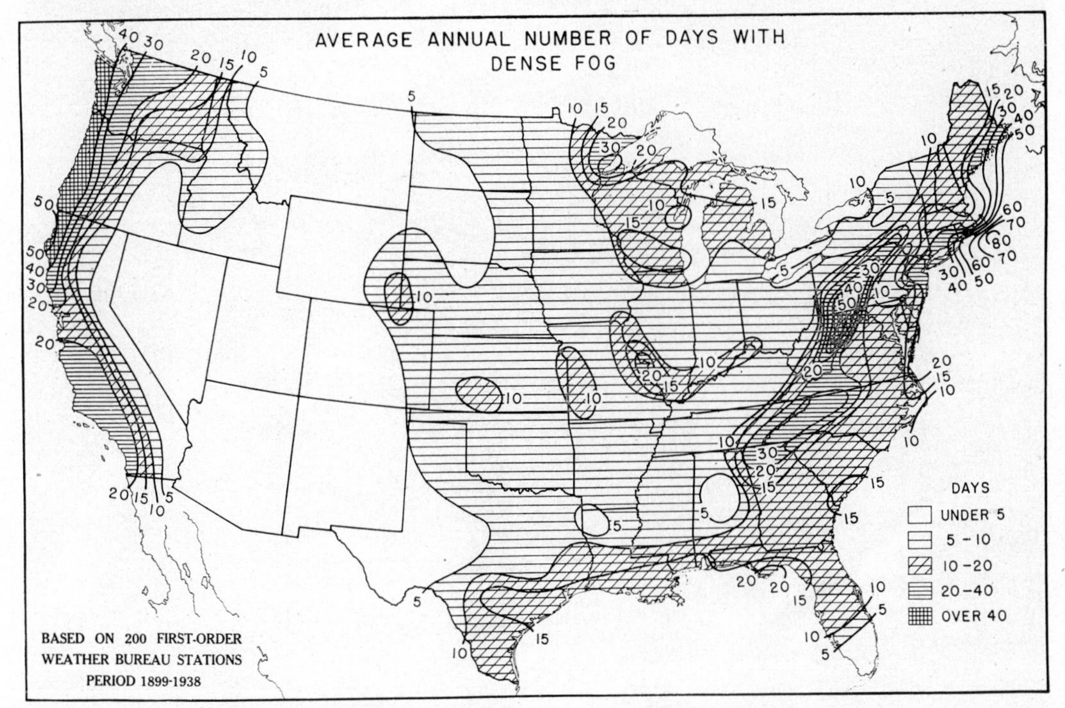

Fig. 6–21. Average annual number of days with dense fog. (*Courtesy U.S. Weather Bureau.*)

heights of 50 feet or more. A good example of arctic sea smoke occurred in Chicago the day after Thanksgiving in 1950. At that time strong westerly and northwesterly winds produced many weird columns swirling upward 30 to 40 feet and extending ½ to ¾ mile offshore into Lake Michigan. The phenomenon was connected with the severe and well-remembered late fall storm which piled deep blankets of snow over the eastern part of the midwest and the adjoining southern states, plummeting temperatures below zero and ensnarling traffic for three days.

When air temperatures are below freezing, the contact of supercooled water droplets in fogs (or clouds) with objects such as poles, wires, buildings, or trees, particularly on mountains, results in a spectacular, and oftentimes fantastic, snow or ice formation on the objects, the coating sometimes aggregating several inches in thickness. This type of formation is called *rime* and is shown in Figure 6–22.

Dew and Frost

Dew as it is explained under the subject of dew point is simply condensation of water vapor on objects which have radiated sufficient heat to reduce the temperature of the air in contact with them to the point at which the air becomes saturated so that a part of its water vapor must condense. If the temperature of the object drops to freezing or lower, then the condensation will be in the form of *frost*. Contrary to a rather common belief, frost is *not* frozen dew. When dew freezes, clear beads of ice are formed; frozen dew is a relatively rare phenomenon.

Frost is frequently called a form of *sublimation* because it results when there is a direct change from water vapor (a gas) into ice (a solid). For the same reason, snow, mentioned in the next paragraph, is often called a form of sublimation. Throughout this book, however, frost is referred to as a form of condensation, and snow as a form of precipitation.

Fig. 6–22. Rime on Mount Washington, N.H. (Courtesy U.S. Weather Bureau.)

Precipitation

Forms of Precipitation

When water droplets in clouds begin to coalesce and form larger drops that become so large they cannot remain suspended in the air, they fall as *rain*. In North America, *mist* is a form of rain and is synonymous with *drizzle*, if sufficiently concentrated. On a warm, sunny day in summer, a strong convective column may cause the formation of ice pellets, usually rather spherical in shape and having fairly distinct concentric layers of ice; such a formation is known as *hail* and is described in greater detail in connection with the subject of thunderstorms later in this chapter. The size of hailstones depends upon the amount of snow or additional rain encountered in its turbulent course. Hailstones as large as baseballs have been authentically reported and photographed (Figure 6–23).

Snow occurs only when the condensing medium has a temperature *well* below freezing, 20° or even lower. Snow is generally in the form of individual crystals, or in flakes that are aggregates of many crystals. Snow crystals are almost always hexagonal in form and of such an amazing variety that it is said no two crystals are ever exactly alike. A few types of crystals are shown in Figure 6–24. *Sleet*, according to the Weather Bureau, is regarded as frozen, or partly frozen, rain, or as raindrops frozen in the form of clear ice particles; it may also be considered a mixture of rain and snow. Occasionally the ground air may be below freezing while a warm, moist wind rides up over it. The forced ascent causes rain that falls into the surface layer of cold air and freezes upon contact with the ground, cold plant forms, wires, buildings, and other structures. Such *ice storms*, or *glaze storms*, may coat objects with ice as much as 2 inches thick in the most extreme cases (Figure 10–19). The extra weight may become so great as to result in tremendous destruction of trees, poles, wires, and even buildings. Less frequently encountered forms of precipitation are graupel, or soft hail, granular snow, snow pellets, and ice needles or spicules.

Precipitation is measured by the depth of the rain, or the water equivalent of snow or sleet, in hundredths of inches or in millimeters. On the average, 10 inches of snow is considered the equivalent of 1 inch of rain; but this varies widely depending largely upon temperature. For example, in the far north during the coldest part of the year, snow comes in the form of tiny, needlelike crystals which form a very compact layer on the ground, so that from 3 to 5 inches of such snow will represent 1 inch of water. On the other hand, snow on the east shores of Lake Michigan or Lake Huron is often so exceedingly fluffy that as much as 18 inches may be required to produce 1 inch of water, and when the temperature on the ground is somewhat above freezing, the snow may be so wet that 5 inches or less will be the equivalent of 1 inch of rain. Precipitation is measured by a *rain gauge* commonly in the form of a cylindrical can or tank 8 inches in diameter (Figure 6–25). Over the opened end of this cylinder is a funnel of the same diameter that empties into an inner cylindrical tank; this inner tank has a cross-sectional area one-tenth that of the receiving part of the funnel. Thus a measurement of 16.4 inches in the small tank means an actual measurement of 1.64 inches. The smaller tank and funnel can be lifted out readily, leaving only the larger 8-inch cylinder. It is then transformed into a snow gauge. Many stations use the *tipping-bucket gauge* (Figure 6–26) or the *weighing gauge* (Figure 6–27). Each of these types measures and records the rainfall electrically.

Fig. 6–23. Hailstones which fell near Meligh, Nebr., during the night of June 13–14, 1950. (Photograph by L. Gillespie, U.S. Weather Bureau.)

Fig. 6–24. Snow crystals. (Photograph by W. A. Bentley, U.S. Weather Bureau.)

Variation of Precipitation with Altitude

Due to the evaporation of raindrops as they fall, precipitation is greatest at the base of the rain cloud that is normally at a higher elevation in summer than in winter. In summer, it is not infrequent for all the rain which starts to fall from the clouds to evaporate completely before reaching the ground. On the average, there is an increase in precipitation on windward mountain slopes up to 6,000 or 7,000 feet, the increase being due to forced ascent of the air as well as the smaller loss by evaporation. The increase does not continue beyond this because of the loss of water vapor; as has already been stated, about one-half the water vapor in the air is below 6,500 feet.

Precipitation Types

Any form of condensation or precipitation can result only through the cooling of moist air. The cooling may be caused in several ways, two of which have already been mentioned: (1) *convectional* precipitation occurs when moist air over heated ground is forced to rise, and (2) *orographic* precipitation occurs when a moisture-laden wind is forced to rise in passing over hills or mountains. Other types of precipitation are: (3) *cyclonic*, and occurring when winds converge or overlap; those caused by (4) *mixing* of masses of moist air of contrasting temperatures, which usually results in but little precipitation, (5) *invasion* of a large mass of cold air, forcing itself under warmer air, and (6) *overriding* of cold, residual air by a warm mass of air, usually from a southerly direction. The last two types are essentially the same as orographic, and they, along with (3), are associated mostly with frontal conditions in cyclonic storms.

Exceptionally Heavy Precipitation

There are instances of remarkably heavy precipitation for periods as short as one minute or over much longer periods of time. Some of these instances are listed in Table 6–2.

A heavy snowfall may not be so devastating as a heavy rainfall because of the immediate floods associated with the latter, but heavy snows can disrupt transportation, maroon people, and cause the loss of human lives as well as considerable economic loss. The 28-inch fall of snow in Washington, D.C., January 27–28, 1922, caused the collapse of a theater roof, killing 96 persons. The removal of the 26-inch fall of snow in New York, December 26–27, 1947, is said to have cost the city 8 million dollars. The marooning of over 200 people for three days in a passenger train in Emigrant Gap, California, in January, 1952 and the great blizzard of March, 1888, when a wind of 70 miles per hour brought 21 inches of snow to New York, are two of the thousands of instances recorded in the annals of the Weather Bureau. Snowfalls unusual because of the latitudinal location or the time of year were: 15 inches at Galveston, Texas, in February, 1899; and 87 inches in 27 hours at Silver Lake, Colorado, in April, 1921. A fall of 96 inches of snow in four days at Vanceboro, Maine, and a total of 884 inches in one year (1906–1907) at Tamarack, California, are records that may well stand for many years. More recently, in late March, 1957, a blizzard swept over the central part of the Great Plains, piling drifts 30 feet high, marooning two trains and thousands of motor vehicles for several days, and causing indirectly more than 40 deaths. A late-season blizzard, April 2–3, 1957, brought up to 3½ feet of snow to the Denver-Pueblo area of Colorado and lesser amounts to Wyoming, Nebraska, and Kansas.

Fig. 6–24 (continued).

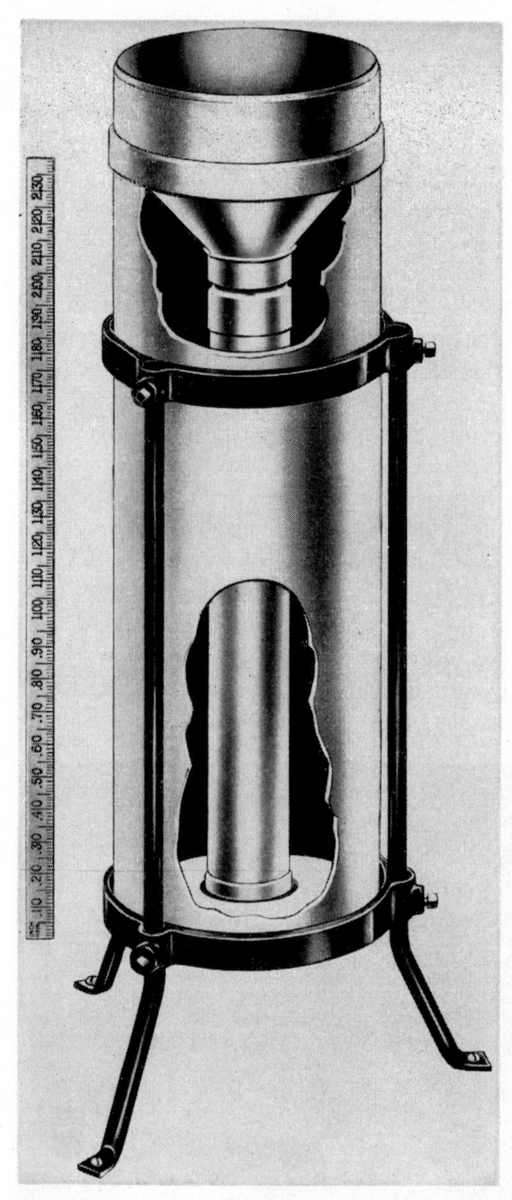

Fig. 6–25. Standard rain gauge. (Courtesy Bendix-Friez.)

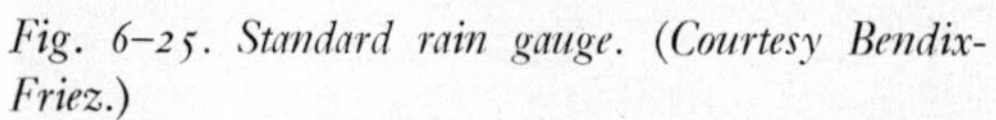

Fig. 6–26. Tipping-bucket rain gauge. (Courtesy Bendix-Friez.)

Fig. 6–27. Weighing rain gauge. (Courtesy Bendix-Friez.)

*Table 6–2. World's Greatest Observed Rainfall**

Duration	Depth, in.	Place	Date
1 min	0.65	Opid's Camp, Calif.	Apr. 5, 1926
5 min	2.48	Portobelo, Pan.	Nov. 29, 1911
8 min	4.96	Füssen, Bavaria	May 25, 1920
15 min	7.80	Plumb Point, Jam.	May 12, 1916
20 min	8.10	Curtea-de-Arges, Rom.	July 7, 1889
40 min	9.25	Guinea, Va.	Aug. 24, 1906
42 min	12.00	Holt, Mo.	June 22, 1947
2 hr, 10 min	19.00	Rockport, W. Va	July 18, 1889
2 hr, 45 min	22.00	D'Hanis, Tex. (17 m. NNW.)	May 31, 1935
4 hr	23.00	Basseterre, St. Kitts I., W.I.	Jan. 12, 1880
4 hr, 30 min	30.8 +	Smethport, Pa.	July 18, 1942
15 hr	34.50	" "	July 17–18, 1942
18 hr	36.40	Thrall, Tex.	Sept. 9, 1921
24 hr	45.99	Baguio, P.I.	July 14–15, 1911
39 hr	62.39	" "	July 14–16, 1911
2 days	65.79	Funkiko, For.	July 19–20, 1913
2 days	57.50	Silver Hill Plantation, Jam.	Nov. 6–7, 1909
2 days, 15 hr	79.12	Baguio, P.I.	July 14–17, 1911
3 days	81.54	Funkiko, For.	July 18–20, 1913
3 days, 15 hr	97.01	Baguio, P.I.	July 14–18, 1911
4 days	101.84	Cherrapunji, India	June 12–15, 1876
6 days	122.50	Silver Hill Plantation, Jam.	Nov. 5–10, 1909
7 days	131.15	Cherrapunji, India	June 24–30, 1931
8 days	135.05	" "	June 24–July 1, 1931
8 days	135.00	Silver Hill Plantation, Jam.	Nov. 4–11, 1909
15 days	188.88	Cherrapunji, India	June 24–July 8, 1931
31 days	366.14	" "	July 1861
2 mo	502.63	Cherrapunji, India	June–July 1861
3 mo	644.44	" "	May–July 1861
4 mo	737.70	" "	Apr.–July 1861
5 mo	803.62	" "	Apr.–Aug. 1861
6 mo	884.03	" "	Apr.–Sept. 1861
11 mo	905.12	" "	Jan.–Nov. 1861
1 yr	1,041.78	Cherrapunji, India	Aug. 1860–July 1861
2 yr	1,605.05	" "	1860 and 1861

* Tabulation compiled by R. C. Elford (*Courtesy U.S. Weather Bureau*).

Thunderstorms

Causes of Thunderstorms

Thunderstorms are so interesting in themselves, so stupendous in their development and destructive effects, and so illustrative of many meteorological principles that they deserve more than passing notice. Thunderstorms are really of two types: *frontal*, or *general*, *thunderstorms*, occurring over wide areas in connection with the passing of a cyclonic disturbance (mentioned later), and *local thunderstorms*, forming as a result of strong local convection (Figure 6–29). The diagram (Figure 6–28) attempts to portray a few of

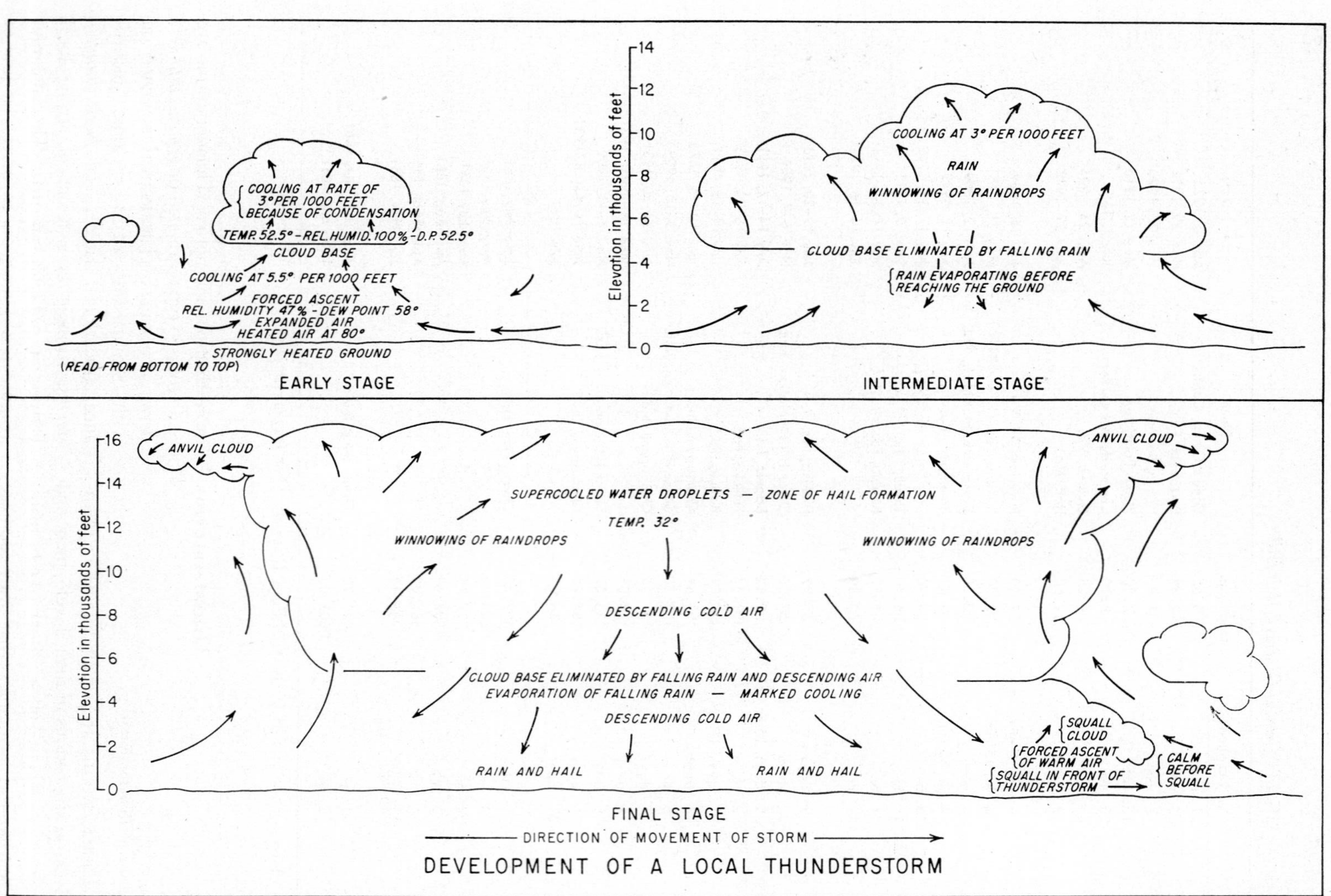

Fig. 6–28. Development of a local thunderstorm.

Fig. 6–29. A well-developed thunderstorm. (Photograph by Butler, U.S. Weather Bureau.)

the phenomena accompanying a well-developed local thunderstorm.

The first essential for a local thunderstorm is strong heating of the ground with consequent heating of the overlying air. Expansion of the heated air and its forced ascent by the cooler air surrounding it result. The rising column of air is cooled at the adiabatic rate of 5.5 degrees per 1,000 feet until the dew point is reached. A cumulus cloud forms at this point and continues to develop as long as the air continues to ascend, although the rate of cooling within the cloud itself is the condensation adiabatic. Of course, the elevation of the base of the cumulus cloud depends upon the relative humidity of the air, or the dew point. Because of the decreased pressure aloft, it is found that the dew point of a rising mass of air decreases approximately 1.1 degrees per 1,000 feet (0.2 degree C per 100 meters). Consequently, the rate at which the dew point is reached in a rising column of air is not the regular adiabatic rate, but roughly 1.1 degrees less than that, or 4.4 degrees per 1,000 feet.

Development of Thunderstorms

Referring to that part of the diagram showing the "*Early Stage* in the Development of a Local Thunderstorm," you will note that the following psychrometric data at the ground are assumed: dry-bulb temperature 80° and wet-bulb temperature 66°, thus making a 14-degree difference between the two readings. The relative humidity and dew point tables in the Appendix will show the relative humidity to be 47 per cent and the dew point 58°. The height of the cumulus-cloud base is easily determined as 5,000 feet (the difference between 80 and 58 divided by 4.4). It is obvious that the relative humidity at the cloud base will be 100 per cent; the dew point at the cloud base, of course, will be the same as the air temperature of 52.5° (found by multiplying 5.5 by the number of thousands of feet and subtracting from 80). Condensation begins to take place at this point (*Intermediate Stage*). With continued ascent, raindrops form and start falling. Some reach the ground; others evaporate before they reach the ground, especially in the forward portion of the thunderstorm. The evaporation causes marked cooling of the surrounding air, which then descends and sweeps down the center and out in front of the storm as a cool squall, sometimes of destructive velocity. Because of the forced ascent of the relatively warm air in front and its mixing with the cold descending air, a *squall*, or *roll cloud*, develops at a rather low altitude and rolls along in front of the thunderstorm (*Final Stage*).

The map, Figure 6–30, shows for the United States the average annual number of days with all types of thunderstorms. It should be noted that thunderstorms are most frequent along the middle and east Gulf Coast, in the lower Mississippi basin, and along the east slopes of the Rocky Mountains southward from Wyoming; they are least frequent in the Pacific Coast states.

Lightning and Thunder

In a well-developed local thunderstorm, many of the raindrops in their descent, while still in the cloud, become too large for cohesion and split in the upward-moving air in the storm. According to C. F. Brooks, the winnowing of larger from smaller fragments leads to a separation of electrical charges; this separation and the differences in potential arising from the differing rates of condensation in progress seem to be responsible for the electrical discharges we call *lightning* (Figure 6–31). When a thunderstorm is below the horizon, the reflection of the lightning flashes against clouds gives rise to what is commonly called *heat*, or *sheet*, *lightning*.

Some authorities say that raindrops are positively charged when growing and negatively charged when evaporating. Others have said, although the accuracy of their view is now somewhat doubtful, that the smaller fragments of raindrops are negatively charged and are forced higher, while the larger nuclei are positively charged and remain in the base of the cloud or descend. Since the earth's surface is negatively charged, a strong difference of potential would develop so that a discharge,

Fig. 6–30. Average annual number of days with thunderstorms. (*Courtesy U.S. Weather Bureau.*)

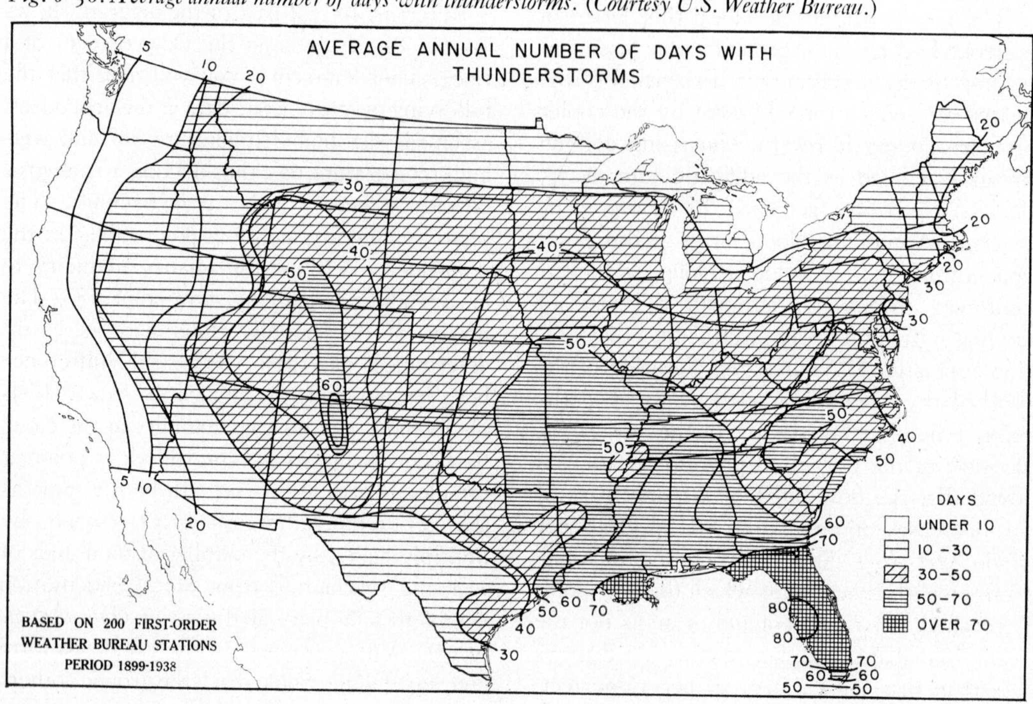

Fig. 6–31. Lightning at Miami Beach. (Photograph by C. S. Watson, U.S. Weather Bureau.)

or flash of lightning, would occur, usually between the cloud and the earth but frequently between two clouds.

H. R. Byers and workers on the project report, *The Thunderstorm,*[1] found that "there were several electric charge centers: a positive charge center in the upper third of the cloud, another smaller positive charge center in and below the base of the cloud where the rainfall was heaviest, and a third, conical-shaped negative charge area extending downward from the center of the cloud to and below the base and generally surrounding the smaller positive charge center. The resulting areal differences in potential cause lightning discharges." Regardless of the cause of the lightning, the sudden expansion of the air and the dissociation of molecules and atoms, caused by the tremendous heating (up to 50,000°) in the path of the lightning, cause *thunder*.

[1] Horace R. Byers et al., "The Thunderstorm," Government Printing Office, Washington, 1949.

Hail

Hail, already mentioned as a form of precipitation, generally occurs only in connection with thunderstorms, although probably only a small percentage of the thunderstorms occurring over the world produces an appreciable amount of hail. There are several related theories concerning the formation of hail; the most reasonable seems to be this. Ice crystals or snowflakes, or clumps of snowflakes, which form above the zone of freezing during a thunderstorm, fall through a stratum of supercooled water droplets (that is, water droplets well below 32°, perhaps as low as 20°). The contact of the ice or snow particles with the supercooled water droplets causes a film of ice to form on the snow or ice pellet. The pellet may continue to fall a considerable distance before it is carried up again by a strong vertical current into the stratum of supercooled water droplets where another film of water covers it. This process may be repeated many times until the pellet can no

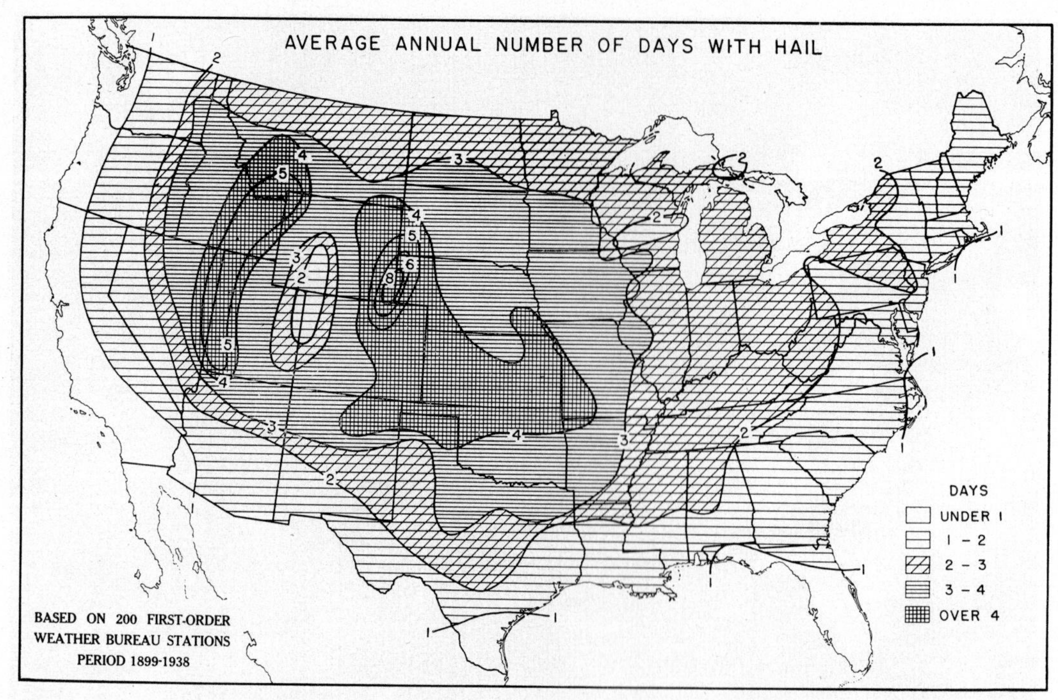

Fig. 6–32. Average annual number of days with hail. (*Courtesy U.S. Weather Bureau.*)

longer be supported by the convective updraft and falls to the ground as hail. It is believed that, during the rapid freezing of the supercooled water, sufficient heat is released through the latent heat of fusion to raise the temperature of the water film to freezing so that a layer of water momentarily covers the ice or hail pellet; thus the distinct, alternate layers of clear and milky ice. It is also possible, and many believe probable, that the hailstone falls low enough to be coated with a film of rain water and then is carried into the zone of freezing—a process which may occur several times. This seems to account for the fact that hailstones of large size are seldom actually spherical; they are more likely to be flat, or hemispherical, a shape which could readily be explained by the force of the updraft brushing the water off one side of the hailstone.

Distribution of Hail

A comparison of the two maps of the United States, Figures 6–30 and 6–32, shows that hail occurs most frequently over the Great Plains, whereas the greatest frequency of thunderstorms is, for the most part, in the southeastern states. The occurrence of hail is obviously related to the height of the zone of freezing. Since this zone is located at high altitudes in the tropics and subtropics, a height to which convection in a thunderstorm does not always reach, it follows that hail will be relatively infrequent in low latitudes, and, although the freezing zone is found at much lower elevations in high latitudes, convection is usually too weak to reach even that low elevation. It seems more likely, therefore, that hail will occur in middle latitudes and particularly over the Great Plains where convection on a hot day can, and does, reach remarkable heights well above the freezing zone. The great size of hailstones in this region of the United States attests to the violence of convective updrafts in a large thunderstorm. It is said these vertical currents frequently attain a speed of 100 miles per hour and sometimes as much as 160 miles per hour. No wonder aircraft avoid thunderstorms!

7

Wind

Nature, Cause, and Force

Air that usually moves parallel to any part of the earth's surface is referred to as *wind*. Vertical or nearly vertical movements of air resulting from turbulence, convection, or any other cause should be referred to as *air currents*. Wind is always the result of a difference in pressure between two areas, the movement of air being from the area of higher pressure toward the area of lower pressure even when the difference between the two is very slight.

The distribution of pressure over a particular area on the earth's surface or at any altitude is best represented on a map by drawing the *isobars*, which are lines drawn through points having the same pressure. Isobars are usually curved, more or less concentric lines, although at times, and especially aloft, they may be nearly straight and parallel. Regardless of their form, they indicate differences in pressure.

If it were not for the rotation of the earth, air would simply flow from an area of high pressure to an area of low pressure, and the line of direction would be at right angles to the isobars. But the deflective effect of the earth's rotation, known as the *Coriolis force*, causes the moving air, or wind, to be deflected from its original course—to the right in the northern hemisphere, to the left in the southern hemisphere. This apparent force, or, more accurately, effect, was stated as a law in the middle of the nineteenth century by Buys Ballot, the Dutch meteorologist; a slightly modified version of that law amounts to this: stand with your back to the wind, stretch out your arms horizontally, and turn 30 degrees to the right; your left hand will point toward the area of low pressure and your right hand toward the area of high pressure. Obviously, at higher altitudes in the free air where ground friction does not exist, the deflective effect is much greater, the direction of the wind being essentially parallel to the isobars. A further explanation of the Coriolis effect is given in the next chapter in connection with the general circulation of the atmosphere.

Normally, the greater the difference in pressure, the greater is the velocity, or speed,

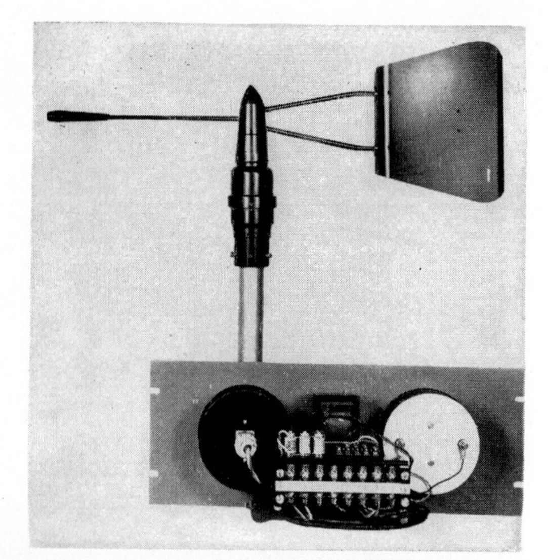

Fig. 7–1. Wind vane. (Courtesy U.S. Weather Bureau.)

Fig. 7–2. Recorder for both wind direction and wind speed. (Courtesy Bendix-Friez.)

of the wind. The difference in pressure in any direction, or for any distance, is referred to as the *pressure gradient.* The pressure gradient is *steep* or *gentle*, depending upon whether the difference in pressure is great or small. Consequently, a strong, or steep, gradient means a strong wind; a weak, or gentle, gradient means a light wind.

Air density is also a factor in wind speed which is inversely proportional to the air density, other conditions remaining the same. This fact accounts, in part, for the high wind speed aloft. For example, if the air density at 18,000 feet is assumed to be one-half that at sea level, then the wind speed at that high elevation would certainly be at least double that at sea level, provided the pressure gradient was the same. Friction on the ground—forests, buildings, the changing relief of the land—and the lack of friction in the free air are important additional factors in accounting for increased wind speeds at high elevations.

Wind Direction

A wind is named according to the direction from which it blows. A wind coming from the west is a west wind; a wind or breeze coming from the sea is a sea breeze; that blowing up a mountain slope from a valley is a valley breeze. In observing wind directions, four, eight, sixteen, and even thirty-two points of the compass may be used. The most common practice, however, is to use eight: the four cardinal points of north, east, south, and west and the four intermediary points of northeast, southeast, southwest, and northwest. A wind which blows more frequently from one direction than from any other is said to be the *prevailing wind* for the place of observation. The term is usually applied on a climatic basis to a season or an entire year, although it may be used to denote prevalence for a shorter length of time.

The direction from which a wind blows is termed *windward;* that toward which it blows, *leeward.* In the case of the Coast and Cascade Ranges of the Pacific Northwest, for example, the west side is not only the prevailing-wind side but also the windward side, the slope on which the air is ascending. It is also the rainy side, since air is cooled sufficiently in its ascent to cause considerable precipitation. The east side of the ranges is the lee, or leeward, side, the side on which the air is descending. Since the air on the leeward side is being heated by compression as it descends, no precipitation falls. The leeward side is also called the "dry side" or the "rain shadow side."

Winds are said to *veer* when their direction changes progressively in a clockwise direction, that is, from east through southeast, south, southwest, and eventually to the west or northwest. Winds *back* when they change in the opposite or counterclockwise direction, from east to northwest by way of north. The terms *veering* and *backing* are applied especially to the changing wind directions during the passing of a low-pressure area, but they are sometimes used in other cases.

The usual method of determining wind directions is by means of a *wind vane* (Figure 7–1). This often consists of two thin pieces of wood placed with their broad surfaces at an angle of 20 or 25 degrees and so mounted on a pole that the device will turn with even a slight wind. More elaborate commercial varieties also have this divided, or flared, tail which aids in keeping the point of the vane nosed *into*, or *toward*, the source of wind at all times, especially when there are gusts and hence frequent changes in direction. Unless they are defective, wind vanes always point *toward the source of wind.* If a wind vane is connected to a self-recording apparatus, it is called a *recording wind vane* (Figure 7–2). There are several varieties available from meteorological instrument companies.

Wind Velocity

The velocity of the wind is measured in miles per hour, knots per hour, feet per second, or meters per second. The relation of these equivalents may be readily determined from the list of equivalents in Figure 3–5. An

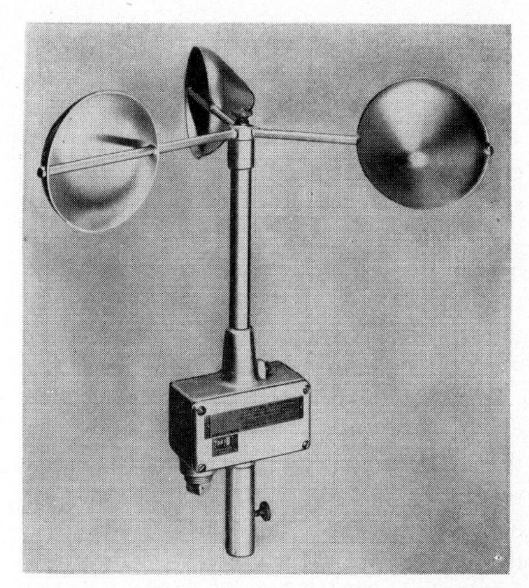

Fig. 7–3. Anemometer. (Courtesy Bendix-Friez.)

anemometer is the instrument used to measure the velocity of the wind (Figure 7–3). The most common form is the cup anemometer which consists of three (less commonly four) cups attached by arms to a vertical rod and this in turn to a system of cog wheels which records the number of revolutions of the rod. Some anemometers translate the record of revolutions to miles per hour on a dial in the building where they are installed (Figure 7–4). A certain multiple of these revolutions may be connected electrically to an *anemograph* that makes a continuous record of the wind mileage (Figure 7–2). The so-called "meteorograph," or *triple register* (sometimes called the "quadruple recorder"), gives a graphic record of both wind direction and velocity as well as a record of sunshine and precipitation (Figure 7–5).

Wind velocities may be estimated quite accurately by using the Beaufort scale given in Table 7–1. Although the Beaufort scale was devised over a hundred years ago for marine observations, it has been remodeled with appropriate descriptions that apply to land observations too. Until a few years ago, the Beaufort scale designations were from 0 (calm) to 12 (hurricane); now the scale has been extended to 17 in order to indicate the measurements of occasional winds of greater velocity than 82 statute miles per hour (mph). Unless he has instruments, the lay observer can scarcely distinguish Beaufort scale number 16 and its destructive force from Beaufort scale number 12, but the weatherman's records will give such details. The World Meteorological Organization, established as an important unit of the United Nations and to which the United States belongs, has adopted the 17-fold Beaufort scale as the basis of its reports. The

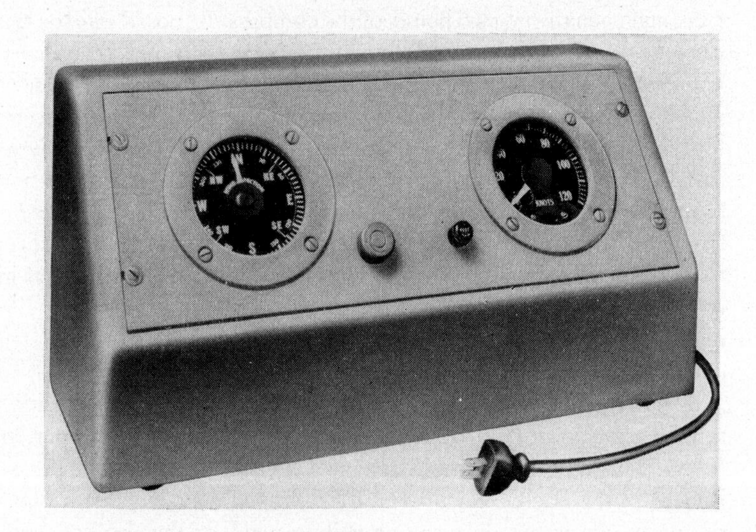

Fig. 7–4. Indicator for both wind direction and wind speed. (Courtesy Bendix-Friez.)

feather-and-flag symbols used by the U.S. Weather Bureau on its daily weather maps depart somewhat from the Beaufort scale in that they give more details, but the scale as given is frequently referred to in weather parlance.

Because of friction with the ground, wind velocities are usually less over land than over the ocean, and for the same reason they increase under average conditions quite uniformly with an increase in altitude. On the whole, they are greater by day than by night. Exceptions to this occur, of course, during the passage of cyclonic storms that are no respecters of the clock.

Convectional Circulation

The phenomenon of simple convection in the atmosphere has already been discussed with thunderstorms, but the disposal of the ascending air was not mentioned. It has been shown, however, that intense heating of the land locally causes the air over that portion of land to become greatly heated and to expand, causing some of the air aloft to flow away, lowering the pressure on the hot area, and raising the pressure around it. The cooler and heavier surrounding air (as well as the cooler, heavier air from aloft) flows under the heated air, forcing the latter to rise and resulting in

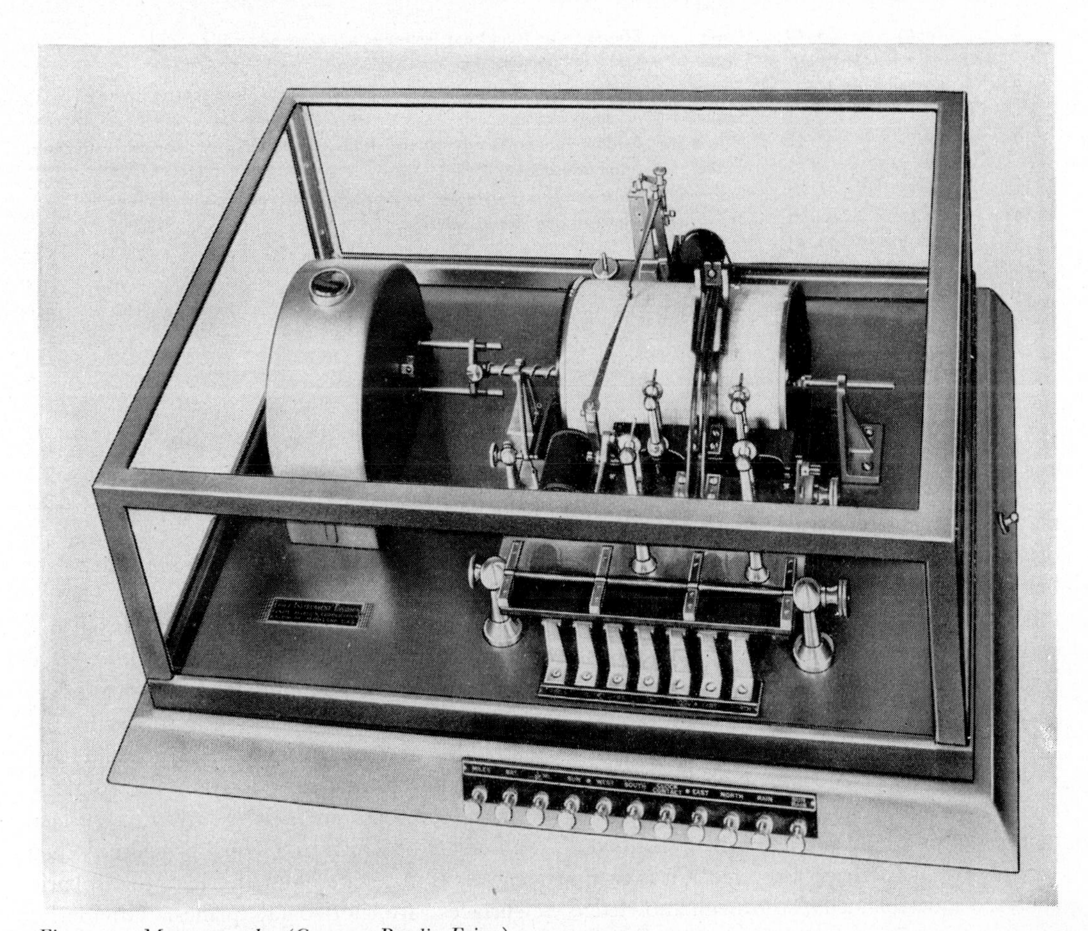

Fig. 7–5. Meteorograph. (Courtesy Bendix-Friez.)

Table 7–1. The Beaufort Scale for Wind

Beaufort scale number	Statute miles per hour (mph)	Knots (nautical miles)	Explanatory terms	Observable description
0	Less than 1	Less than 1	Calm	Calm; smoke rises vertically
1	1–3	1–3	Light air	Direction of wind shown by smoke drift but not by wind vanes
2	4–7	4–6	Light breeze	Wind felt on face; leaves rustle; ordinary vane moved by wind
3	8–12	7–10	Gentle breeze	Leaves and small twigs in constant motion; wind extends light flag
4	13–18	11–16	Moderate breeze	Raises dust and loose paper; small branches are moved
5	19–24	17–21	Fresh breeze	Small trees in leaf begin to sway; wavelets form on inland water
6	25–31	22–27	Strong breeze	Large branches in motion; whistling heard in telegraph wires; umbrellas used with difficulty
7	32–38	28–33	Moderate gale	Whole trees in motion; inconvenience felt when walking against wind
8	39–46	34–40	Fresh gale	Breaks twigs from trees; impedes progress generally
9	47–54	41–47	Strong gale	Slight structural damage occurs; chimneys and slates carried away
10	55–63	48–55	Whole gale	Seldom experienced inland; trees uprooted; considerable structural damage done
11	64–72	56–63	Storm	Very rarely experienced; accompanied by widespread damage
12	73–82	64–71	Hurricane	Violence and destruction
13	83–92	72–80	"	" " "
14	93–103	81–89	"	" " "
15	104–114	90–99	"	" " "
16	115–125	100–108	"	" " "
17	126–136	109–118	"	" " "

local convection. The same sort of circulation may occur on a larger scale, affecting hundreds of square miles, a whole continent, or even a belt stretching around the entire earth. Before this larger scale circulation is taken up, it is well to consider the local result of this differential heating.

The Sea Breeze

The sea breeze is a common phenomenon of the seashore in summer, and because it is quite local in occurrence and rather shallow in vertical extent (perhaps 1,500 feet), it is easily analyzed and explained. A thorough knowledge of the conditions which give rise to it aids greatly in understanding the general circulation of the atmosphere that is discussed in the next chapter.

The essential features of the sea breeze are shown in the diagram in Figure 7–6. The air over the land and sea is assumed to be in a *state of equilibrium*, that is, at any given altitude there is no more air over the land than over the sea. This condition is shown by the straight horizontal lines, partly solid and partly dotted, across the diagram. Each of these lines represents an *isobaric surface*, a surface all points of which have equal pressure. When the air is in a state of equilibrium, the isobaric surfaces are horizontal planes. Obviously, under any condition there could be an infinite number of isobaric surfaces. In the diagram,

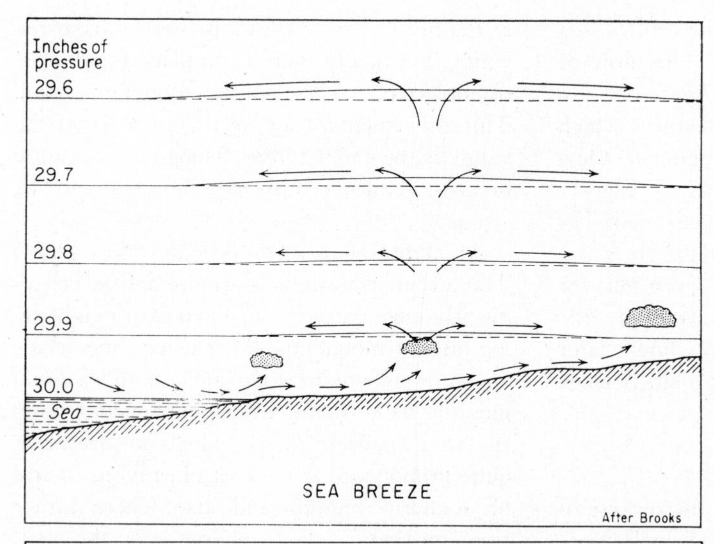

Fig. 7–6. The sea breeze.

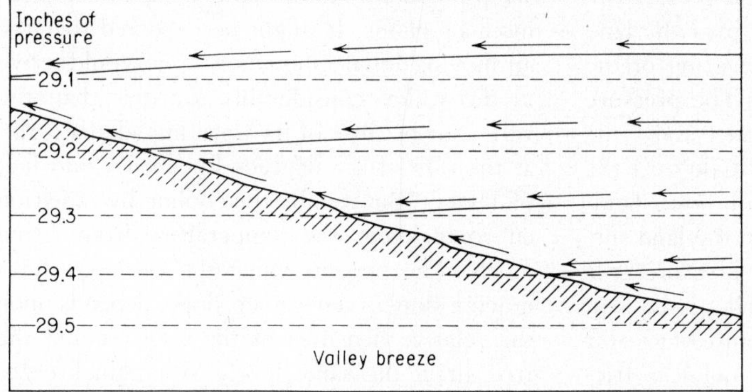

Fig. 7–7. The valley breeze.

pressure at sea level is taken at 30.0 inches; the interval between each isobaric surface represented is taken as 0.1 inch. There are six fundamental aspects which need to be considered.

1. On a sunny morning the temperature of the land increases more rapidly than that of the water.

2. As a consequence, the air over the land expands more, and all the isobaric surfaces are raised. The upper surfaces are raised more than the lower surfaces because each layer of air absorbs some radiation from the land, and that layer expands in its own right.

3. The pressure at a few hundred feet above the land is seen to be greater than the pressure at the same altitude over the sea. This is shown by the curve in the solid lines representing the isobaric surfaces. Such an inclination of an isobaric surface is said to be an *isobaric slope*. No air has been added or taken away; therefore, the dotted line represents slightly greater pressure than the solid line. Even taking into account the lesser density of the air near the top, the increasing divergence of the solid and dotted lines indicates at the higher altitudes a greater difference in pressure between the air over the land and the air over the sea.

4. Because of this difference in pressure, the air for some distance above the land begins to flow from the area of higher pressure toward the sea.

5. The consequent accumulation of air over the sea means greater weight, or greater pressure, on the sea surface at the same time that the pressure over the land is becoming less.

6. The result is that air resting on the sea surface moves toward the land. This flow of air at the surface is called a *sea breeze.*

The sea breeze is one of the features which attract people to the seashore in summer. On a hot day, the cooling sea breeze is eagerly awaited; a sudden drop of 15 degrees is not unusual with the advent of the breeze. It is also frequently experienced inland—even one or two hundred miles inland if carried by the general wind—although it arrives hours later and perhaps appreciably warmed by its passage over land.

The Land Breeze

The land breeze is essentially the reverse of the sea breeze. At night the land cools more rapidly than the water. The air over the land contracts, with a consequent lowering of the isobaric surfaces over the land. The pressure thus becomes less at a short distance above the ground than it is at the same altitude over the water. As a result, the upper air flows landward, increasing the pressure on the land surface. With this difference in pressure, the surface air moves from the land to the sea, giving rise to the *land breeze.* Because of greater daytime temperature contrasts and less friction, sea breezes are usually more pronounced than land breezes.

Mountain and Valley Breezes

Mountain and valley breezes are very similar to land and sea breezes. Although the valley corresponds to the land in the matter of heating, the air does not move into the valley but up the mountain slope. The reason will be readily understood by studying the diagram (Figure 7–7). During a sunny day the valley floor becomes heated; air over it expands; the isobaric surfaces are raised; at corresponding elevations in the upper air, pressure becomes greater over the valley than on the valley walls and on the mountain slopes at the head of the valley; air moves from the valley, but finally there is no place for this air to move to, except up the mountain slope. This movement of air up the slope from the valley is the *valley breeze.* Along coasts, valley breezes reinforced by sea breezes can be strong.

At night the reverse process takes place. The air in the valley contracts and is consequently augmented by air from over neighboring hills or mountains. Also when the surface air on the mountain slope cools, it slides down into the valleys, the whole movement giving rise to a *mountain breeze.* This cool breeze is quite pronounced at the foot of canyons in arid or semiarid regions and also where rather steep mountain valleys emerge upon the piedmont or plains. It might be expected that the air moving down a mountain slope would arrive in the valley considerably warmer than the valley air because of its warming adiabatically at the rate of 5.5 degrees for each 1,000 feet of descent. But usually the cooling by radiation en route keeps the temperature from rising much. The rate at which the cool air slides down a slope, even a steep slope, depends upon the relative densities of the cool air and the free air at the same level. Mountain breezes are usually chilling, and frosts may occur when the valley temperatures are otherwise critically near the freezing point. Yet there are numerous instances where mountain breezes are sufficiently warmed in their descent to prevent what, without them, might have been severe frost. Along coasts, mountain breezes can become quite strong when reinforced by land breezes. Mountain winds and similar *gravity winds* are frequently referred to as *fall winds*, or *katabatic winds*, the latter stemming from the Greek word meaning "to go down."

There are other local winds that depend upon more widespread differences in pressure. These will be discussed in Chapter 10.

8

General Circulation of the Atmosphere

Circulation on a Nonrotating Globe

Let us suppose that the earth had a homogeneous surface: a surface either all water or all land of uniform character and altitude. Let us suppose further that the earth did not rotate but that the direct rays of the sun nevertheless reached all points on the equator at all seasons. Under such conditions the air movement would be very simple. The air near the equator would expand, would overflow, and the cooler air of higher latitudes would crowd under it, forcing it aloft. At or near the earth's surface, the winds in the northern hemisphere would be north winds; those in the southern hemisphere would be south winds. At higher altitudes, the directions of the winds would be the opposite of those at the surface.

None of the above conditions obtains. Land, water, and ocean currents, mountain barriers and altitude, and the changing seasons—whereby the zone of greatest heating shifts north and south of the equator—are sufficient in themselves to complicate the situation. Add to these the rotation of the earth that causes any moving fluid, air or water, to be deflected from the course in which it starts to move, and the resulting conditions of circulation become extremely complicated.

Deflective Effect of the Earth's Rotation

In the northern hemisphere any mass of moving air is deflected to the right; in the southern hemisphere it is deflected to the left. If air north of the equator starts to move directly southward toward the equator as a north wind, it soon assumes a direction from the east, becoming a northeast wind; a wind from the southeast becomes a south wind. Because of friction with the ground, moving air is usually deflected less than 45 degrees from the original direction. The maximum deflection is never more than 90 degrees. This fact is easily understood when we compare moving air with water flowing down a flat cone. This comparison is legitimate since air

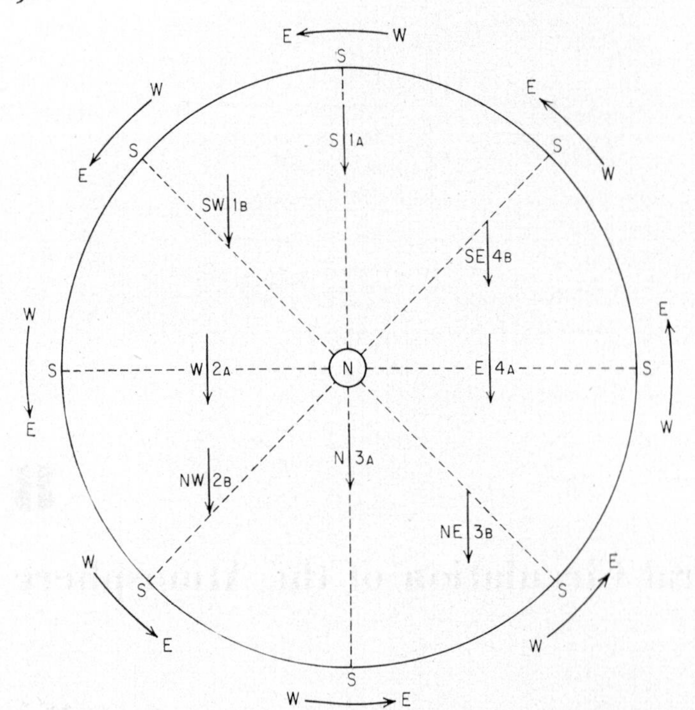

Fig. 8–1. Deflective effect of the earth's rotation at the north pole.

flows from one place to another because of the existence of a pressure gradient, that is, because of the isobaric slope. If we started a tiny stream of water down the cone and if we rotated the cone in a counterclockwise direction—just as the earth rotates when viewed from the north pole—we would find that the stream of water would be deflected to the right. It is obvious that the deflection could be no more than 90 degrees because if it were more, the water would have to start flowing up the slope of the cone. It is the same with moving air: if the air stream were deflected more than 90 degrees, it would have to move back "up" the isobaric slope. Stated another way, the air would have to move from an area of low pressure to an area of high pressure, which is impossible.

Deflective Effect at the North Pole

Perhaps the deflective effect of the earth's rotation, known as the Coriolis force, can best be explained here by assuming that a flat disc is tangent to the earth at the north pole and rotates at the same rate as the earth and in the same direction. This disc is represented in the diagram (Figure 8–1). The arrows parallel with the circumference of the disc indicate the direction of rotation. All directions from the north pole are south, as shown in the diagram. A fundamental law of motion is that any body which starts to move in a given direction will keep that direction while it continues to move unless it is acted upon by some external force. On the basis of this principle, a mass of air which starts to move directly north, as illustrated by the arrow in 1*A* of the diagram, continues to move in the same direction as shown by the arrow 1*B* which is seen to be parallel with 1*A*. That is, with respect to space the direction has not changed. The earth, however, has turned through 45 degrees so that the moving air, or we may call it "wind," has also changed 45 degrees with respect to directions on the earth. Thus the wind which was a south wind at the start, 1*A*, has become a southwest wind at 1*B*; it has been deflected to the right of its initial course. Similarly, a wind which is west, as at 2*A*, becomes a northwest wind by the time the earth has rotated 45 degrees, as shown at 2*B*; and a wind which is northeast at 3*B* becomes an east wind at 4*A*. If we were to imagine the disc tangent to the earth at the south pole, we would find that the moving air, or wind, would be deflected to the left of its original course. If it were possible for a wind to be deflected more than 90 degrees, it follows that, theoretically, the total deflection would be 360 degrees in 24 hours.

Imagine again the disc is tangent to the earth at the equator; it is easy to see there would be no deflection whatever over a period of 24 hours, or over any longer period, because there would be no rotation of the disc. Between the poles and the equator, the amount of possible deflection in 24 hours would depend upon the latitude and would be greater in high latitudes than in low. Since air cannot be deflected more than 90 degrees, it is evident that this amount of deflection at the north pole could take place in a quarter turn of the disc, or in a period of 6 hours. This would be at the rate of 15 degrees in 1 hour. It has been found that the amount of deflection in any given period at any latitude varies with the sine of the latitude.[1] Thus, to find out the maximum deflection in a period of 1 hour at any latitude, you have only to multiply 15 degrees by the sine of that latitude (see table of sines in the Appendix). For example, at the 60th parallel, the maximum deflection in 1 hour would be 0.866×15, or approximately 13 degrees; at the 40th parallel, it would be approximately 10 degrees; at the 30th parallel, exactly one-half, or 7½ degrees; at the equator, zero.

This condition of deflection on a rotating body was demonstrated with a pendulum by Foucault in Paris in 1851; a full description of that experiment may be found in any encyclopedia. In essence, the experiment showed that the pendulum, although it started swinging in a north-south direction, gradually assumed a northeast-southwest direction—not that the pendulum changed its course with respect to space, but the marked line on the earth (in this case, a meridian) turned.

Again, deflection may easily be demonstrated with a globe (preferably a slated globe) and a piece of chalk. Rotate the globe in a counterclockwise direction as viewed from a point above the north pole—that is, from west to east. While the globe is rotating, draw the chalk vertically downward over the northern hemisphere of the globe. The chalk mark will be a curved line running from northeast to southwest. The chalk in your hand did not change its initial direction, but the turning of the globe under the chalk caused an apparent change in the direction of its movement.

To go further into the explanation of deflection would involve considerable mathematics and physics and is beyond the scope of this book. For our purpose, the amount of deflection and a detailed and complete explanation are not so significant as the *fact* of deflection. As we will soon see, deflection is the fundamental reason for the so-called "wind belts" and "pressure belts" and is involved, indirectly at least, in the development of air masses and storms.

General Planetary Circulation

Again, if the earth were a homogeneous surface and if there were no seasonal changes in the insolation received at various latitudes, the rotation of the earth would cause the wind belts to be as illustrated in Figure 8–2.

This condition of greatest heating and expansion at the equator gives rise to overflow and the low-pressure belt called the *doldrums*. Cooler, denser air moves in from each side, forcing aloft the heated equatorial air. The doldrum belt is, therefore, a region of calms

[1] Draw a radius through one end of the arc of a circle, and from the other end of the arc drop a perpendicular upon that radius. The ratio of the perpendicular to the radius is called a *sine*. Therefore, the greater the arc of the circle (that is, the higher the latitude), the greater will be this ratio or sine.

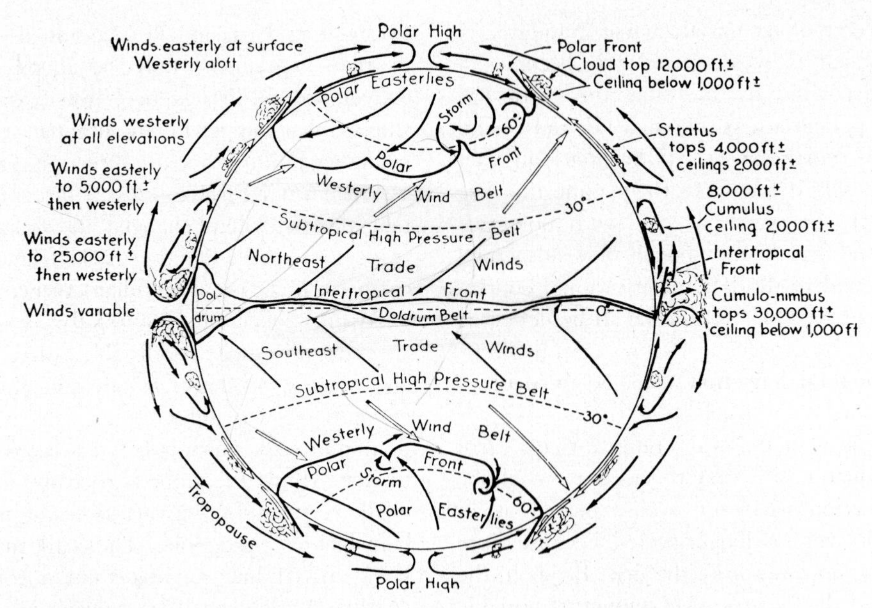

Fig. 8–2. General planetary circulation. (*Courtesy W. L. Donn.*)

or variable winds and occasional squalls and of much cloudiness and rainfall. These inflowing winds, instead of being north winds in the northern hemisphere and south winds in the southern hemisphere, are deflected to the right and left respectively and become the *northeast trades* and the *southeast trades*. They begin approximately at 25 degrees north and south latitude. Why do they not start from the poles? The answer is probably as follows: The expanded air at the equator spreads out aloft on both sides and moves toward the poles. Since there is less friction at higher elevations, the winds are readily deflected to the right and left in the northern and southern hemispheres. By the time they have reached a latitude of 20 or 30 degrees, the deflection has turned them in an eastward direction. This results apparently in an accumulation of air over latitudes 20 to 30 degrees north and south, which means high pressure on the earth's surface at those latitudes or a little beyond. These high-pressure belts are known as the *horse latitudes*, regions of descending air and, therefore, of calms or variable winds and of clear skies. Another factor in the accumulation of air over subtropical latitudes is mentioned in connection with jet streams on the next page.

Correspondingly, near the poles[1] the shrinkage of air due to cooling favors the inflow of air aloft from warmer latitudes. Low subpolar pressures and high polar pressures are favored by this transfer of air. In accordance with the pressure distribution, viz., the *subpolar low-pressure belts*, at approximately 60 degrees from the equator, and the *polar high-pressure* caps, winds flow out from the polar regions and are deflected to the right or left so that they become northeasterly or easterly in the north-polar region and southeasterly or easterly in the south-polar region. These winds are generally referred to simply as the *polar easterlies*. The subpolar low- and the subtropical high-pressure belts cause a surface movement of air poleward that becomes, because of deflection, southwesterly in the northern hemisphere and northwesterly in the southern hemisphere. These winds are commonly called

[1] Not actually at the geographical north pole but rather in Siberia where the cold is most intense in winter.

the *prevailing westerlies*. They are usually deflected more than the trade winds. The latitudes in which they occur in the southern hemisphere are often referred to as the "roaring forties," the "howling fifties," or even the "screeching sixties" because of the frequent gales in those oceanic southern latitudes; the lack of interfering land bodies and the strong pressure gradient account for the high velocities.

At altitudes from 6,000 to 10,000 feet, the pressure distribution seems to be relatively simple, compared with that at the ground. There appears to be high pressure at high altitudes over subtropical latitudes because of the vast importation of air from below, and low pressure at high altitudes over the poles because of the enormous compression of the air at the surface caused by the intense cold.

Such a pressure distribution results in a general poleward movement of air, although the deflective effect of rotation gives the winds a pronounced west–east movement. In addition, but for reasons not well understood, pressure at high altitudes over the equator tends to be uniform although slightly lower than over latitudes immediately to the north and south, resulting in an east–west movement of air over the equatorial region. Owing to lack of friction, to lower air density, and to the presence of a relatively strong pressure gradient, the westerly winds attain enormously high speeds in the upper troposphere and at the tropopause.

Jet Streams

One highly variable and illusive feature of the prevailing westerlies near the tropopause is the occurrence of those powerful rivers of air known as "jet streams," already referred to in Chapter 3. They are becoming increasingly significant to the student who is interested in upper-air currents and their possible effects upon surface-weather conditions; certainly they are of real importance to high-flying airplanes.

Although the existence of these narrow, high-speed wind belts was noted by Bjerknes as early as 1933, they were not actually encountered until bombers in World War II, attempting to fly westward at altitudes above 30,000 feet, found themselves flying backward in the face of such strong head winds. This led to intensive exploration and study, with the result that these enormous streams of air, aptly termed *jet streams*, have been observed and have been charted in considerable detail.

Jet streams are erratic as to speed, direction, and extent, constantly shifting in altitude as well as in latitude but always moving in a meandering course from west to east. Speeds from 150 miles per hour to as much as 400 miles per hour have been noted. Normally they occur close to the tropopause, sometimes actually puncturing it, yet they are frequently reported as low as 20,000 feet. They are often sharp and clear-cut, confined like a river within its banks, particularly on their equatorward margins; the piling up of air on the right, or south, margins in the northern hemisphere (on the left, or north, margins in the southern hemisphere) is believed to be a factor in the great accumulation of air over subtropical latitudes, contributing to the horse latitude belts of high pressure.

A fairly typical jet-stream system for January 18, 1957, at approximately the 30,000-foot level, is shown in Figure 8–3; here wind speed in knots per hour is shown by *isotachs*, lines of equal wind speed. The arrows indicate the main axis of the several components, or tongues. The average position, as well as the strength of the jet stream for January and for July, is shown in Figure 8–4.

Though obscure in many details, the origin, development, and eventual breaking up of jet streams are definitely associated with the contact of dry, cold polar air with warmer and more humid tropical air, very much like the frontal systems at the earth's surface; and like the surface wind and pressure belts, they change with the seasons.

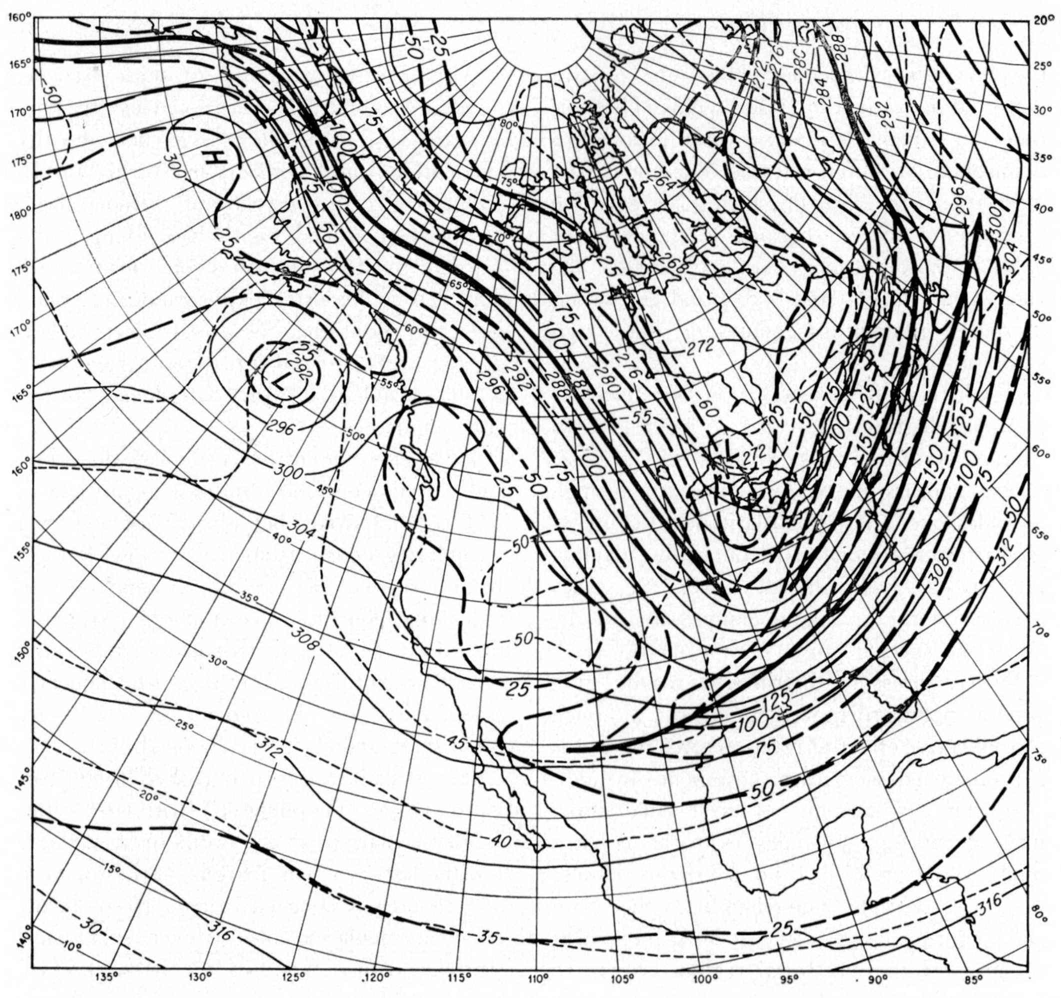

Fig. 8–3. Typical jet stream, 300-mb level, January 18, 1957.

So far the jet stream is used by flyers only in winter because it is then within range of most flights; at this season it is found from 30 to 40 degrees north latitude and at elevations of 4 to 6 miles. In summer it is beyond all except experimental flights; at this season it is at least 7 miles in elevation and around 50 degrees north of the equator. The same sort of seasonal shifting occurs in the southern hemisphere. Obviously, the advantage of the jet stream depends upon the direction in which one is flying. For example, an Air Force C97 cut 7½ hours from its normal flying time from Tokyo to Honolulu by using the jet stream as a tail wind; but if one is flying westward against it, he may "stand still with engines roaring at full throttle." It should also be noted that in the northern hemisphere at all seasons jet streams seem to be farthest north over water and farthest south over land.

There seems to be no doubt that the jet stream will become of greater importance to flying and certainly of greater importance to our knowledge of surface weather as its changing and whimsical moods are recorded, charted, and studied. There is already sufficient evidence to indicate that eddies from jet streams reach down and influence the attendant weather conditions of cyclonic storms: rain or snow, thunderstorms and tornadoes, cold waves and

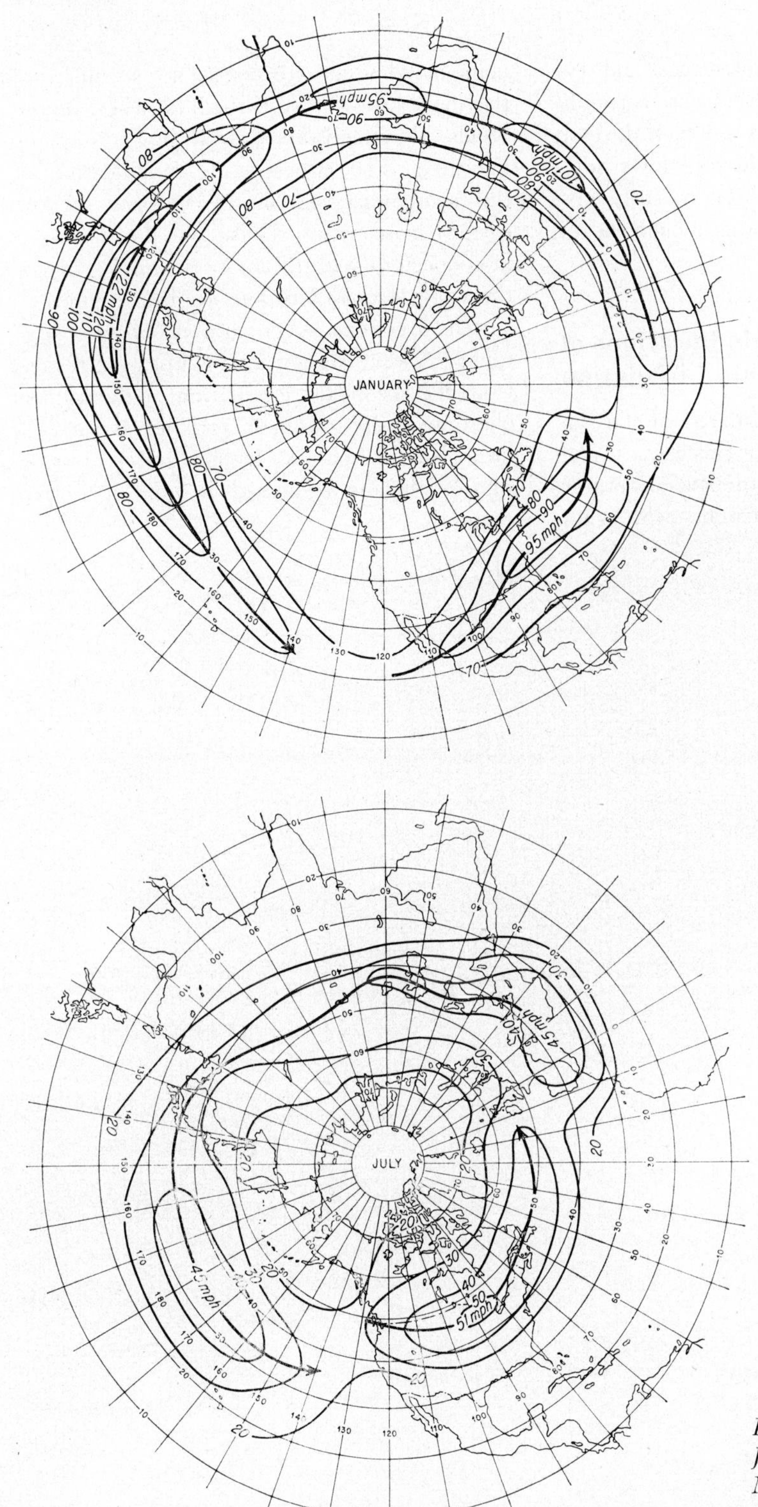

Fig. 8–4. Average January and July jet streams. (Courtesy Jerome Namias.)

blizzards, and probably hurricanes and typhoons. These surface conditions are certainly part and parcel of the storms which develop aloft, whether they are cold-core lows (discussed in Chapter 9) or those recently discovered giant tropical storms which never seem to reach the surface.

Modifications of Planetary Circulation

Because of the apparent march of the sun north and south of the equator twice a year and because of the presence of large land and water bodies as well as ocean currents, the wind belts including jet streams are not so simple as they have just been portrayed. Nowhere do the winds blow constantly from the directions indicated. The largest scale modification of this system is seen in the *monsoons* of the eastern hemisphere. These will be discussed in connection with the more detailed study of the wind belts in Chapter 15. Local winds of varied directions occur as a result of topography. A very significant modification results from the secondary circulation of the atmosphere affecting the characteristics and movements of air masses, fronts, and storms, a subject that is discussed in the next two chapters.

9

Air Masses and Fronts

High and Low Pressure

It has already been pointed out that air is not static—it is in a state of almost constant motion, moving vertically or horizontally depending upon its temperature and humidity and thus upon its weight or pressure with respect to the corresponding conditions of surrounding bodies or masses of air. We have also seen that these differences in weight, or density, manifest themselves by differences in pressure at the ground (and at any height above the ground). If sea-level pressure is low, there must be less air pressing down than if the pressure is high. In other words, low pressure at the ground means an *accumulation* of air aloft. The circumstances involving this withdrawal or accumulation of air are determined in part, particularly in extratropical latitudes, by the thermal and moisture conditions of large masses of air and in part by the heating or cooling of the ground underlying those masses of air.

Air Masses

Air masses are large bodies of air having their own distinctive characteristics; they are three-dimensional. An air mass is said to be *cold* when it is colder than the surface over which it rests or is moving; an air mass is said to be *warm* when it is warmer than the surface over which it rests or is moving. A cold air mass may become a warm air mass without a change in temperature if it passes over a surface having a temperature lower than its own; or a warm air mass may become a cold air mass without change in temperature if it moves over a surface having a temperature higher than its own. Thus a warm air mass moving from cold land to a warm body of water could be quickly converted into a cold air mass even though warmed appreciably by contact with the warm water. Furthermore, a cold air mass by day may become a warm air mass at night because of the rapid cooling of the ground by radiation after sunset. Air masses, however, tend to retain their identity as far as temperature and moisture are concerned because air masses of different densities, like oil and water, do not mix readily.

Cold air masses tend to be turbulent because of convection caused by the warming of the air in contact with the ground; thus cumulus clouds with their billowing domes or turreted tops and flat bases are a characteristic of cold air masses, although not necessarily present at all times. Frequently, the vertical development of clouds may be sufficient to cause local thunderstorms and squalls for short periods. Turbulence results in the bumpiness so well known to the air traveler; this bumpiness and the cumulus clouds are identifying features of cold air masses.

Warm air masses have little or no vertical movement because of the cooling of the air in contact with the ground. Hence the air becomes stratified, and inversions of temperature and fogs are characteristic. Unlike cold air masses, the lack of vertical development in warm air masses causes any dust or smoke to be concentrated near the ground, thereby causing low visibility. Any clouds are the stratus type. Smooth flying would be found above the cloud layer, but landing would be difficult because of the low ceiling, low visibility, and possible fog, smog, or smaze.

Characteristics of Air Masses

Meteorologists throughout the world have come to recognize certain well-defined masses of air that have certain qualities as to weather characteristics, origin, and movement. They are classified first as to origin and are further subdivided as to the nature of the surface over which they originate; they are usually classified as to whether they are cold or warm. To designate them simply and easily, they are given letter designations: *A* for arctic, *P* for polar, *T* for tropical, *E* for equatorial, and *S* for superior (in case it is not in contact with the ground but is descending from aloft); *c* for continental and *m* for maritime; *w* for warm and *k* for cold (really for the German word *kalt*). Instead of the symbols *c* and *m*, further refinement sometimes substitutes for *c* the small letter *p* for Pacific; and for *m*, the small letter *a* for Atlantic and *g* for Gulf.

For simplification and comparison, the sources and characteristics of the various air masses that are most likely to have a bearing on United States weather are summarized in Table 9–1.

Familiarity with the meaning of the symbols furnishes the student with a good concept of any air mass to which a symbol is applied. Thus the symbol *pTw* would indicate the air had its origin over the subtropical portion of the north Pacific and is therefore relatively warm and humid; since it has the symbol *w*, indicating the air is warmer than the land over which it is moving, it would probably represent a winter condition. Again, *cPk* in any season would indicate a cool, or cold, dry mass of air moving in from Canada over the warmer surface of the United States.

Since air masses of different temperatures do not readily mix, there is necessarily a boundary zone between them; this boundary zone is likely to be wavy rather than clear-cut like a plane. Because cold air is heavier than warm air, the latter will be lifted with resultant expansion and adiabatic cooling; this will probably cause cumulus clouds and perhaps rain. The boundary between the two masses is known as a *front*. And since cold air is usually moving toward the southeast and warm air toward the northeast, the general movement is easterly. The front of the cold air moving over the ground formerly occupied by the warm air is said to be a *cold front*, while the warm air moving over ground originally occupied by the cold air retreating toward the east is said to be a *warm front*.

Movement of Air Masses

From the foregoing, it appears that cold fronts (with the cold air behind them) and warm fronts (with the warm air behind them)

Table 9–1. Sources and Characteristics of the Principal Air Masses

Type	Symbol	Origin	Temperature and Moisture Characteristics at Source: Winter	Summer
Polar continental	cP	Alaska, Canada	very cold, dry	cool, dry
Polar maritime	mP	Northern Atlantic or Pacific	moderately cold, humid	cool, humid
Polar Pacific	pP	Northern Pacific		
Polar Atlantic	aP	Northern Atlantic		
Tropical continental	cT	Southwestern U.S., northern Mexico	(not usually found in the U.S.)	hot, dry
Tropical maritime	mT	Subtropical waters	warm, humid	hot, humid
Tropical Pacific	pT	Subtropical North Pacific		
Tropical Atlantic	aT	Subtropical North Atlantic		
Tropical Gulf	gT	Gulf of Mexico and Caribbean		

tend to follow each other with considerable regularity; this is true for the most part. Now the question arises: Why do they do this? From our study of wind and pressure belts in the preceding chapter, the student will recall the position as well as the reasons for the polar easterlies and the prevailing westerlies. There may be occasions when a large accumulation of cold air will force the polar easterlies far south of the 60th parallel, and there may also be occasions when the prevailing westerlies will be pushed well beyond subpolar latitudes. The polar easterlies (or perhaps northeasterlies) represent a cold air mass; they fan out irregularly in pronglike bursts and not in regular and symmetrical fashion. The prevailing westerlies (or southwesterlies) represent a warm air mass; they tend to flow poleward between the prongs of cold air. When these masses (or winds) are moving in opposite directions but parallel to each other, the drag along the surface of discontinuity between the two winds, plus the differences in density, set up a wavelike motion. This early stage in the development of wave fronts is shown in the diagrams *A* and *B* of Figure 9–1. The crest of the wave is the point (or surface) where the cold front merges into the warm front, diagram *C*. Low atmospheric pressure develops at this crest because the warm, humid, and therefore light air rides up over the cold, dry, and heavy air while the latter pushes southward, replacing some of the warm air. Thus there is developed a definite *Low*, or storm center, often referred to as a "cyclonic depression," or simply as a *cyclone*. A further development is shown in diagram *D* where there is a definite counterclockwise circulation of air, a well-defined sector of warm air being pinched and made smaller by the resistance of the cold air to the right and by the onward force of the cold air to the left; here we have a well-marked warm front and cold front. A still further development is shown in diagrams *E* and *F* in which the cold air has moved so rapidly it overtakes the warm front, pushing the warm front aloft and *occluding* it, or closing it off, from the surface, thus filling up the area of low atmospheric pressure.

Types of Occlusion

There are two types of occlusion; they are easily understood by reference to the two diagrams in Figure 9–2. If the cold air which overtakes the warm air is colder than the retreating cold air, it will push the warm front upward so that the latter does not appear at the ground but aloft, forming an *upper warm front*. This process is known as the *cold-front type* of occlusion because the advancing cold air does the displacing; it is illustrated by diagram *A*. It is the most common type of occlusion and occurs in most instances east of the Rocky Mountains where the retreating cold air mass has had a chance to warm up; it

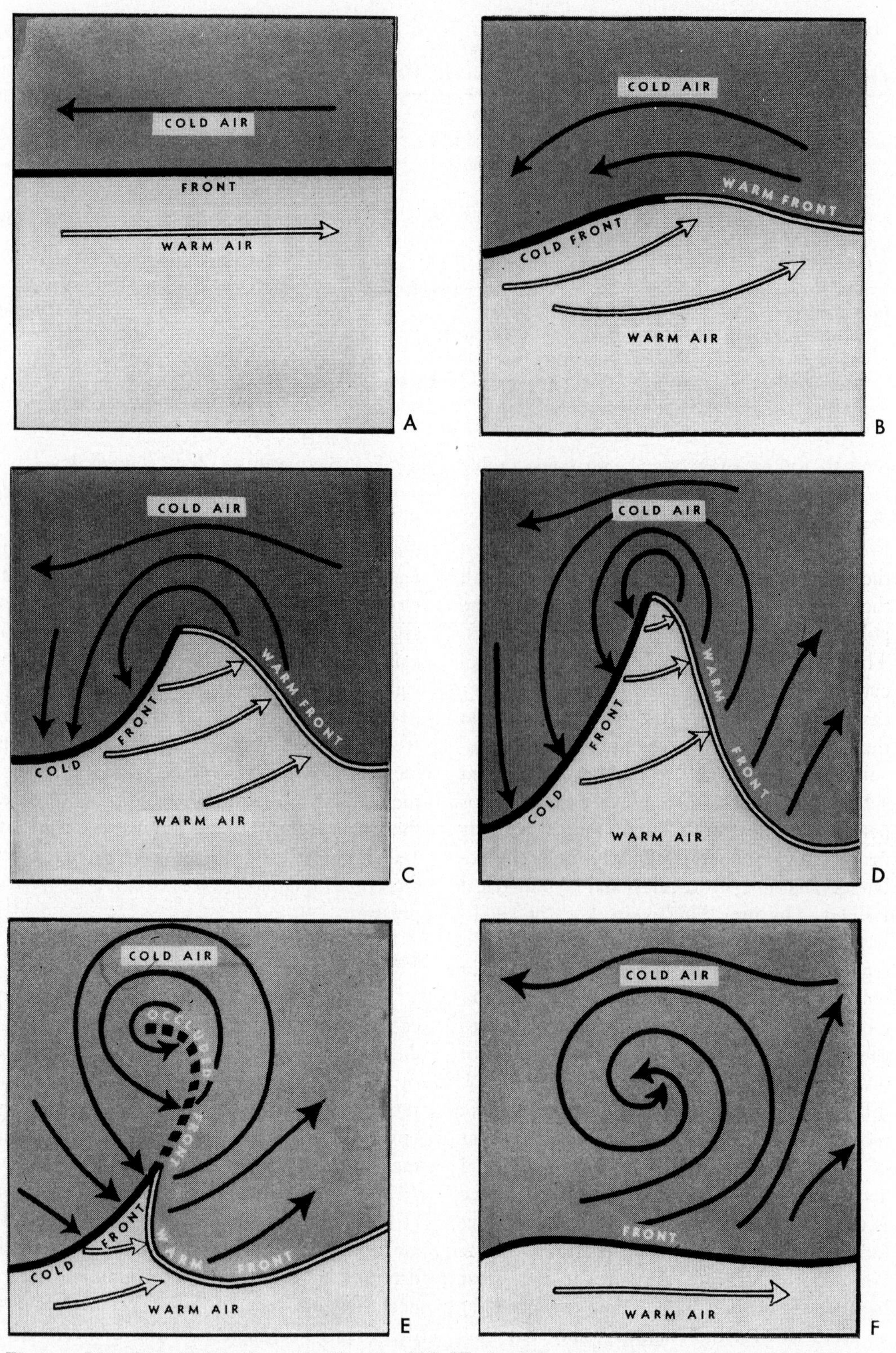

Fig. 9–1. Life cycle of a frontal system. (*Courtesy U.S. Weather Bureau.*)

is, therefore, warmer than the advancing cold air mass from the northwest, particularly the *cP* type. In rarer instances, however, the retreating cold air mass will be colder than the advancing cold air mass so that the latter is forced to override the residual, or retreating, colder air mass. The actual cold front is, therefore, held aloft and is called an *upper cold front.* This process is known as a *warm-front type* of occlusion because it is the retreating edge of a former cold air mass which does the displacing; it is illustrated by diagram *B.* This type is most likely to occur when the retreating cold air becomes increasingly cold by radiation, as over the dry western or Great Basin states, and when the advancing cold air mass is the *mP* type. In either type of occlusion, the warm air mass is forced upward, resulting in cooling, condensation, and probable precipitation. The only essential difference in surface-weather conditions is that the west-east weather band is wider in the warm-front type just as it is wider along a surface warm front; and the precipitation is generally lighter than with the cold-front type of occlusion.

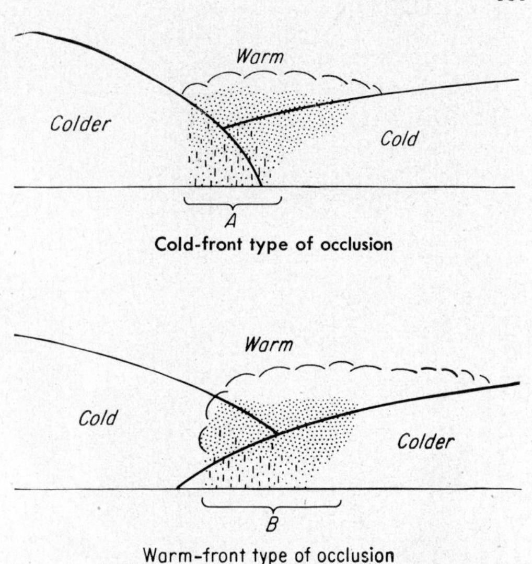

Fig. 9–2. Occlusions. Note the wider weather band in the warm-front type of occlusion.

Life Cycle of a Frontal System

The whole process—from the time the prong of polar air encounters the northward flow of warm air to the final, complete occlusion and filling up of the area of low pressure—is known as the *life cycle* of a frontal system according to the *wave theory* as originally developed by Vilhelm Bjerknes, the Norwegian meteorologist. It should be pointed out here, however, that not all meteorologists are agreed that low-pressure areas or cyclones *always* develop as waves on polar fronts because, as they say, the polar-front theory does not accurately describe conditions in the upper air; and they add that cyclones can develop without polar fronts. All this hinges on the theory that jet streams must have a significant relationship to the development of surface storms for it seems to be fairly well established that cyclonic activity and precipitation are generally concentrated along the equatorward margins of jet streams and that, when a pronounced jet stream lies over a distinct frontal zone, strong cyclonic activity almost invariably develops.

A well-developed frontal system may be found on almost any daily weather map, particularly in winter. An ideal one without isobars (lines of equal pressure), with the insert showing isobars, is given in Figure 9–3. Note that the cold air masses, because of their coldness and low water-vapor content, are shown as well-defined areas of high pressure marked *HIGH* on the map; such areas are generally referred to as *anticyclones.* The succession of *Highs* and *Lows* (that is, waves of cold air and of warm air) across the country gives rise to the succession of weather changes so typical of middle latitudes, as in the United States, Canada, and Europe.

Characteristics of a Low-pressure Area

Some of the detailed and usual characteristics of a low-pressure area are indicated by the diagram of an "ideal cyclone," Figure

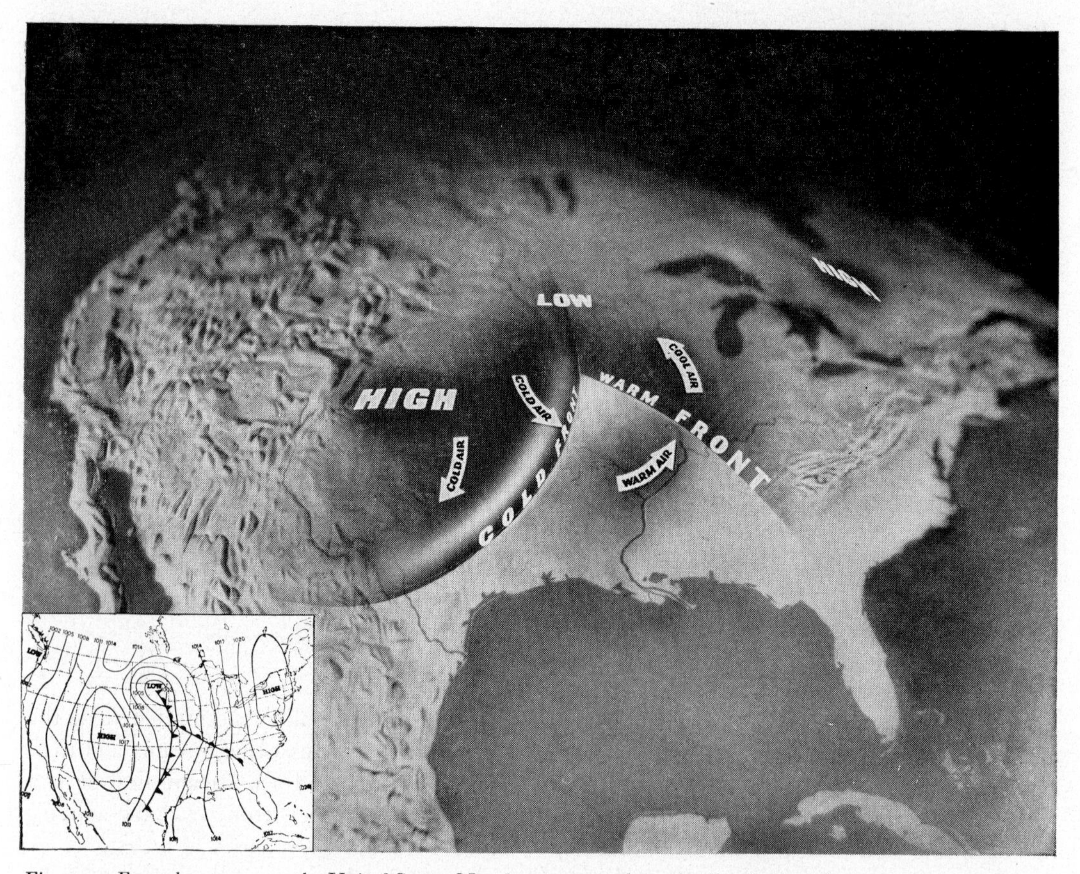

Fig. 9–3. Frontal system over the United States. Note inset map to show the distribution of pressure. (Courtesy U.S. Weather Bureau.)

9–4. This diagram is a slightly modified version of the original made by Bjerknes. No isobars are shown, merely the circulation of the air. The portion of the diagram labeled *A* at the top of page 103 represents a west-east vertical section through the upper portion of diagram *B*. It should be noted that, as the cyclone or storm center moves eastward, no place in the path of the northern portion of it will experience any exceptionally warm weather. Instead, as illustrated in the vertical section, the air will be relatively cool at first, followed by even colder air as the storm passes. At first the sky becomes covered with cirrus clouds, thickening to cirro-stratus, alto-stratus, and finally nimbus, as some of the air of the warm sector rides over the cold air. Precipitation is at first light but becomes heavier where the cold air is most shallow. With the passing of the storm, precipitation gradually ceases, and the nimbus clouds give way rather rapidly to strato-cumulus and alto-stratus in low levels and to cirro-stratus in high levels, which soon disappear, leaving the sky clear.

The weather is very different for places lying in the path of the southern portion of the storm. This is illustrated in the west-east vertical section *C*. As in section *A*, cirrus clouds appear first, hours and sometimes days before, but they are followed eventually by cirro-stratus and alto-stratus; rain or snow soon follows as a result of the forced ascent of the warm air over the cold residual air. With the advent of the warm winds, precipitation ceases rather abruptly, the sky clears, and there are several hours of relatively warm, sunny weather, perhaps with occasional

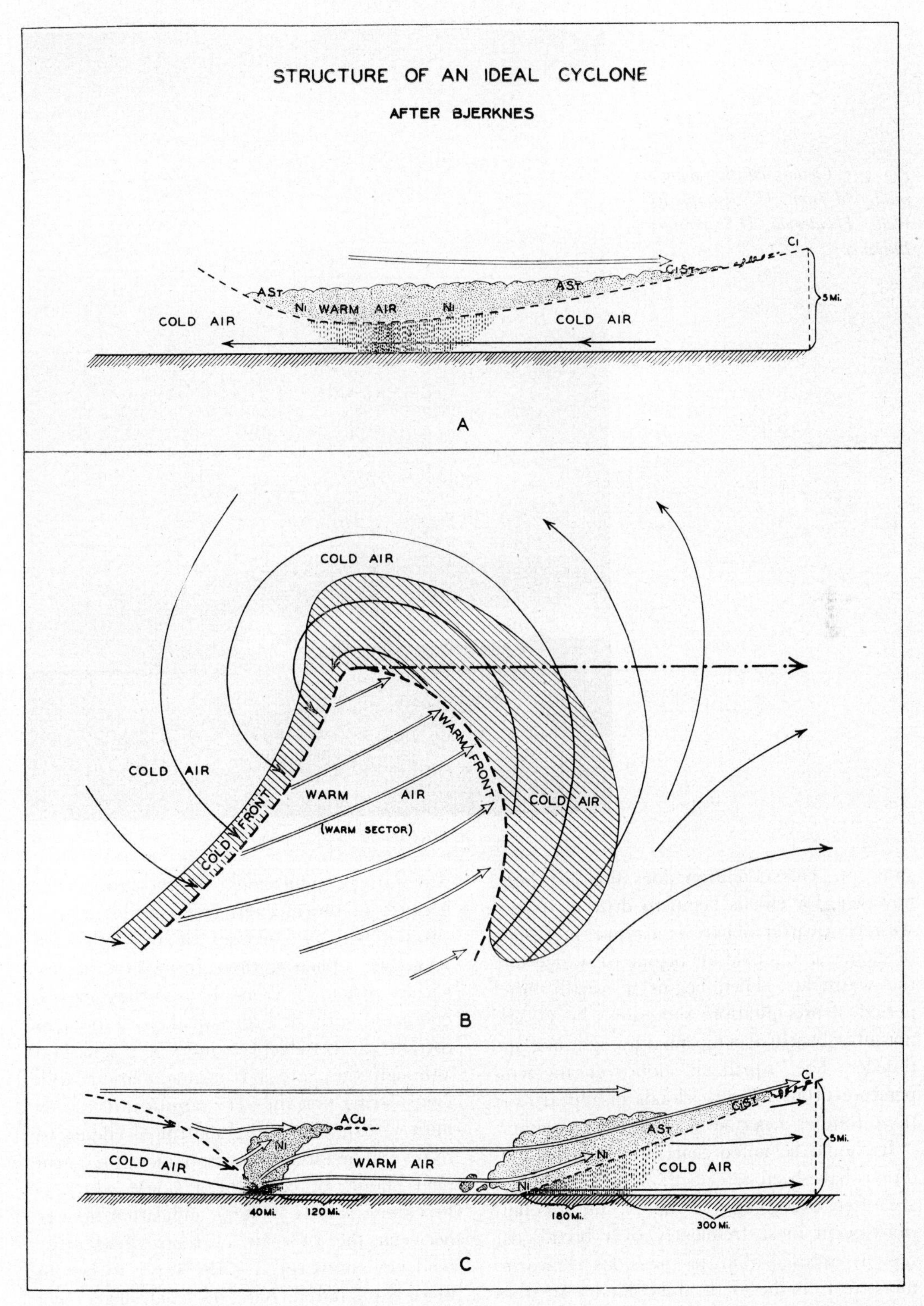

Fig. 9–4. Structure of an ideal cyclone. (After Bjerknes.)

Fig. 9–5. Cloud formation along a mild cold front. (Photograph by Walter Henderson, U.S. Weather Bureau.)

showers. This condition does not last long, however, for clouds begin to drift in from a westerly quarter. There is a rapid thickening of clouds as the cold air begins to wedge into the warm air. Then begins a usually brief period of precipitation, sometimes heavy and not infrequently accompanied by lightning and thunder. The "squall" is soon over, the temperature drops rapidly, clouds disappear, and there follows a period of cold, clear weather.

It should be noted that only "ideal" conditions have been described. Actual conditions are often exceedingly complex. Ideal conditions occur most frequently over broad, flat regions exposed alike to invasions of warm, moist air from the south, and cold, dry air from the north. The meeting of the cold *polar* air with the warm *tropical* air gives rise to eddies, waves, or undulations in the atmosphere because of the motions and contrasting densities of the two masses of air. While the waves are similar to those formed on the sea's surface when the wind blows, they may be compared with the eddies made by a mountain torrent as it suddenly enters a placid pool, although they are on a much larger scale. Considering first the very largest of these, we must visualize the development of eddies 100 to several hundred miles in diameter, requiring many hours or days for a single whirl. At first there is only a slight undulation or wave between the two air currents. There is a tendency, however, for the wave to become more pronounced, until the cold current perhaps curves completely around the mass of warm air, forcing the latter and its warm front aloft and resulting in an occlusion such as we referred to earlier.

Characteristics of Fronts

Let us review and note some definite characteristics of fronts and of the weather accompanying them. As viewed from the south, cold fronts are on the left of the observer, warm fronts on the right. The cold front is relatively steep, and it is likely to be tongue-shaped rather than like a simple inclined plane because of the frictional drag as the air moves over the irregular ground surface. Frequently the tip of the tongue of cold air may be hundreds or even several thousand feet above the ground and may extend as much as 100 or 200 miles ahead of the front on the ground. The cold front has a narrow weather band (from west to east) with great vertical development of cumulus or cumulo-nimbus clouds. It is marked by rather violent weather with possible lightning, thunder, squalls, rain or snow, and hail; tornadoes may also be a feature. Cold fronts do not necessarily march across the country like an ocean wave. Rough terrain, even cities or forests, may cause the front to lag behind locally while other parts of the front are advancing more rapidly over smoother terrain. The actual front, therefore, may be a very irregular line on the earth's surface, although it is generally shown as a distinctly regular line on the weather map. At times the vanguard of cold air may advance so far as to be cut off completely from the main air mass, eventually losing its identity. A cold front with associated clouds is shown in Figure 9–5.

A warm front has a gentle slope because it is the trailing edge of a former cold air mass; it is a *warm* front because the ground is still cold and there is a frictional drag as the trailing edge moves over the terrain. There is only a gradual lifting of the warm air over the cold air mass; therefore the warm front has a wide weather band from west to east. There is little vertical development of clouds so the stratus type predominates. Drizzles and fogs are common. In winter, ice or glaze storms are likely.

Before we can go into a more complete explanation of the various weather phenomena in connection with fronts, it is necessary to go into further detail concerning weather maps or charts—how they are made and what they are used for. Charts of this sort are called *synoptic* (from the Greek meaning "viewed together") *charts* because they give a comprehensive picture of weather conditions pertaining to fair weather, cold weather, thunderstorm weather, hurricane weather, or any other kind of weather. Synoptic weather and specific types of storms, descriptive as well as prognostic, are taken up in the next chapter.

10

Synoptic Weather, Storms, and Local Winds

Weather Observations and Maps

Reference has already been made to the fact that the United States is one of many countries whose meteorological services are part of the World Meteorological Organization. The purpose of each of the more than 14,000 weather stations belonging to this organization is to make daily, or even more frequent, observations of the various weather elements. Some of the stations are located in the arctic within 800 miles of the north pole. All make ground observations, but many make upper-air observations by one or more of the following means: (1) by radiosonde, which makes direct recordings of pressure, temperature, and humidity at various elevations; (2) by pilot balloons, the data from which permit the calculation of wind speed and direction at various elevations; and (3) by aircraft-reconnaissance reports. Since January, 1958, all regular weather stations in the United States make ground observations at 1:00 A.M., 7:00 A.M., 1:00 P.M., and 7:00 P.M.; they also make observations at other hours as the situation requires. All observations, whether made in the United States or in other countries, are transmitted in an abbreviated numerical code to certain central offices by radio, telegraph, or teletype. In order to analyze the observations for use in making weather forecasts, the observations are plotted on maps; most are surface maps, many are upper-air charts. The Daily Weather Maps, reproduced in Figures 10–1 to 10–5 were made from observations taken at 1:30 A.M. E.S.T. At frequent intervals, valuable information of various types is printed on the reverse side of the weather map; as a whole, the information on weather maps constitutes a good course in weather principles and processes.

In the case of surface or ground maps as used in the United States, all, or almost all, of the observations sent in by an individual station are plotted around the station which is indicated on the map by a small circle. Instrumental observations, such as certain phases of pressure, temperature, and moisture, are indicated by numerals, while other observations are indicated by symbols in ac-

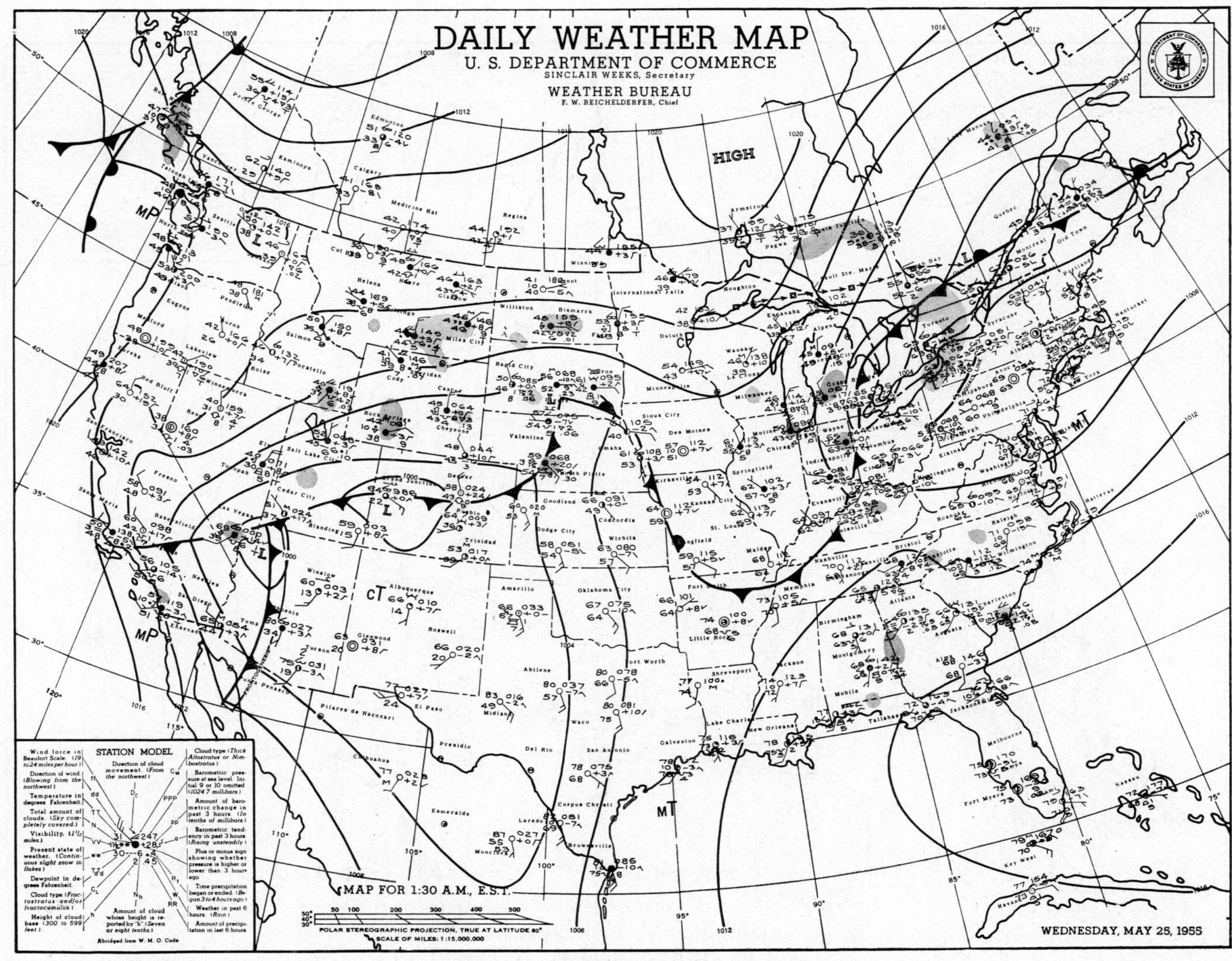

Fig. 10–1. Daily Weather Map, May 25, 1955. (Courtesy U.S. Weather Bureau.)

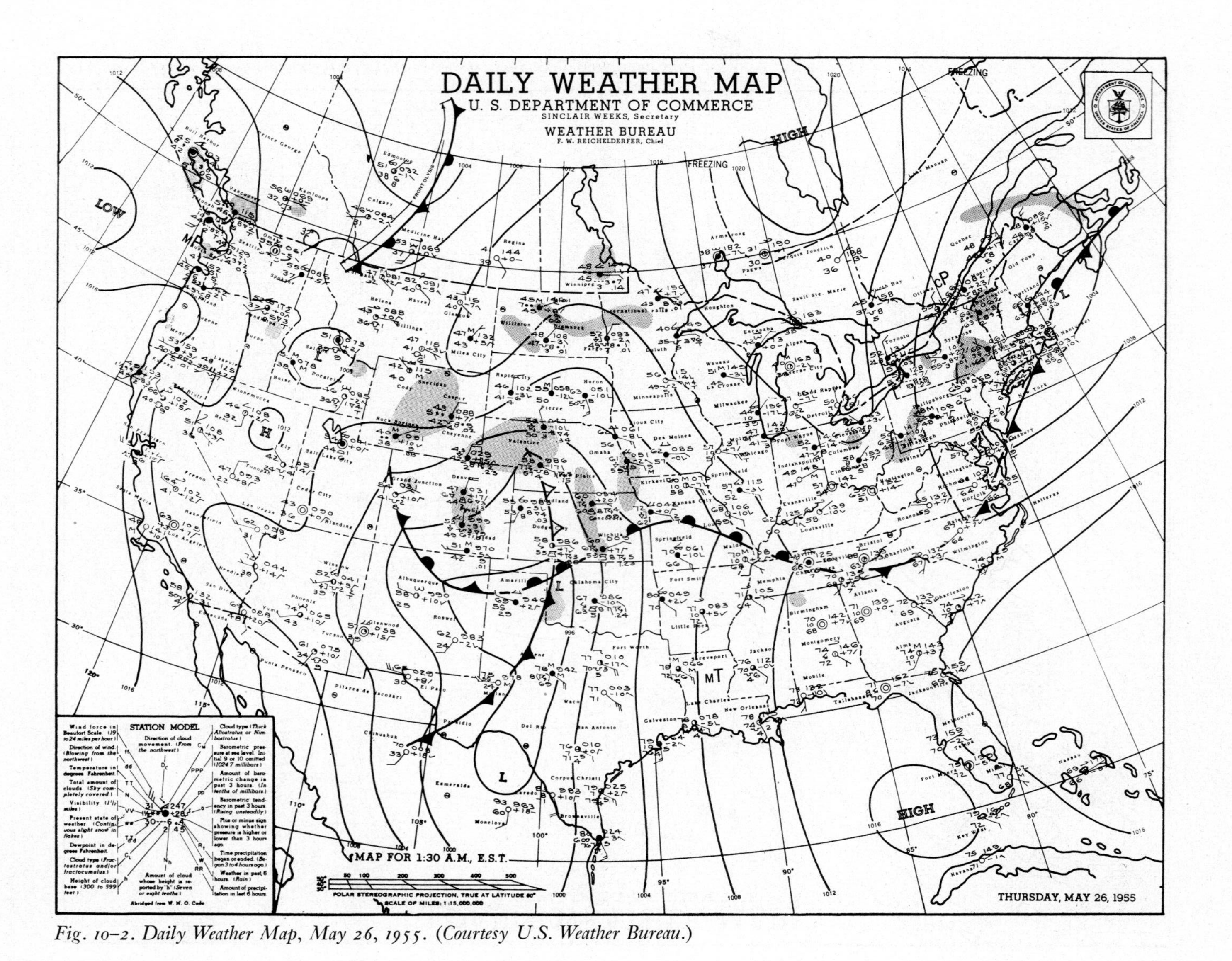

Fig. 10–2. *Daily Weather Map, May 26, 1955. (Courtesy U.S. Weather Bureau.)*

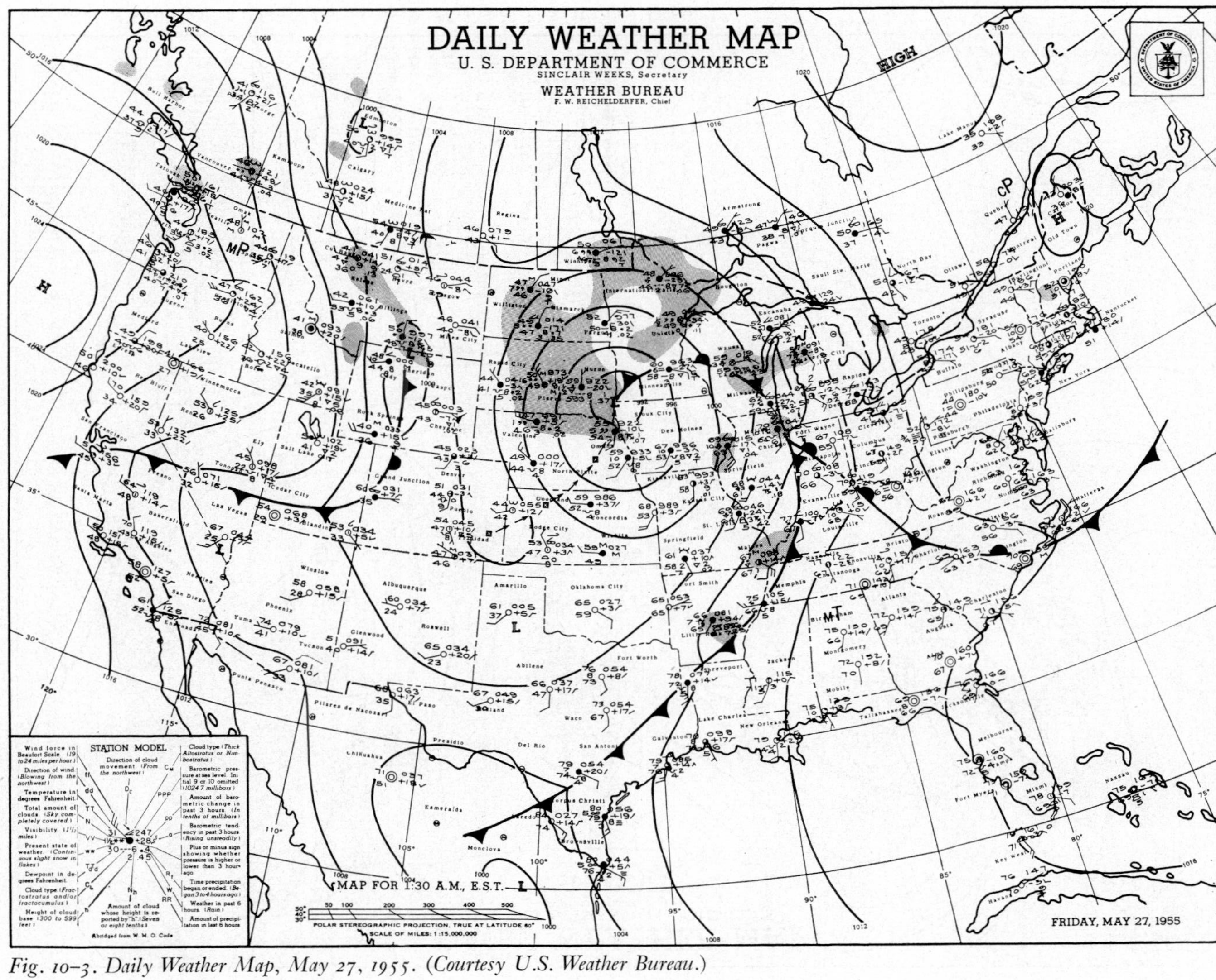

Fig. 10–3. *Daily Weather Map, May 27, 1955. (Courtesy U.S. Weather Bureau.)*

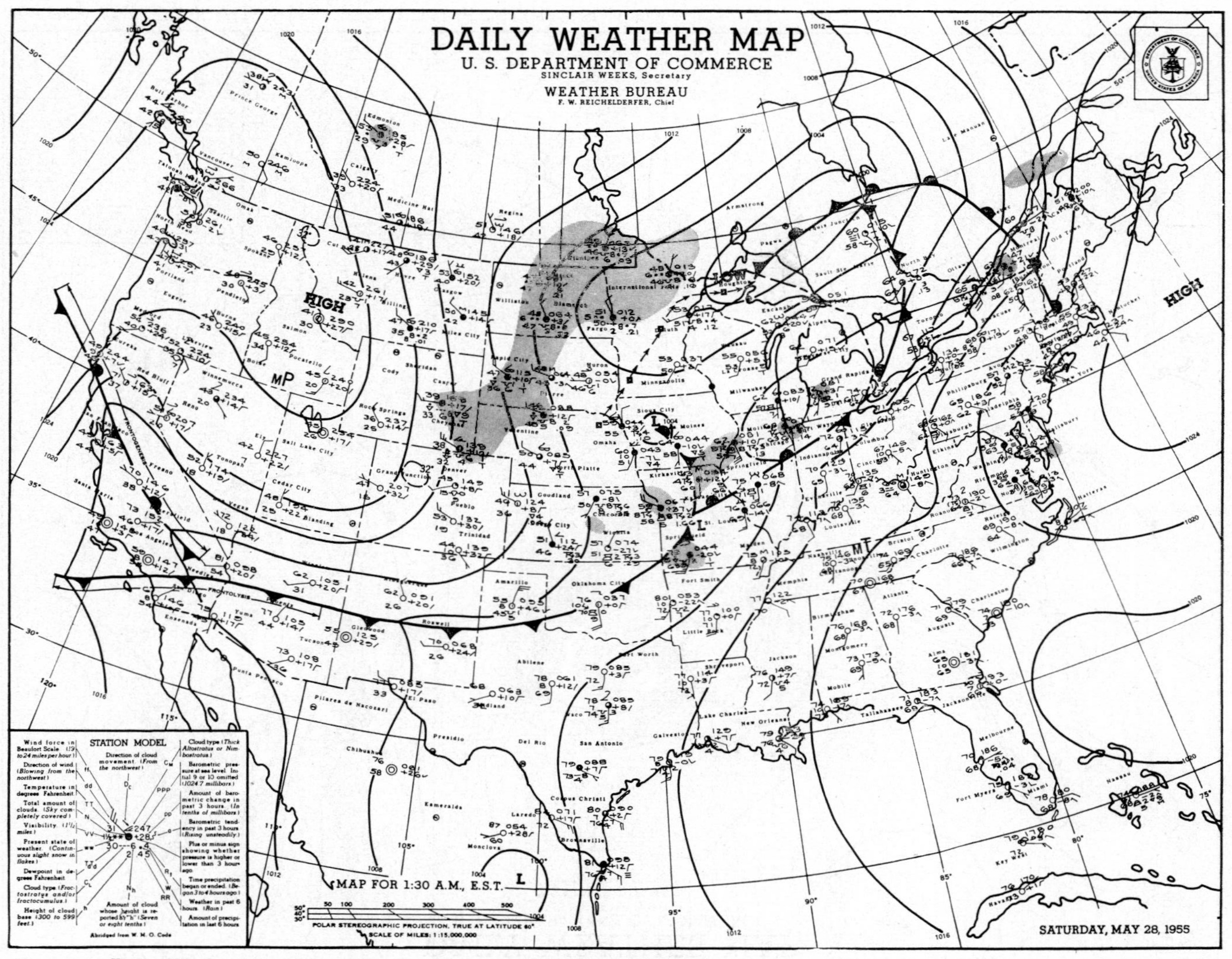

Fig. 10–4. Daily Weather Map, May 28, 1955. (Courtesy U.S. Weather Bureau.)

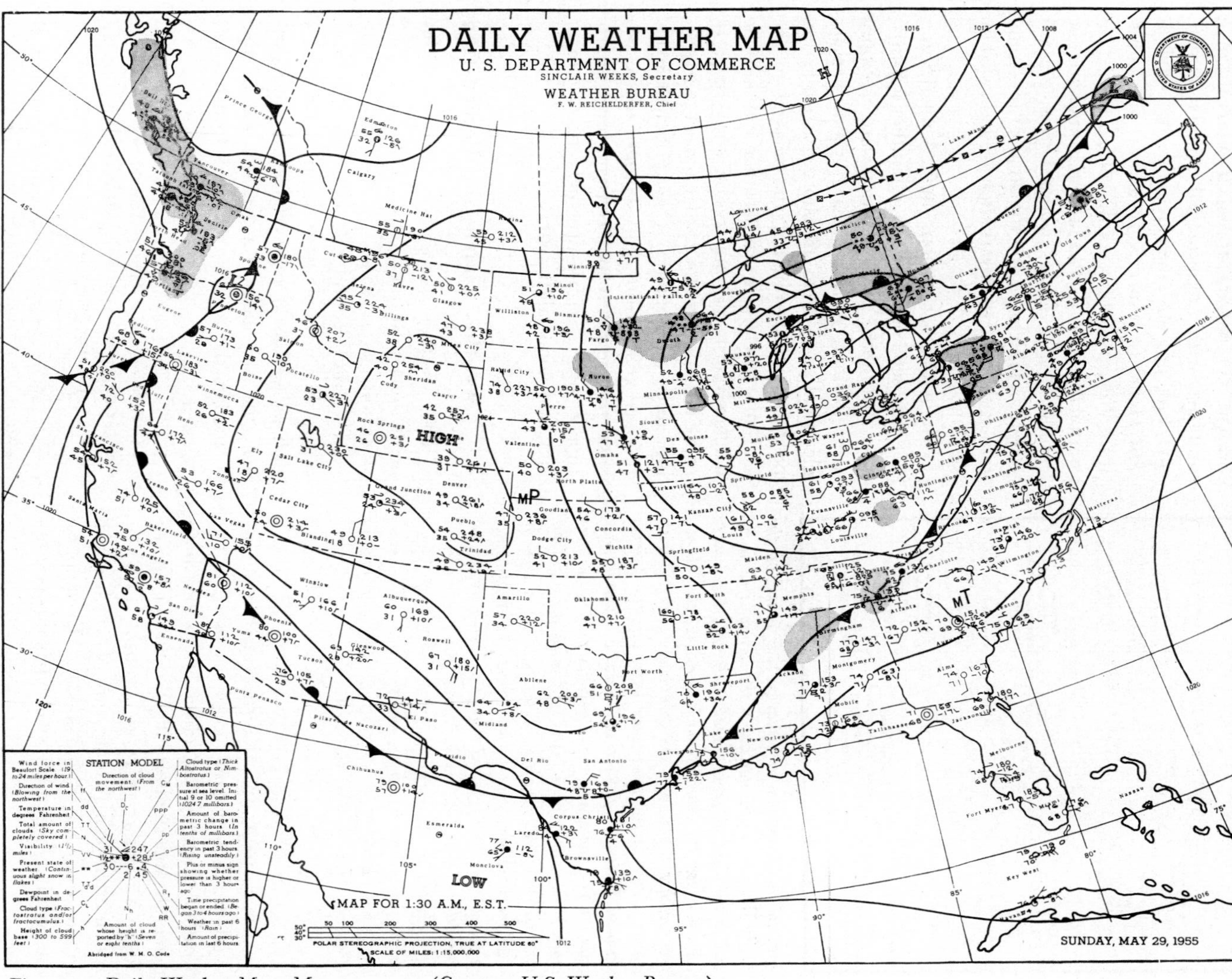

Fig. 10–5. Daily Weather Map, May 29, 1955. (Courtesy U.S. Weather Bureau.)

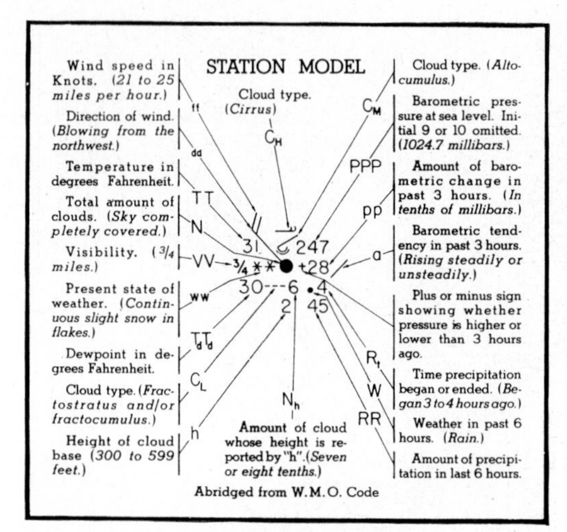

Fig. 10–6. Station model (enlarged) as found on the Daily Weather Map. Starting at the station circle itself and then proceeding from the top of the model and reading clockwise, the symbols and figures give the following weather conditions: sky completely covered with clouds, dense cirrus above and alto-cumulus below; sea-level barometer 1024.7 mb, now 2.8 mb higher than it was 3 hours ago, with tendency to increase; precipitation began 3 to 4 hours ago, rain having been intermittent during the past 6 hours, with a total of 0.45 inches during that period; seven- or eight-tenths of the sky between 300 and 600 feet covered by fracto-cumulus or fracto-stratus clouds; dew point 30°; present precipitation in the form of continuous fall of snowflakes; visibility ¾ mile; temperature 31°; wind northwest, Beaufort 4 (21 to 25 miles per hour). (Courtesy U.S. Weather Bureau.)

cordance with the international code of symbols. The station model, as used on the weather map, gives observations in both numerals and symbols; it is shown in Figure 10–6. Altogether there are probably close to 200 international symbols, any of which could appear around a station model.[1] To simplify the weather map, only two types of lines are drawn: (1) isobars drawn at intervals of 4 millibars with the principal centers of high and low pressure indicated by the symbols *HIGH* and *L*, respectively, and (2) any significant *fronts*, whether warm, cold, occluded, or stationary. Air masses are indicated by the combination of letters referred to in the last chapter, Table 9–1, and the areas of precipitation at the time of observation are indicated by shading.

Major weather stations[2] in the United States not only send out but receive weather information constantly. Some stations are equipped with *facsimile machines* that reproduce maps electrically. A map is photoelectrically reproduced in Washington and relayed to the facsimile machine at the weather station. The machine burns the impression on a sheet of paper coated with zinc oxide, a process requiring about 20 minutes. The principal maps received are the four daily synoptic maps, six-hour prognostic maps, constant-level charts (upper air), constant-pressure charts (upper air), and winds-aloft charts.

Weather Charts and Forecasting

The weather forecaster in making his predictions must interpret not only the ground conditions as revealed by the surface map (and as revealed by his own local observations), but he must also identify air masses aloft as well as at the surface. He must estimate the movement of air masses and be able to predict the location of their fronts hours

[1] For a complete list of symbols, see the reverse side of the Sunday Daily Weather Map or the "Manual for Synoptic Code," 1955 edition, both published by the United States Department of Commerce but obtained from the Superintendent of Documents, Washington 25, D.C.

[2] In addition to the regular Weather Bureau stations, there are some 4,000 or more cooperative stations in the United States that make observations only once a day, usually near sunset. These minor stations keep records of temperature, precipitation, and unusual conditions or phenomena which occur from time to time. Their observations are largely for climatological purposes and are printed regularly in the monthly and annual summaries of *Climatological Data* published by the Weather Bureau. Educational institutions of certain types are able to obtain these data without charge.

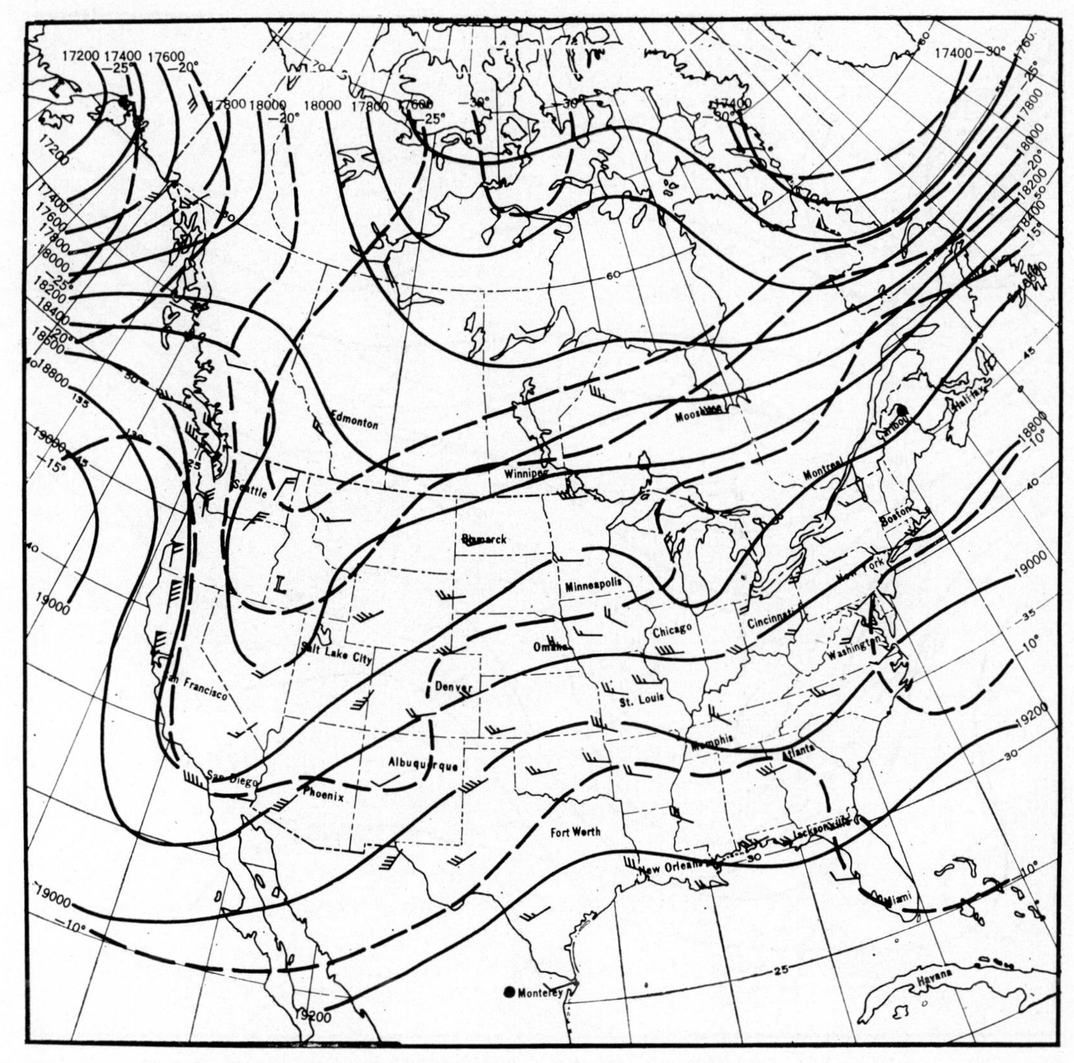

Fig. 10–7. 500-millibar chart, May 25, 1955. (*Courtesy U.S. Weather Bureau.*)

or even days hence. He must know the height of each air mass and the slope of its front: Is the slope steep or gentle? Is the degree of slope constant? Is it like a wedge or tongue, or is it like a simple inclined plane? He must have a complete knowledge of lapse rates in the particular air mass: Is the air stable (small lapse rates) with possible inversions of temperature, or is the air highly unstable (large lapse rates)? Are the winds strong enough to cause mixing along air-mass surfaces?

The forecaster knows that with increasing height, abrupt and minute changes in pressure, temperature, and wind as caused by surface irregularities tend to disappear, and the broader and more general features of circulation become evident; he knows also that these general conditions aloft govern surface conditions and air-mass movements. He knows that a change in the height of an air mass by uplift or subsidence does not necessarily mean a change in the identifying characteristics of the air mass; the change in height may result in adiabatic cooling or warming and consequently in changes in relative humidity, but unless there is condensation, it is the same air mass.

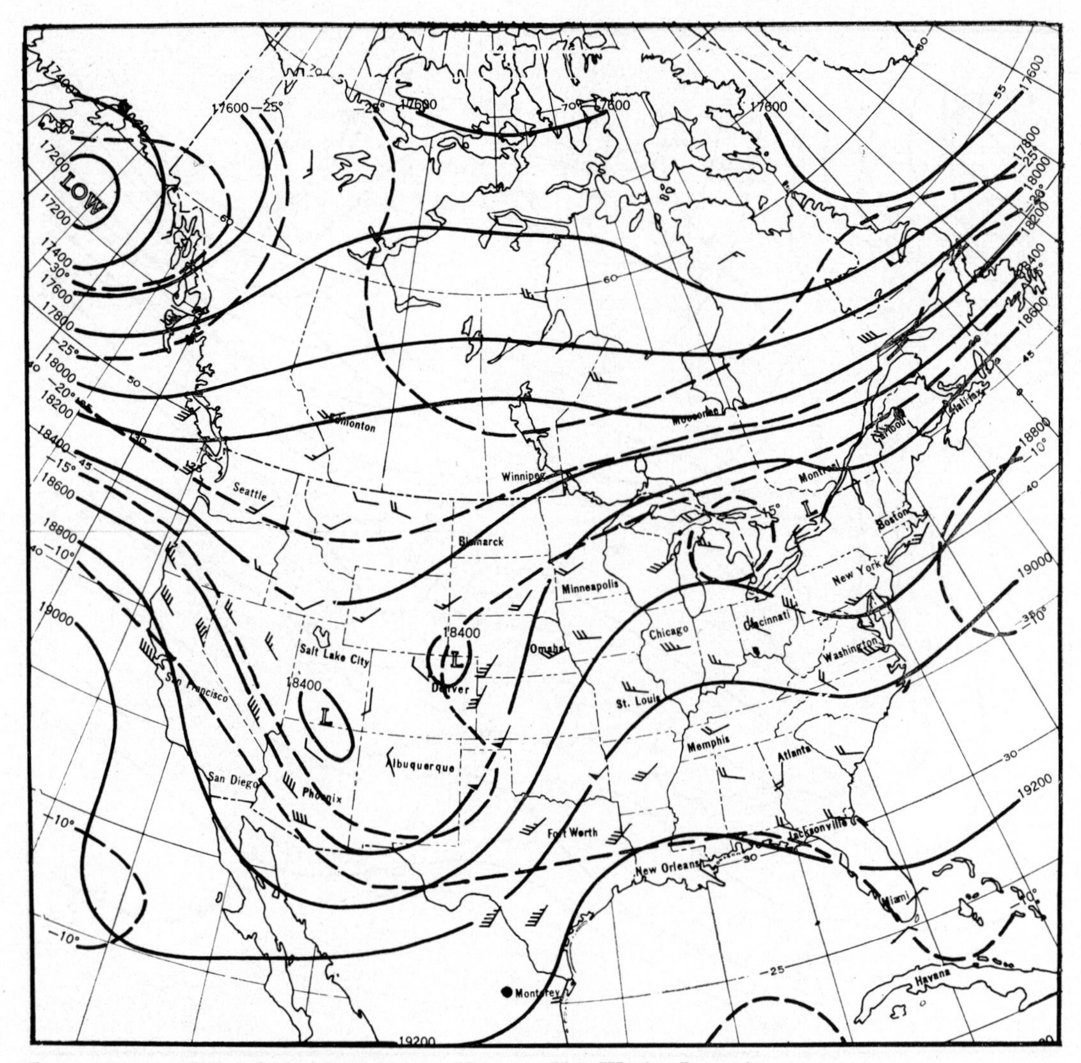

Fig. 10–8. 500-millibar chart, May 26, 1955. (Courtesy U.S. Weather Bureau.)

Therefore, in order to make an accurate forecast of weather conditions, the forecaster must have access to and must study a variety of charts and graphs. Cross sections and profiles must be made across the country from the data obtained from radiosonde flights; *isopleths* (charts showing two or more variables) must be constructed. He must have access to *constant-level charts* in which differences in pressure, temperature, and humidity are shown at any particular level. Of still more importance, he needs to have available and to study *constant-pressure charts*, as in Figures 10–7 to 10–10 inclusive, sometimes called "isobaric surface charts"; these charts show differences in elevation, temperature, and humidity. The lines connecting points of equal elevation on a constant-pressure chart are like contours on a topographic map: they might be closed, indicating definite depressions or hills; they might merely show troughs (valleys) and ridges. The pattern of the succession of troughs and ridges has a direct relationship to their speed of movement, which in turn is reflected in the movement of air masses and fronts on the ground. Since the movement of these troughs

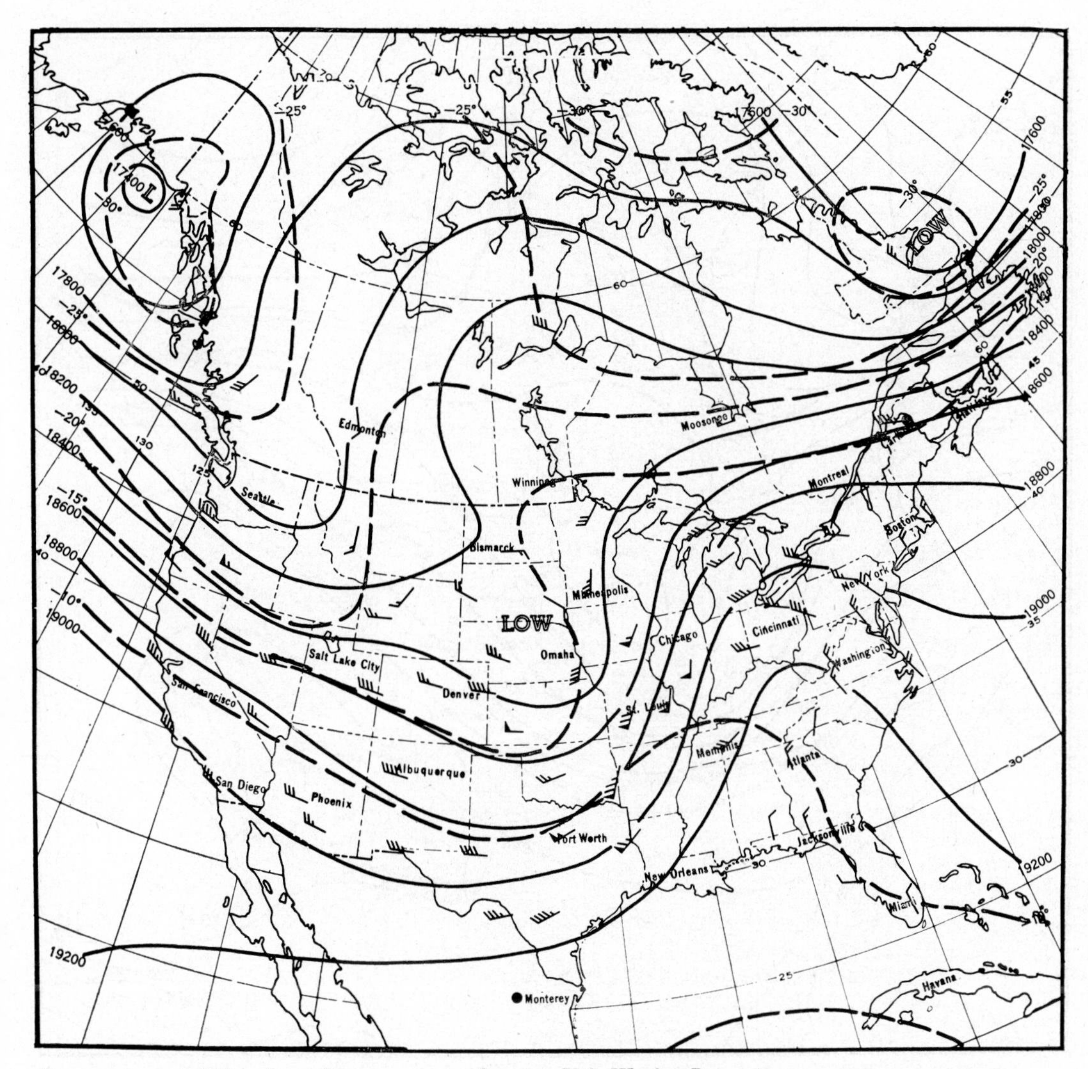

Fig. 10–9. 500-millibar chart, May 27, 1955. (Courtesy U.S. Weather Bureau.)

and ridges can be determined with mathematical accuracy for several days in advance, it follows that the forecaster can make accurate predictions of surface, or ground, conditions for several days in advance. Constant-pressure charts may be made, and are made, for several different pressure levels. There are several other upper-air charts that the trained and experienced forecaster uses with considerable facility, but a simple explanation of them is not easy, and a complete explanation of them is far beyond the scope of this book.

There are also some fairly recent technical developments which have added greatly to the science of weather forecasting. One of these is the use of radar, which enables the forecaster to follow and predict the movement of certain types of storms. Actual condensation in the form of ice crystals or water droplets seems to reflect radar pulses, thus enabling the forecaster to predict the intensity and to determine the movement, with respect to both speed and direction, of such disturbances as thunderstorms, tornadoes, hurricanes, and frontal storms in general. Distances, however, are limited and do not usually exceed 200

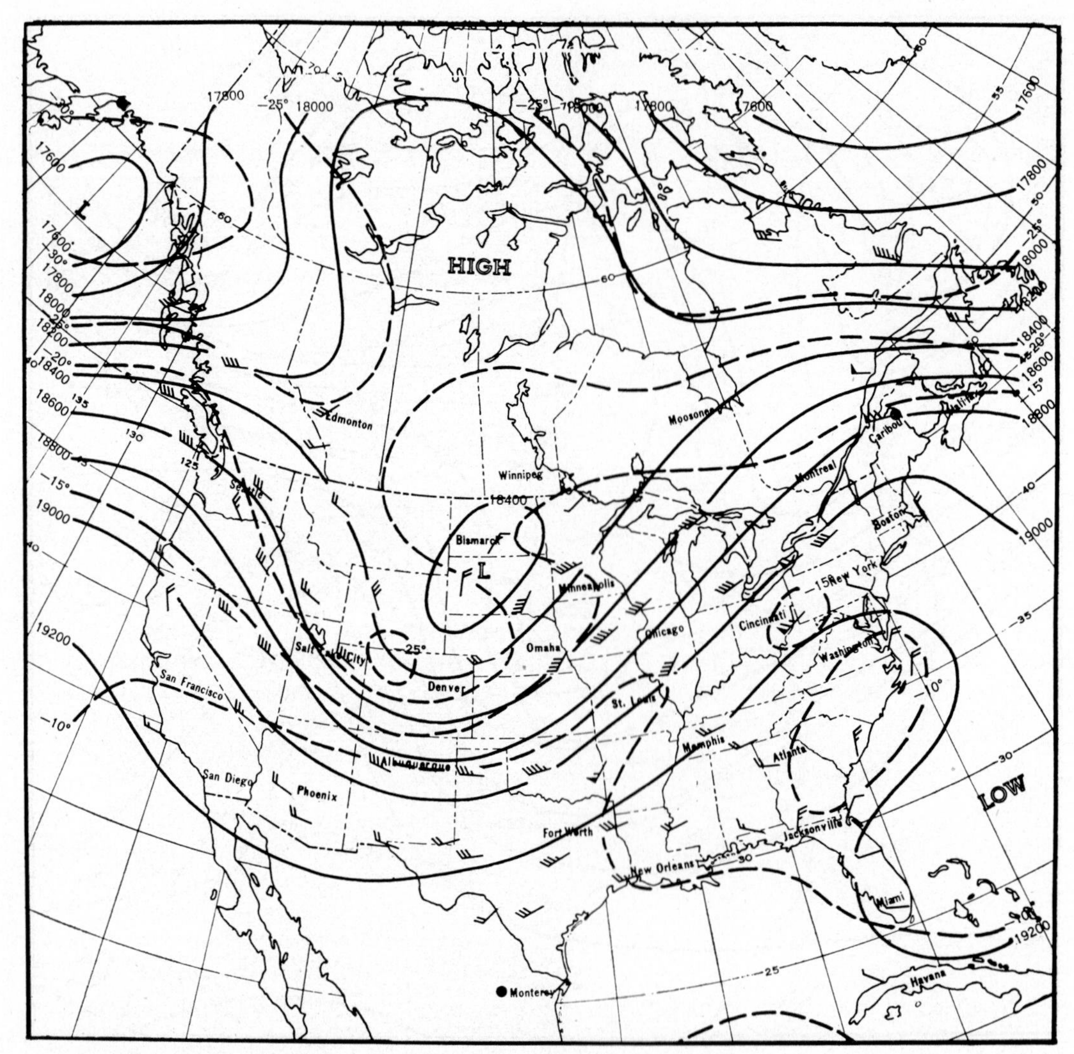

Fig. 10–10. 500-millibar chart, May 28, 1955. (Courtesy U.S. Weather Bureau.)

miles; and the closer the storm, the more accurate the determination of its intensity and extent. Snow is rarely detected beyond 50 miles, but thunderstorms may be detected as far as 300 miles away.

Some of the equipment used in a radar detection system is shown in Figures 10–11 to 10–13. The equipment consists essentially of an antenna, a transmitter-receiver, a control unit, and an indicator. The standard antenna of the Weather Bureau is a rotating, concave metal disc 6 feet in diameter (Figure 10–11), mounted on top of a high building or tower and protected by a domed, cylindrical cover (Figure 10–13). The antenna mounted in an airplane is of the same type but is only about 2 feet in diameter. The indicator is something like a TV screen (Figure 10–12). An actual storm as seen on the indicator at three ranges—150 miles, 50 miles, and 20 miles—is shown in Figure 10–14.

According to the U.S. Weather Bureau,[1] "Radar can observe the location, extent and

[1] From a memorandum, *Information on the Weather Bureau Radar Program*, Dec. 20, 1954.

Fig. 10–11. Antenna and pedestal for receiving radar pulses; the antenna is 6 feet in diameter. (Courtesy U.S. Weather Bureau.)

direction, and speed of movement of the precipitation associated with storms; the 'clear' areas between storms, the height of the tops of severe storms, the height of the melting level, whether the precipitation is rain or snow, something of the wind speed and direction aloft; the stability of the atmosphere. From the nature and movement of the echoes, deductions can be made about the pressure and wind fields in the storm area and the extent of the turbulence. Associated phenomena, such as cold fronts, squall lines, hail, tornadoes, and hurricanes, may therefore be indirectly observed and their movements tracked but not always positively identified." Some of the fields of service in which radar observations are useful, aside from short-period forecasting and the detection and tracking of hurricanes and tornadoes, are: storm-avoidance information for pilots, flash-flood forecasting, and fire-weather forecasting (from detection of thunderstorms with their accompanying lightning). Although radar equipment, including its installations, is expensive, its use is being greatly expanded both on the ground and in aircraft.

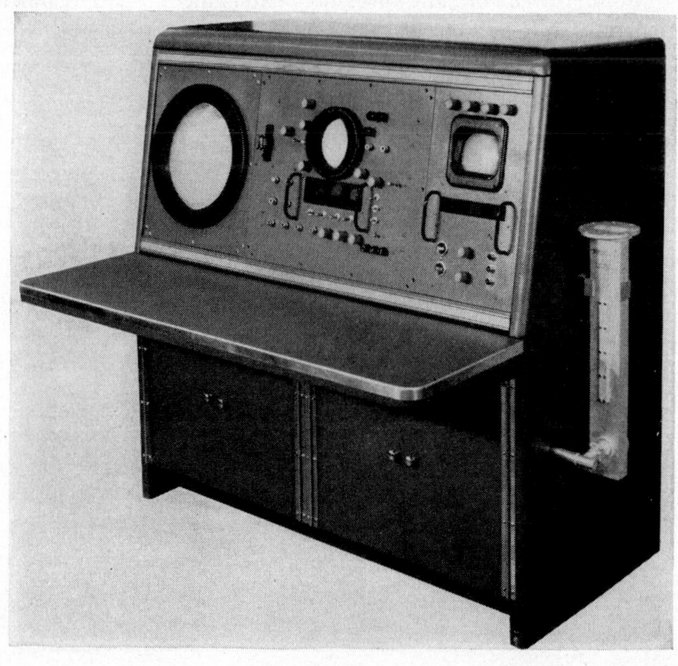

Fig. 10–12. Console for reception of radar pulses. (Courtesy U.S. Weather Bureau.)

Some General Characteristics of Highs and Lows

Obviously the average student does not have available all the charts and diagrams which the professional Weather Bureau forecaster uses. Furthermore, the student would not have the ability to use some of those charts intelligently. It is possible, however, for any individual to subscribe for the Daily Weather Map from Washington, and it is possible for some people to see the daily facsimile map at a local weather bureau station. It seems desirable, therefore, to give some attention and study to the types of weather conditions which the surface maps show.

Pressure Patterns

In the first place, the patterns formed by the isobars are significant. The isobars enclosing the *Lows* are frequently circular, as on the map for May 27 (Figure 10–3); sometimes they are elliptical, as on May 28. Often they are V-shaped, with the point of the V extending toward the southwest; at times this V-shaped depression may be sufficiently elongated to form a *trough* of low pressure extending generally NE–SW across the entire United States. On the map for May 29, the 1016-mb isobar forms a distinct V-shaped depression in the eastern United States, with the point of the V near Galveston, Texas. When the isobars are circular, there is a decided tendency for the winds to blow uniformly and spirally toward the center of the *Low* in a counterclockwise manner. Because of this tendency for the winds to blow toward the center and obliquely across the isobars, we utilize the modified version of Buys Ballot's law: stand with your back to the wind with arms outstretched and turn 30 degrees to the right; your left hand will point in the direction of the *Low*, and your right hand in the direction of the *High*. When the *Low* is elongated into a trough, or V-shaped depression, the winds to the left (west) tend to come from a northwesterly quarter while those on the right (east) are likely to come from a direction almost opposite, or from a southerly quarter. *Lows*, however, do not always conform to the pattern as outlined above, as evidenced by the maps for May 25 and 26.

The anticyclones, or *Highs*, are larger in area and often more uniform in shape, being in general roughly circular. Winds tend to flow spirally outward from *Highs* in a clockwise manner. This type of circulation is quite evident over the western United States on the weather maps for May 28 and 29 (Figures 10–4 and 10–5).

Fig. 10–13. A standard 6-foot antenna housed (A) and mounted on tower. (Courtesy U.S. Weather Bureau.)

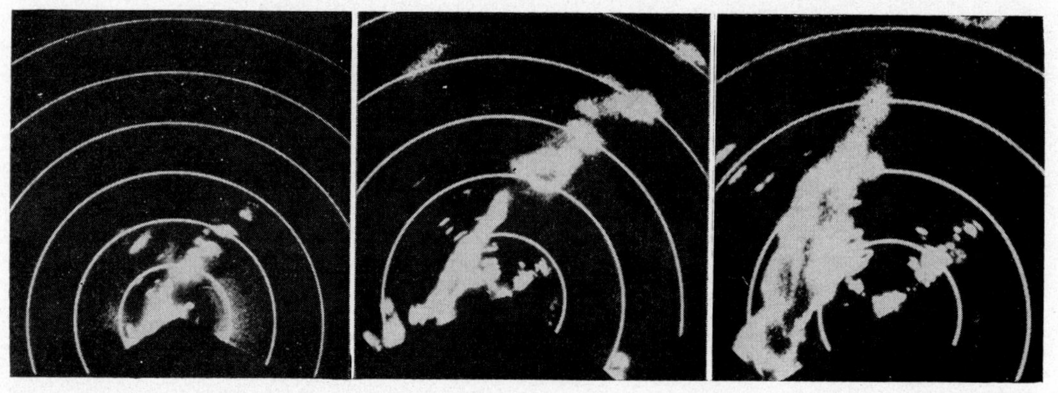

Fig. 10–14. Typical storm pattern as shown on radar indicator. (Courtesy Bendix-Friez.)

Gradient Winds

Another significant rule, or law, which applies to winds not obstructed by features of the terrain is that the speed of the wind is inversely proportional to the pressure gradient (the perpendicular distance between isobars) so that the closer the isobars are to one another, the stronger will be the wind. As the student already knows, because winds are deflected by reason of the earth's rotation (Coriolis force) they do not blow perpendicularly to the isobars, so that eventually, if the winds blow long enough and if we neglect the retarding influence of friction, they could be blowing at the maximum deflection of 90 degrees and thus parallel to the isobars. Winds which attain this maximum deflection are termed *gradient winds*. Obviously, at some distance above the earth's surface, say 1,500 feet or more, all winds would probably be gradient winds.

Core Highs and Lows

Weather reports frequently refer to *cold lows* and *warm highs*, seemingly paradoxical conditions. Actually, conditions giving rise to such expressions are common and are very significant weatherwise. The diagrams of Figure 10–15 illustrate four distinct types of pressure distribution as represented by isobaric surfaces. (It would be well for the student at this point to review the steps in the formation of a sea breeze, Figure 7–6.) Diagram *a* shows the pressure distribution vertically, in the formation of a cold-core low; just as water flows down sloping terrain, so air flows down an isobaric slope, indicated by arrows in the diagram. Cold-core lows may be at a sufficiently low elevation to be noticed on the ground, but more often they are located far aloft, beyond even the 500-mb level and up to 30,000 feet. It is this occurrence at a considerable height which gives the forecaster trouble: he doesnot see the condition on the ground, and he finds that clouds and precipitation develop unexpectedly.

Diagram *b* shows a warm-core high so common in summer in those parts of the United States where the subtropical belt of high pressure, well developed over the oceans, extends over the land in spite of the strong heating of the land. This condition is particularly marked over the southeastern states and southern California.

Warm-core lows (diagram *c*) and cold-core highs (diagram *d*) are the normal surface *Lows* and *Highs* and are clearly indicated on surface charts. The *Highs* are relatively shallow as compared with the *Lows*. In the case of the warm-core low, note the characteristic removal of air aloft to cause the low pressure at the surface; and in the case of the cold-core high, note the characteristic accumulation of air aloft to cause the high pressure on the surface.

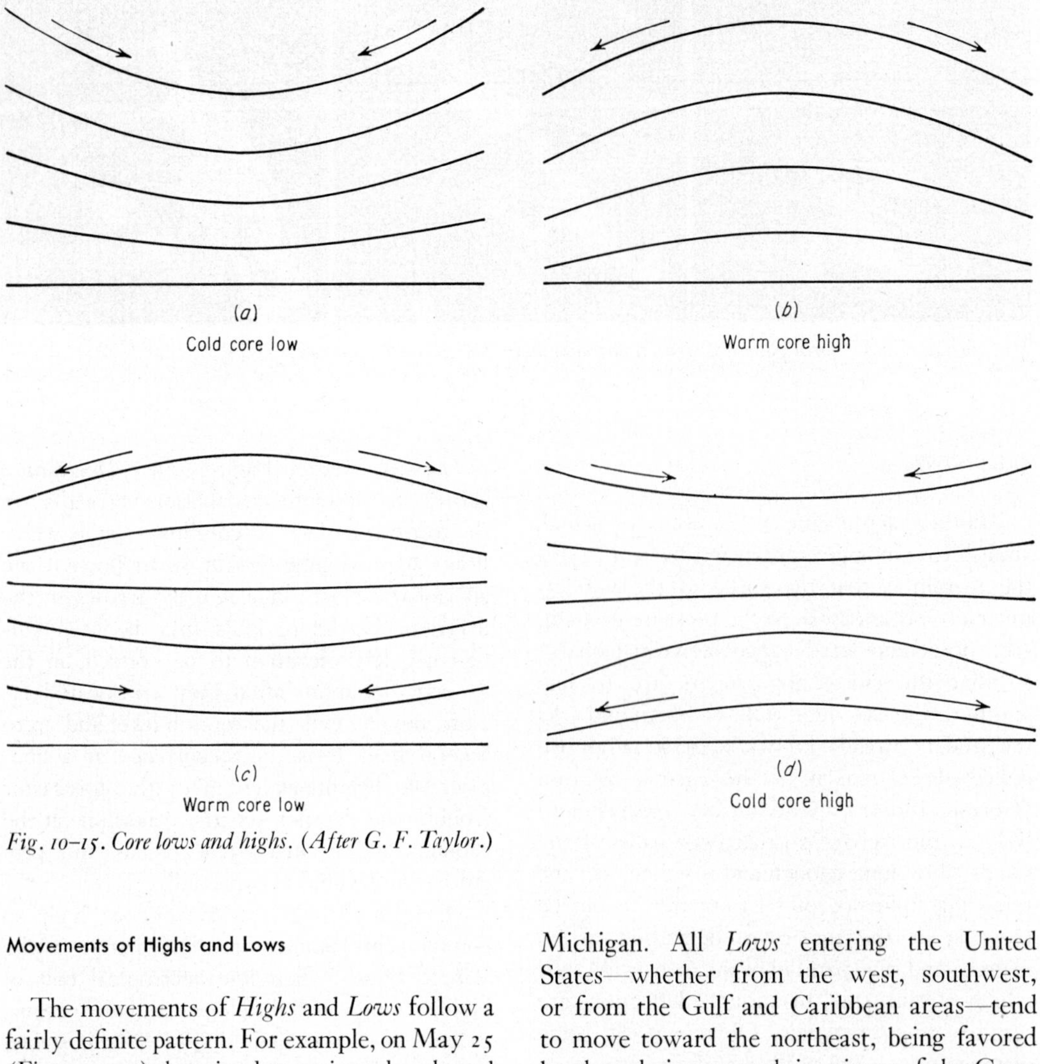

(a)
Cold core low

(b)
Warm core high

(c)
Warm core low

(d)
Cold core high

Fig. 10–15. Core lows and highs. (After G. F. Taylor.)

Movements of Highs and Lows

The movements of *Highs* and *Lows* follow a fairly definite pattern. For example, on May 25 (Figure 10–1) there is a large triangular-shaped *Low* over the southern plateau states with two small but distinct centers over northwestern Arizona and western Colorado; on May 26 the *Low* has moved eastward to the high plains of Texas; on May 27 it has moved northeastward to southwestern Minnesota; by May 28 it is still farther northeast over Lake Superior—but note that a small secondary *Low* has formed over Iowa; by May 29 the original *Low* has moved over the Gulf of St. Lawrence, while the secondary *Low* has become greatly intensified and, like its parent, has moved northeastward over northern Lake Michigan. All *Lows* entering the United States—whether from the west, southwest, or from the Gulf and Caribbean areas—tend to move toward the northeast, being favored by the relative warmth in winter of the Great Lakes and by the warm waters of the Gulf Stream; they eventually join the *subpolar low* of the North Atlantic.

The *High* over the Pacific Northwest on May 27 has moved eastward and by May 28, with its center over Wyoming, has covered most of the Rocky Mountain states. In fact, this same *High* moved southeastward to the eastern states where it actually remained until June 5, finally settling over the subtropical western Atlantic to join the *horse latitude belt* of high pressure.

Charted Air Masses and Fronts

All these *Highs* and *Lows* are simply manifestations of cold and warm air masses. The *High* over the Pacific Northwest on May 27 is an *mP* (polar maritime) air mass. It has all the characteristics of a cold air mass: clear skies; moderately low temperatures, considering the time of year; and low dew points, indicating relatively dry air. Radiation through the cloudless skies at night keeps the air from warming up very much in spite of the bright sunshine during the day. The 500-mb pressure-level charts for May 25–28 show the winds to be strong (as much as 40 to 45 miles per hour, Beaufort 8) and well below freezing. It should be also noted that at the high level of 18,000 or 19,000 feet the winds sweep southeastward, eastward, and finally northeastward in a great wedge of cold air, probably the lower portion of a jet stream.

Warm air masses are clearly evident on each of the surface-weather maps. This air in most cases is indicated by *mT* (tropical maritime). In general the air is moving from a southerly quarter, temperatures are relatively high, and the water-vapor content must likewise be high, especially in the southeastern states, as shown by the high dew points. Although the relative humidity may be high, the skies are clear or partly cloudy except near the warm front. Winds are moderately light, generally under 10 miles per hour.

Cold fronts are clearly indicated on all maps. Note that the front is indicated by the line where temperatures change definitely from relatively high to relatively low and that the front tends to follow the line where southerly winds of the *mT* air mass give way to westerly winds of the *mP* air mass. This latter condition is not the case everywhere, however, because the cold air mass with its more rapid movement may sweep far ahead, becoming a southerly wind, as on May 27, and overtaking the warm front, thus causing an occlusion extending in a great arc from the center of the *Low* to the Chicago area. Other well-marked occlusions may be pointed out on the map for May 28 where the occlusion extends from the center of the *Low* in Iowa southward to central Missouri and from the center of the *Low* over Lake Superior northeastward to James Bay; also on the map for May 29 an occlusion extends from eastern Wisconsin to a point north of Lake Huron. There are other occlusions clearly indicated on the maps by the characteristic symbol, especially in the western states. Do not, however, confuse a *stationary front* with an occluded front; on each of the weather maps, May 25–29, stationary fronts are shown either in the eastern United States or in the contiguous portions of southern Canada. A front is shown as stationary when little or no movement is indicated at the time the map is prepared.

Weather Maps and Charts for North America

Every day the Weather Bureau publishes two maps of North America as a part of the Daily Weather Map. One, already referred to, is the 500-mb constant-pressure chart, Figures 10–7 to 10–10; the other is the surface map—shown in Figure 10–16—giving the isobaric and frontal pattern which existed 12 hours previous to the time of the regular weather map, that is, at 1:30 P.M. of the preceding day. The student should refer to each of these two groups of charts as he studies the weather maps of the corresponding dates.

The Constant-pressure Chart

In connection with the 500-mb chart, it should be noted that the solid lines are really contours drawn through points having the same elevation. It should be noted, too, that each of these charts bears out the statement made in Chapter 3 that at approximately 3½ miles (18,500 feet) above sea level, pressure is about one-half (approximately 500 millibars) that at sea level. Almost invariably, these charts show that the 500-mb pressure level is lower in high latitudes than in low latitudes

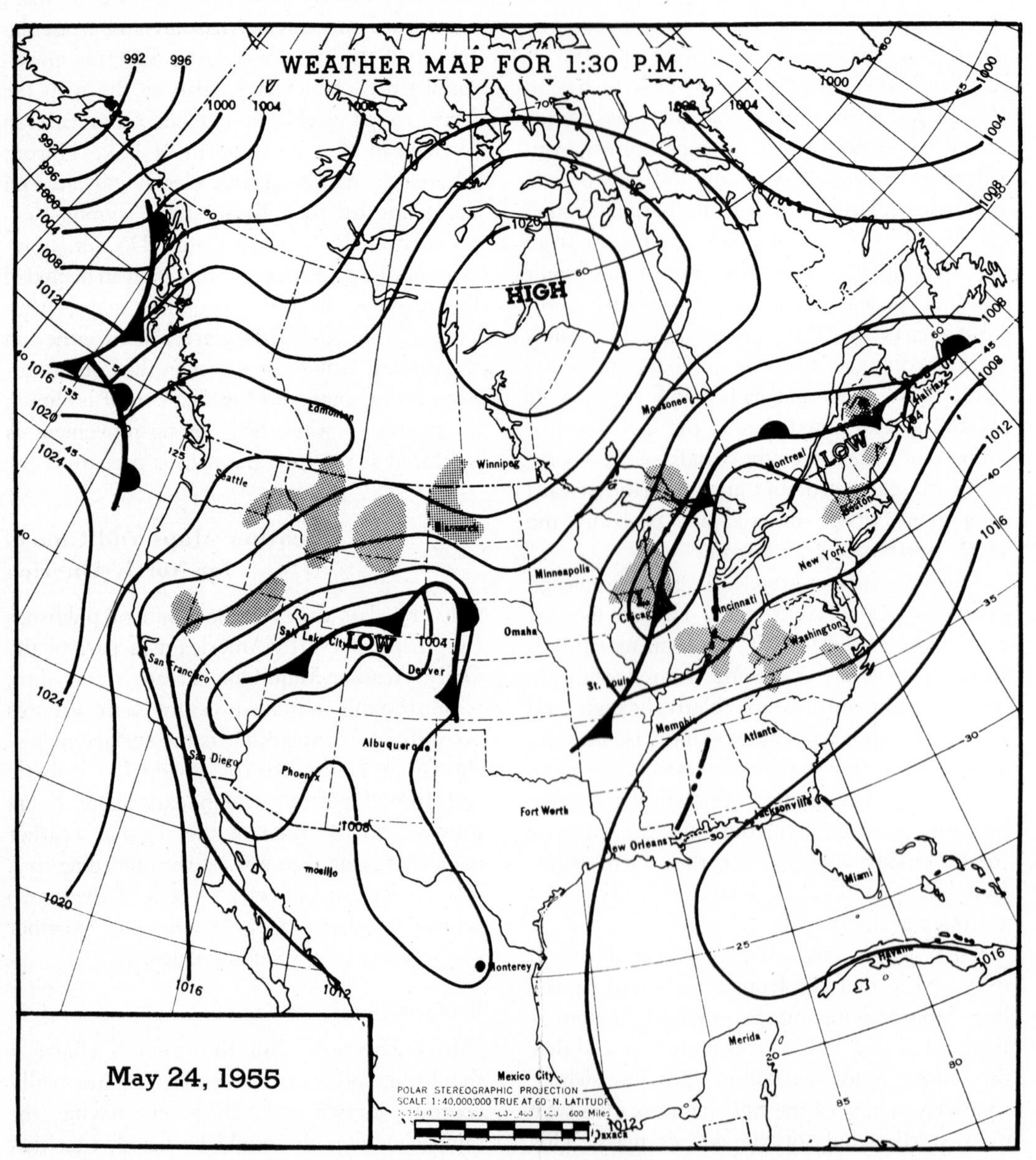

Fig. 10–16. Weather maps, 1:30 P.M., May 24–27, 1955. (Courtesy U.S. Weather Bureau.)

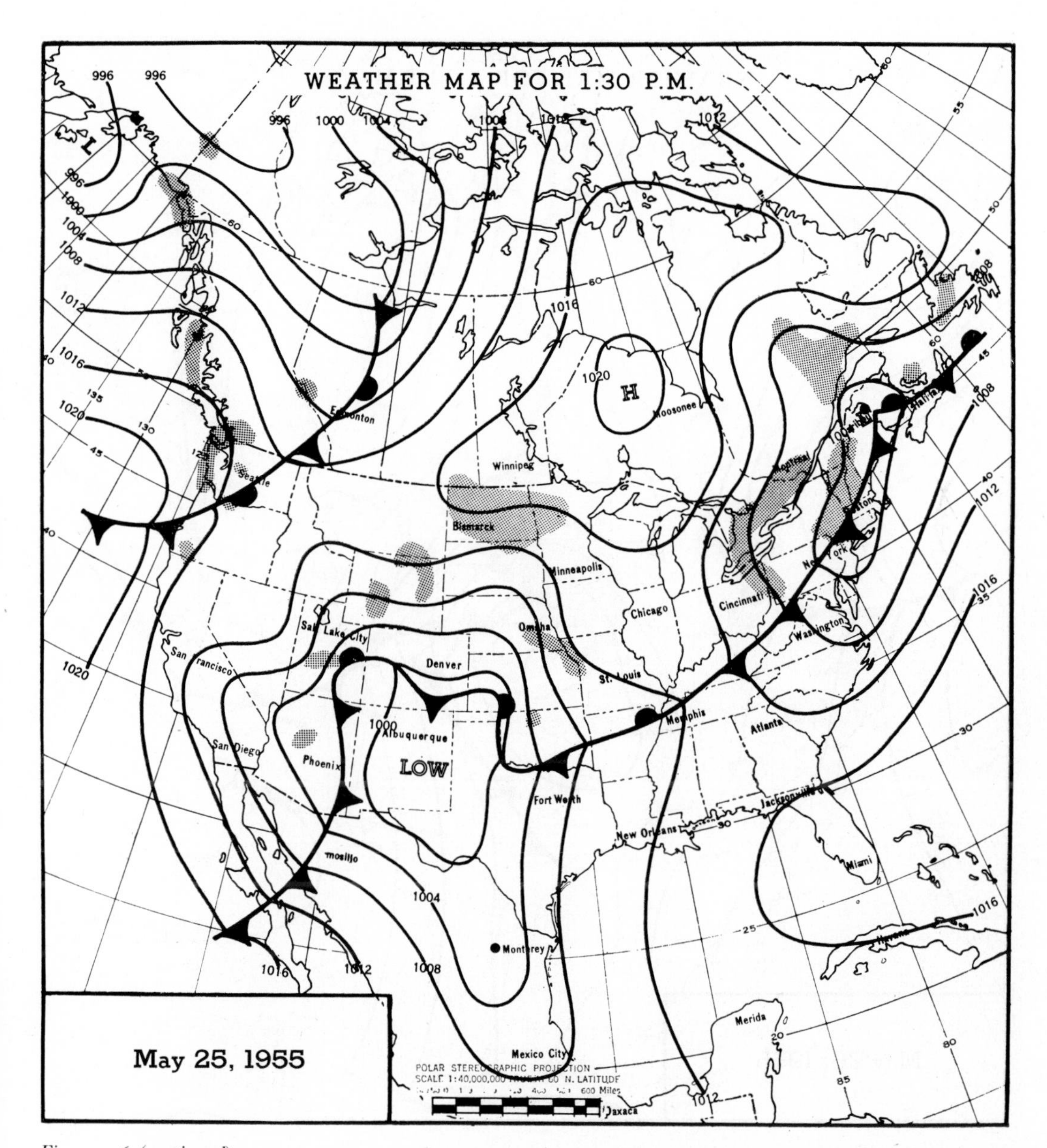

Fig. 10–16 (continued).

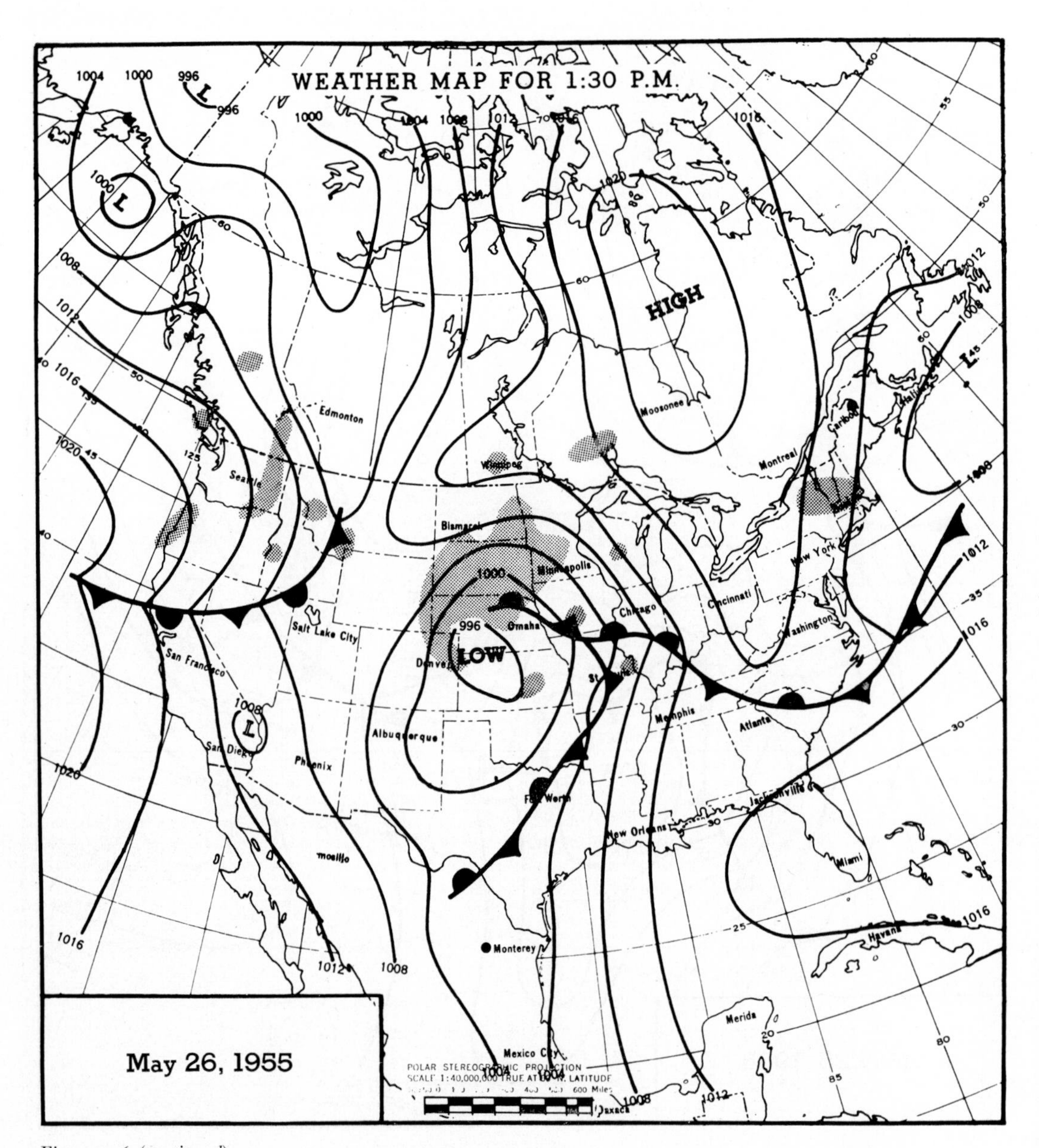

Fig. 10–16 (continued).

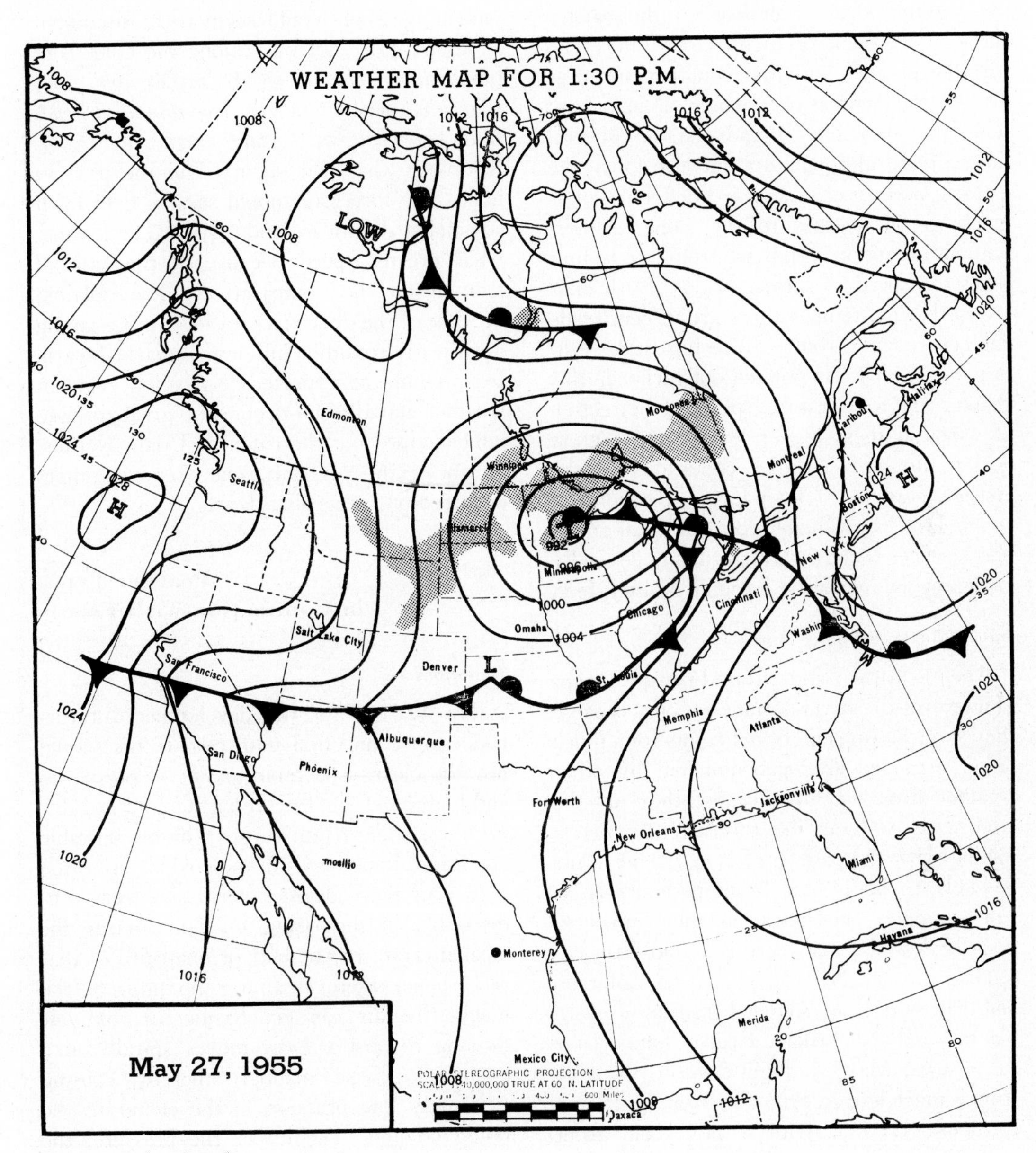

Fig. 10–16 (*continued*).

because of the greater density of the cold, dry air in high latitudes. Depressions—areas where the 500-mb level is closest to the earth's surface—are found on each chart. Winds tend to blow parallel with the contours and thus toward the depressions, which is in accord with the principle, already stated several times, that, although air moves down the isobaric slope, it can move at a deflection of as much as 90 degrees to the slope. Almost without exception, winds are from two to four times the speed of surface winds. Dash lines are isotherms; temperatures are in centigrade degrees, ranging from −10 to less than −30° (14 to −22°F). As pointed out earlier in this chapter, the temperature, speed, and direction of winds at these higher levels have a great deal to do with weather conditions on the ground; certainly a knowledge of conditions at this level is invaluable to modern air-transport services because through-flights generally cruise at an elevation around 18,000 feet.

The 1:30 P.M. Weather Map

The isobaric- and frontal-pattern maps (Figure 10–16) are important not only because they show storm centers and fronts for a much larger area and are unencumbered by other weather data, but also because they aid the student as well as the forecaster in determining what the pattern of isobars and fronts is likely to be 36, 48, or more hours later. For example, one notes on the 1:30 P.M. map for May 24 that there is a cold front extending through Nevada, Utah, and Colorado and that there is an occluded front in western Kansas and Nebraska. Twelve hours later (1:30 A.M., May 25, Figure 10–1), the cold front is much longer, extending from California to South Dakota, with a new cold front extending north and south in Arizona. The *Low* has increased in extent and intensity. Winds throughout Kansas, Oklahoma, and Texas are southeasterly with *mT* characteristics. Light rains are occurring over the mountain states with heavier rains over the plains states along the cold front. With the pronounced contrast in wind direction, temperature, and humidity along the cold front, violent weather would seem to be imminent at least locally not only along the cold front but *well in advance of the surface cold front since the cold air above the ground extends well beyond it in a giant tongue, or wedge.* Knowing what had occurred in the past 12 hours, the forecaster could see the cold front advancing southward and eastward at a somewhat greater speed because of the lack of friction on the comparatively level plains; because of the great mass of *mT* air streaming northward from the Gulf, a well-marked warm front would be expected to develop. Violent weather actually did occur: the great tornado which wiped out the town of Udall, Kansas, on the 25th was only one; several others occurred on succeeding days.

Weather Types in Connection with Fronts

Tornadoes

The reasons for the development of tornadoes in connection with cold fronts should now be clear. The rapid ascent of the warm and humid *mT* air forced by the cold and dry *mP* (and often *cP*) air causes a highly unstable condition. Lapse rates in the cold air are often 6 to 8 or more degrees per 1,000 feet; condensation in the rising *mT* air retards the adiabatic rate to one-half or one-third of that rate; hence there is almost no limit to the height the air can go. Rising air, like air flowing toward a *Low*, moves spirally in a counterclockwise manner, thereby causing extremely low pressure in the center of the rising column. The lower the pressure, the stronger the winds; and the stronger the winds, the greater the gyratory action in the updraft and the more intense the low pressure becomes. The lowering pressure cools the air rapidly to below the dew point; as a result, a cloud develops in conformity with this chimney of low pressure; hence, the characteristic funnel-

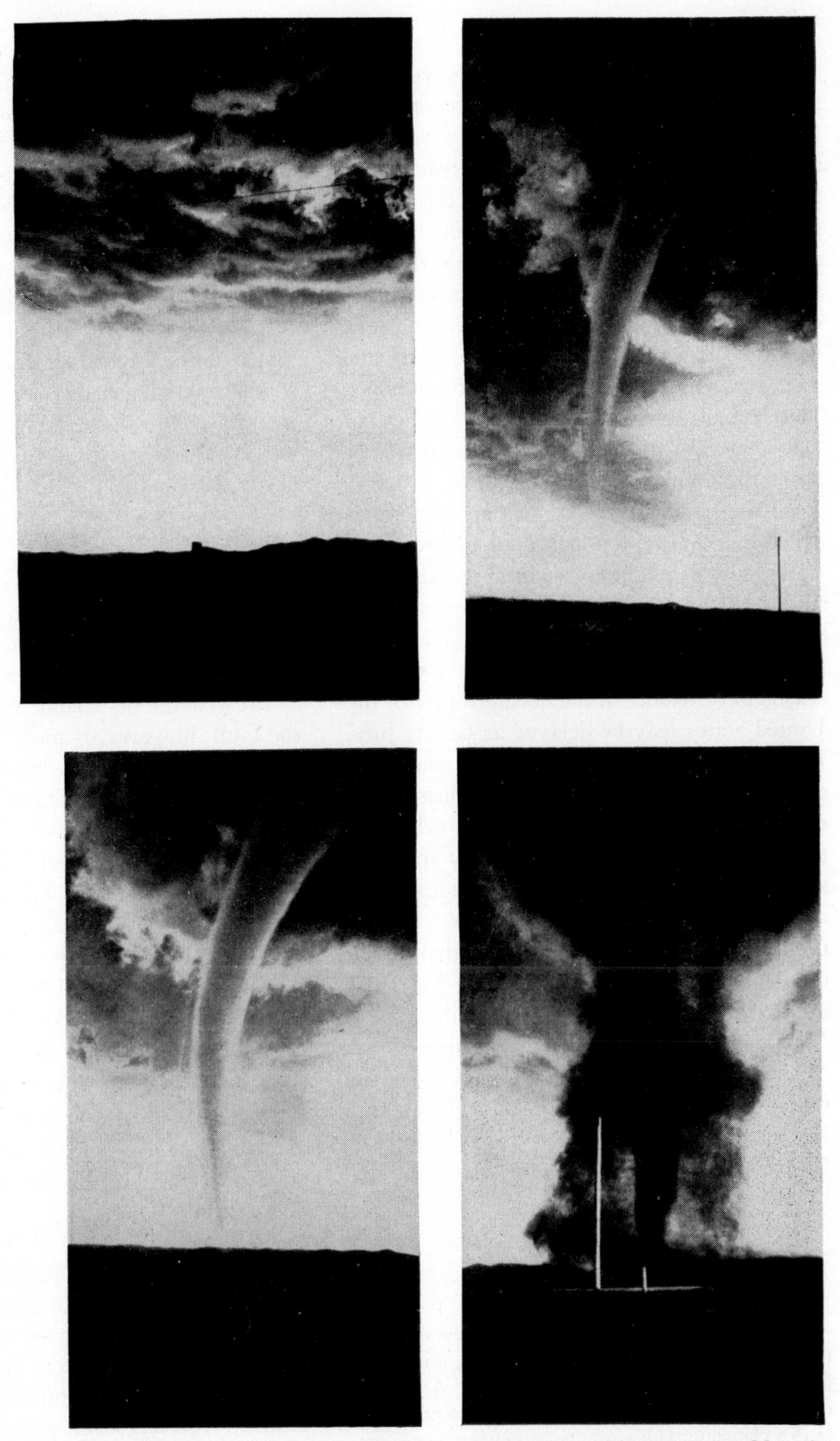

Fig. 10–17. Sequence in the development of a tornado. (Photograph by Mrs. O. Homer, Courtesy U.S. Weather Bureau.)

shaped cloud (Figure 10–17).[1] The very low pressure (as low as 23 inches recorded at Minneapolis in August, 1904) causes buildings to explode when the funnel cloud reaches the ground, and the terrific velocities of the wind—perhaps as great as 500 miles per hour—usually prostrate every standing object in the tornado's path. Fortunately these storms do not commonly devastate an area more than 1/4 mile in width or more than 25 miles in length, and their average life span is only about 8 minutes. Waterspouts, which occur over bodies of water, are essentially the same as tornadoes but are seldom as intense.

Tornadoes, or *twisters* as they are sometimes called, are most pronounced in the southern states; they begin in late February and extend into April or May. As the sun moves northward in spring, the area of greatest contrast in air masses also moves northward so that the season of tornadoes in the northern part of the United States may be delayed as late as July. Because tornadoes are so local in their occurrence, it is not easy to forecast just where they may strike, but it is comparatively easy to forecast the likely occurrence of them. A study of weather maps will give an idea of the conditions which favor tornadoes. Besides the Udall tornado of May 25, 1955, there were others occurring from May 26 to May 29: in the vicinity of Blackwell, Oklahoma, on May 26; in several areas in Oklahoma and Missouri on May 27; in Osage County in Oklahoma, parts of Michigan, west-central Indiana, and Clintonville, Wisconsin, all on May 28; in Jacksonville, North Carolina, and in South Bay, Florida, on May 29. Although these storms are most common in the middle west from the Gulf of Mexico practically to the Canadian border, other parts of the country are subject to their visitations. In Massachusetts in 1953, the top of the thunderhead in the Worcester tornado (Figure 10–18) is said to have reached a height of 14 miles; this tornado was the result of the same condition which brought the devastating tornado to southern Michigan the day before. Tornadoes are not confined to the United States; they have occurred in France, China, and as far north as Scandinavia.

[1] There are other theories concerning the origin of tornadoes. One is that cold, dry air aloft, usually *cP*, overrides *mT* air, thereby forming a "lid." A penetration of the lid by the *mT* air causes a violent updraft as in a chimney, with a resultant gyratory movement. Some suggest that it is the *cP* air which penetrates the lid, forcing the *mT* air violently upward. In any case, the spiral or gyratory motion is characteristic.

Other Frontal Storms

In winter an invasion of cold air from Canada may give rise to a *cold wave*. At times the cold air may penetrate far south to the Gulf, causing freezing temperatures as far as Brownsville or Miami; their penetration to the Gulf, however, is more likely to occur in early rather than late winter. On the western plains of the United States and Canada, the first advance of the cold wave is usually accompanied by snow and high winds. Gradually the snow ceases, but the strong winds continue so that the snow on the ground is caught up and whitens the air. The whole phenomenon is referred to as a *blizzard*. After a blizzard, the average depth of the snow on the ground may vary from a few inches to a few feet. The characteristic features of blizzards are severe cold, strong winds, and snow-filled air rather than a great amount of snow, although local drifts may block surface transportation.

A feature of warm fronts that is often severe enough to tie up transportation and cause heavy property loss, especially to open wire and cable lines, is the *ice*, or *glaze*, *storm* so frequent in the eastern half of the United States from the Gulf states northward. If the cold air retreats over ground that is below freezing and if the warm air overriding the cold air is well above freezing (or even below freezing but with supercooled water droplets), condensation at the cloud level will result in rain. But when the rain strikes the ground, or objects on the ground, it freezes and coats the objects—pavement, trees, telephone wires—with ice, sometimes as much as ½ inch or more in thickness (Figure 10–19).

The well-known *northeaster* of the North Atlantic seaboard that brings heavy rain or snow is the result of a continuous importation of warm, moist air over the cool surface air well north of the center of a *Low*. Here the surface winds will be easterly, but the air aloft will be southerly, bringing in large volumes of *gT* or *aT* air. This condition may last for hours or even for days. The heavy rain, extending from Virginia to Maine, that accompanied the two hurricanes of August, 1955, was of this type. Northeasters, however, are most typical of the winter season.

Tropical Cyclones

Hurricanes and Typhoons

Low-pressure areas which originate in the tropics, particularly in the area of the doldrums where they are well beyond the equator, and which generally travel westward with the prevailing winds of the area are called *tropical cyclones*. They are known to occur in all tropical seas near the equator, except in the Atlantic south of the equator. In the region of

Fig. 10–18. Destruction by a tornado. (Photograph by the Worcester Telegram Gazette, Courtesy U.S. Weather Bureau.)

Fig. 10–19. Effects of an ice storm. (Courtesy U.S. Weather Bureau.)

the West Indies, Mexico, and Central America, and in Australia also, they are referred to as *hurricanes;* in the region of the western Pacific, from Indochina to Japan, they are called *typhoons*, while in the Indian Ocean they are referred to simply as *cyclones*. Wherever they occur, they differ only in name. On a weather map, the isobars are more nearly circular than in extratropical cyclones, and the isobars are closer together, which means the pressure gradient is much steeper than in cyclones of middle latitudes. This steepness of the pressure gradient, with almost gradient winds, accounts for the high wind velocities and the consequent destructiveness of tropical cyclones. Some unusually low pressures have been observed, as in the Florida Keys hurricane of September 2, 1935, with a reading of 27.02 inches, although an unofficial aneroid reading of 26.35 inches was reported; and the Luzon, Philippine Islands, typhoon of August 18, 1927, that was reported as having a low of 26.185 inches.

Occasionally a hurricane may originate in subtropical waters of the North Pacific off the coast of Mexico. In August, 1951, a mild one passed over Lower California, bringing nearly 1 inch of rain to San Diego which is normally rainless in August; another one in September, 1939, brought approximately 2.5 inches to the same city.

Characteristics of Tropical Cyclones

As pointed out in Chapters 3 and 6, warm air with a high moisture content (found over tropical waters) is much lighter than warm air with a low moisture content. Warm, moist air, therefore, favors low pressure and easy convection. Even a small drop in pressure starts the air moving in a circular manner and eventually in a spinning or gyratory manner. If there is condensation of moisture in the convection column, the updraft of air is speeded up and the gyratory motion increased. The circulation of air around an area of tropical low pressure may be compared to the water draining from a laundry tub or lavatory that has a drain in the center of it: at first there is only a slight circular motion of the water, but in a short time a definite funnel-like depression, or vortex, develops, with the water circulating more rapidly around the tub until all the water is drained. The development of a hurricane is based somewhat upon the same sort of conditions; comparison with a basin of water is legitimate because the pressure of the water is uniformly decreasing from the margins of the tub toward the vortex, just as the pressure of the air tends to decrease uniformly toward the vortex of the hurricane; the concentric pattern of the isobars is an indication of this uniform decrease of pressure. This may seem to be an oversimplification of a very complex subject, for there are obviously other conditions which have an influence upon the origin of these violent tropical storms. Just as jet streams seem to have an important part in the development and intensity of extratropical cyclones, so jet streams may have a definite relationship to hurricanes. A wave or tongue of polar air at high altitudes may at times sweep far south over the land and reach down to comparatively low altitudes, pulling or dragging lower air along with it as the axis turns and sweeps northeastward; this is meant as a possibility, not necessarily a reality.

Just as a spinning top tends to move, so the spinning motion of the air in a hurricane seems to cause the storm center to move. This motion is probably attributable, in part at least, to the general air movement in the area where hurricanes develop: the trades tend to push them westward in the region of the Caribbean and Gulf until they reach the belt of prevailing westerlies where they are pushed eastward. This is in conformity with the air circulation as shown by the map of July pressure and winds (Figure 15–3). Furthermore, typhoons off the east coast of Asia seem generally to move northward in conformity with the flow of air in the summer monsoon of that area (Figure 15–3).

The pronounced intensity of tropical cyclones must be attributed in part to the fact that they originate over tropical waters where there is abundant moisture to supply energy and little friction with the surface to dissipate the storms; this is substantially what was stated in the first paragraph of this section. In addition, however, it should be noted that the regions of occurrence are largely confined to areas of converging and more or less conflicting winds—trades against westerlies. The fact that they are most numerous in late summer and early fall, when the temperature of ocean waters has reached its seasonal maximum and when the cooling of the continental interiors farther north is beginning or proceeding rapidly, suggests that the formation of all tropical cyclones is not caused alone by weather conditions of purely tropical origin.

The Weather Map for September 11, 1954, (Figure 10–20) shows a well-developed hurricane of moderate intensity centered near Cape Hatteras, North Carolina. This storm was first noted east of the Bahama Islands on September 7. Obviously, its movement was slow, at least at first, for it progressed only about 1,500 miles in 4 days—roughly 15 miles per hour. As it moved northward along the Carolina coast, it increased somewhat in speed. The pattern of concentric circles formed by the isobars is typical, but the spiral movement of air around the center is shown only in part because most of the storm was located over water where complete observations of wind were not possible. To indicate how the winds change at any given point as the storm approaches and departs, three successive positions of this particular hurricane are mapped in

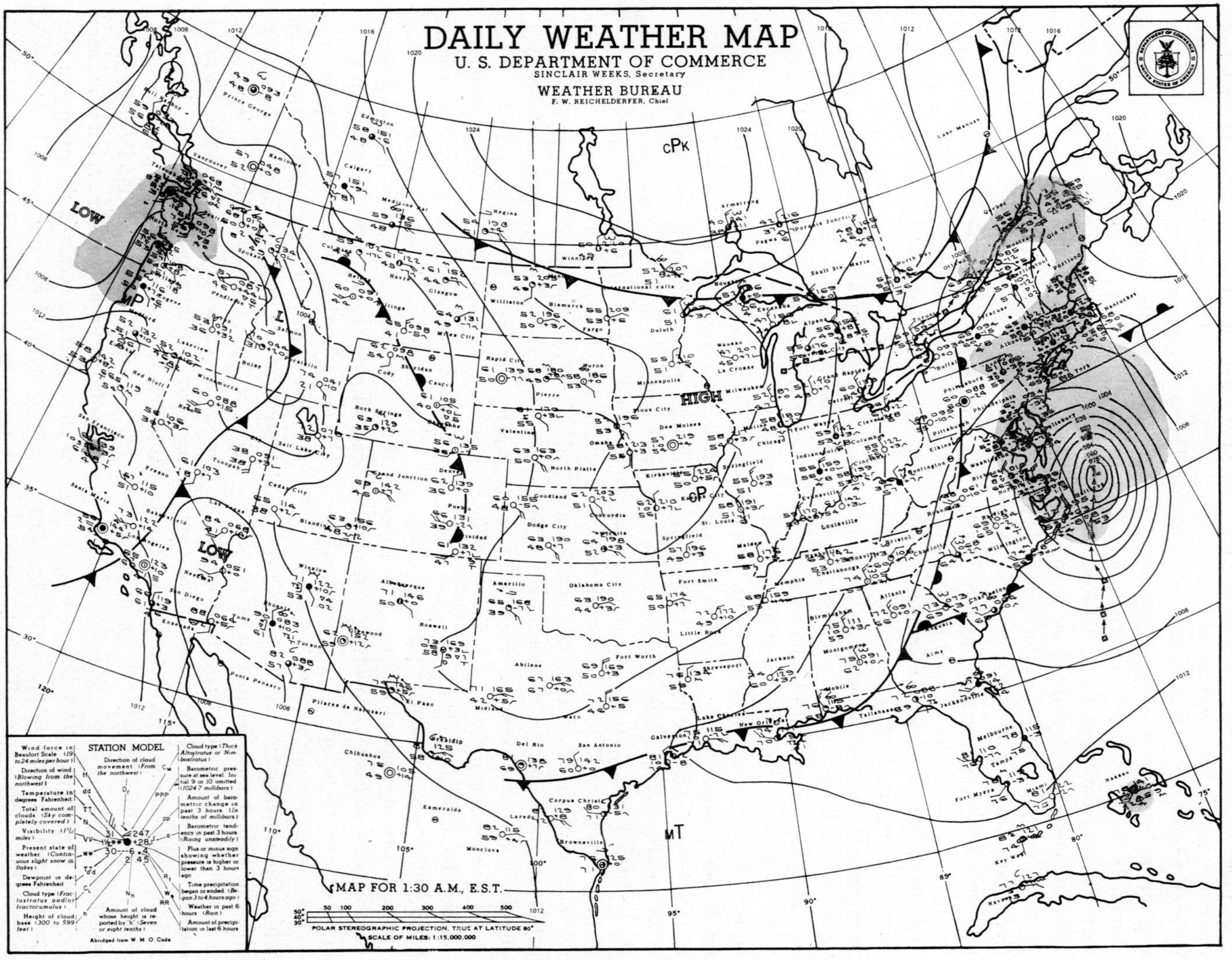

Fig. 10–20. Daily Weather Map, September 11, 1954. (Courtesy U.S. Weather Bureau.)

Figure 10–21, and the given point is Wilmington, South Carolina. Note that in diagram *A*, when the storm center was south of Wilmington, the wind was from the northeast with a Beaufort speed of 4 (approximately 15 miles per hour). In diagram *B*, the wind had changed to northwest with a Beaufort speed of 6 (about 27 miles per hour). In diagram C, when the storm center was definitely north of Wilmington, the wind had changed to west with a Beaufort speed of 5 (about 21 miles per hour). It should be noted that, because the storm center passed to the east of Wilmington, the winds backed, the direction changing in a counterclockwise manner; if the storm center had passed to the west of Wilmington, the winds would have veered, the direction changing in a clockwise manner, probably from northeast through southeast to southwest.

The distribution of the weather elements in hurricanes and typhoons is somewhat different from that in extratropical cyclones. In general, the winds are strongest in the right rear sector (southeast quadrant), and they tend to move almost perpendicularly to the direction of the storm's movement. Otherwise the winds tend to blow as gradient winds—almost parallel to the isobars—at least during the period in which the storms are over water. On land, as noted at Wilmington, the winds are not gradient winds because they blow at a distinct angle to the isobars. Precipitation seems generally to be greatest in the southeast quadrant, although at a distance of 50 to 150 miles from the center. In the left rear quadrant, where the winds are northerly or northwesterly, rainfall is light. Of course there are no distinct cold fronts or warm fronts while the storms are over tropical waters. Frontal characteristics develop, however, as the storms move over land and to the north and east.

An interesting feature of almost all well-developed hurricanes or typhoons is the *calm central eye*, often an area as much as 20 or 25 miles in diameter. In it the air appears to be descending; therefore, skies are frequently quite clear. This calm, clear weather in the center of the storm may be genuinely deceiving. For example, when the center of the Miami hurricane of September, 1926, was still out at sea to the southeast, northwesterly winds up to 100 miles per hour caused extensive damage along the east coast and along the shores of Lake Okeechobee. When the center of the storm was over Miami, the winds subsided, the sun came out, and many people went about their accustomed tasks. This fine weather lasted perhaps one hour. Then, as the

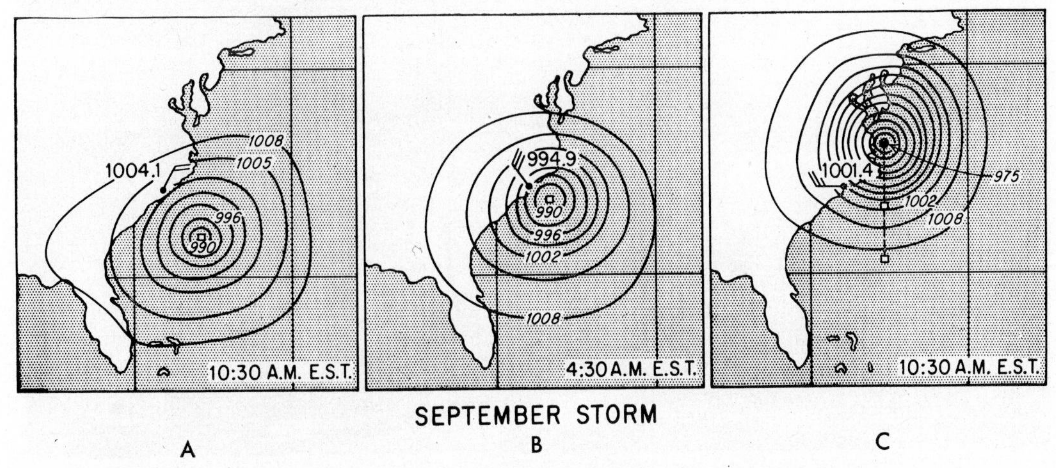

Fig. 10–21. Successive positions of the Wilmington hurricane, September, 1954. (Courtesy U.S. Weather Bureau.)

center of the storm passed inland and toward the northwest, violent winds of still greater speed and accompanied by rain swept in from the southwest, resulting in considerable destruction of life and property even though the Weather Bureau had given warning to prepare for what was to come.

Typhoons occur over a much longer period than hurricanes do, but they are practically unknown in late winter and early spring; with either type, September is the month of greatest frequency. As soon as one of these storms moves over land, it begins to weaken, and as it moves into the zone of prevailing westerlies, it gradually takes on the characteristics of any middle-latitude *Low*. Cyclones that have their origin in hurricanes from the West Indian region have been traced northeastward along the Atlantic seaboard into Canada and even as far north as Greenland. Such storms are usually more intense than the ordinary ones of middle latitudes, causing stronger winds and heavier precipitation.

Destruction by Tropical Cyclones

In addition to floods, much of the destruction in coastal towns and cities visited by hurricanes or typhoons is due to the great avalanches of sea water piled up and driven onshore by the force of the wind (Figure 10–22). The Galveston hurricane of September, 1900, in which 6,000 lives were lost, and the Swatow, China, typhoon of August, 1922, in which 40,000 lost their lives, were cases of this sort. Other destructive tropical cyclones may be noted. One of the best known occurred in the year 1281 and destroyed the vast fleet of 3,500 vessels assembled by Emperor Kublai Khan for the invasion of Japan; of the 100,000 who embarked on that fateful expedition, only 3 returned. The Florida hurricane of September, 1928, resulted in 2,000 deaths, chiefly from drowning. More recently, in December of 1944, a typhoon in the East China Sea sank three United States naval craft, destroyed 164 planes, and killed 790 men. During the

Fig. 10–22. Fishing draggers washed ashore by a New England hurricane. (Photograph by the New Bedford Standard Times, Courtesy U.S. Weather Bureau.)

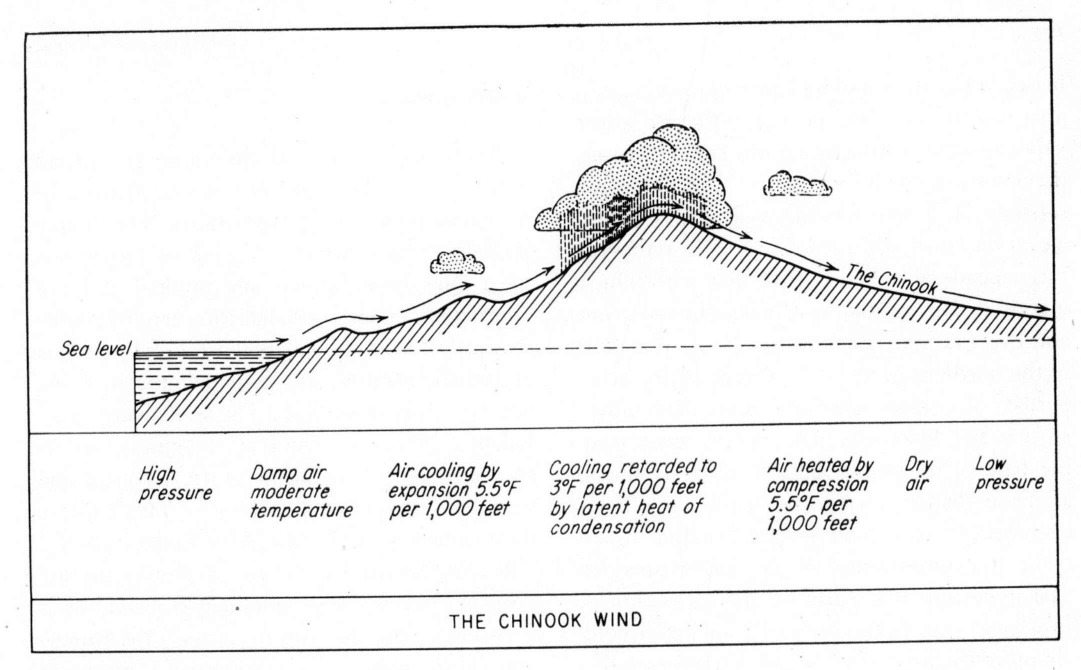

Fig. 10–23. The chinook.

last two decades, West Indian hurricanes have been more frequent along the Atlantic seaboard, causing great destruction of property not only by wind but also by floods. There are many instances of excessive rains, in some cases as much as 15 inches in a single storm.

Forecasting Tropical Cyclones

The approach of a tropical cyclone may frequently be detected by the streamers of cirrus clouds radiating outward in all directions and gradually thickening into cirrostratus, alto-stratus, and denser cloud types. Their advance can be gauged somewhat by the decreasing frequency of sea swells from a normal of about eight per minute to as few as five per minute, a phenomenon not easily explained. Our knowledge of tropical cyclones is increasing each year, not only through direct surface observations but also through airplane reconnaissance; presumably further research in connection with jet streams will add to our knowledge of causes and therefore to our ability to forecast.

Local Winds

In almost every part of the world, winds of a certain intensity, from a certain direction, or with certain characteristics of temperature and humidity have received specific names. In fact, there are probably several hundred with local names. Whatever their characteristics or names may connote, they are caused, like all winds, by differences in air pressure.

Foehns and Chinooks

One of the most interesting of local winds is the *foehn* (pronounced like "*fern*" with the *r* left out) which was first observed and studied in the Alps. It is a warm, dry wind which descends the slopes and valleys of the mountains. Foehns which occur on the Great Plains of North America as well as those which occur in the interior valleys of the cordillera are given the name *chinook*. The cause of their warmth and dryness is shown by the diagram in Figure 10–23, but essential features are (1) a considerable elevation of land lying between (2) an area of high pressure and (3) an area of low pressure and (4) sufficient water vapor in the air moving up the slope to cause precipitation on the windward side. Thus air starting at 2,000 feet arrives at the same elevation on the leeward side with a temperature considerably higher than that with which it started, disregarding loss of heat by radiation and evaporation. Chinooks are most common in the northern parts of the Great Plains after periods of cold weather and often after snowstorms and blizzards. They may occur near the base of mountains but are more frequent at some distance from the base; they have been known to extend several hundred miles from the mountains, as on the plains of Saskatchewan. If a warm air mass (low-pressure area) covers the Great Plains and a cold air mass (high-pressure area) settles over the Great Basin, conditions are just right for a strong chinook. They will blow out of the mountain valleys onto the plains, raising the temperature as much as 30 or 40 degrees in a few hours and causing any snow cover on the plains to evaporate as if by magic. The economic importance of the chinook lies chiefly in its rapid removal of snow, thereby permitting the open-range stock to feed. The failure of a chinook to arrive when it is needed may cause great loss of sheep and cattle through starvation and sometimes through suffocation if the animals crowd under some protecting ledge to avoid the snow and cold.

Mountain Waves

Another type of local wind associated with mountains is the *mountain wave*, first noted by glider pilots during the 1930s. The strong updrafts on the windward side of mountains have long been known and studied and are quite easily understood, but the currents on the leeward side, although generally following a systematic pattern, are quite complicated and not so easily analyzed. These leeward, turbulent currents are the true mountain waves; they are highly significant to the aeronaut and very baffling and dangerous if he fails to note their existence or to study their characteristics. The diagram in Figure 10–24 shows the air flow in an average mountain-wave condition. As soon as the flow of air crosses the mountain crest, downdrafts predominate, attaining speeds of more than 30 miles per hour. Turbulence with strong updrafts begins some 5 to 10 miles from the mountain crest; this is the primary wave. Farther leeward, a second downdraft followed by an updraft occurs; this is a secondary wave, and it is usually less intense than the first. There are instances of as many as six or more distinct waves being encountered in succession but with decreasing intensity as they gain distance from the mountain crest.

Strong mountain waves occur quite frequently to the lee of the Sierra Nevada range, and they have been especially well observed

Fig. 10–24. Diagram of a mountain wave. (Adapted from a U.S. Weather Bureau chart.)

and studied near Bishop, California. Their best means of identification are provided by characteristic cloud forms, unless the air happens to be so dry that clouds cannot form: the *cap cloud* covering the crest of the mountain; the *roll cloud*, an irregular or ragged cloud of the cumulus type, occurring to the lee of the crest and identifying the strong updrafts and zone of turbulence; the *lenticular clouds*, developing at considerable elevations—as much as 35,000 feet—and serving to identify the top of the wave. The distance between successive bands of lenticular clouds measures the distance between successive waves; these distances are usually 10 to 15 miles apart but may be as much as 20 miles apart. Normally, wave action becomes more intense with stronger winds, yet winds of more than 100 knots in the free air may eliminate systematic wave patterns but cause excessive turbulence. Aviators, naturally, avoid areas where mountain waves are indicated; certainly, if they happen to be in the area, they avoid the ragged and irregular lenticular and roll clouds. On the other hand, gliders are known to utilize mountain waves; and there is one instance of a glider, under such impetus, having sailed from the Sierras eastward to St. Louis. Strong winds blowing perpendicularly against a mountain range are a good indication that mountain waves of considerable intensity are likely to occur to leeward.

Other Local Winds

The *bora* descending in winter from the Istrian Plateau to the northeast shore of the Adriatic, the *mistral* descending from the Auvergne Plateau to the Rhone Valley, and the *etesian* descending from the Rhodope Plateau to the Aegean are all strong winds that blow from areas of high pressure over central Europe toward a passing *Low* of the Mediterranean Sea and are cold largely by contrast because they are necessarily somewhat warmed adiabatically in their descent. Small ships at sea are often seriously damaged by these winds.

Other winds—some of them hot and dry, others cold and dust-laden, still others cold and raw—occur under a variety of conditions. Among the cold air-mass type may be named: the *pampero*, a squally, dust-laden, and sometimes rainy wind out of the south that blows across the plains of central Argentina; the *buran* of eastern Russia and the *purga* out of the Russian tundra, northerly or easterly winds of great violence at times and always extremely cold, the former usually bearing dust and the latter snow; the well-known *norther* of Texas. Successive invasions of the buran over western Europe in February, 1956, brought—even to Spain and North Africa—the coldest weather experienced in a century. There are certain warm air-mass types that are hot and dry: the *simoom* of the Sahara; the *harmattan*, a northeast wind out of the Sahara blowing to the Guinea Coast, sometimes chilling because of great evaporation; the *brickfielder* of Australia. The *black roller* of our own Great Plains is a strong, dust-laden, continental type of air mass, either *cT* from the southwest or *cP* from the northwest. The ravages of this wind are too well known to need further comment (Figure 10–25).

Foehnlike *fall* (descending) *winds* are common throughout the world whenever a mass of air over an upland area sweeps down depressions or valleys to the lowlands. They may be warm or distinctly cold, depending upon the temperature with which they started or the temperature of the area they enter; they are always strong and relatively dry. Such a wind is the *Santa Ana*, so common in southern California, blowing out of the east or northeast as a very warm and desiccating wind. It occurs when an accumulation of cold air, filling the Great Basin to overflowing, spills through the mountain depressions and sweeps out onto the lowlands. It constitutes a distinct fire hazard and often results in considerable financial loss to orchardists. Similar winds, but not so warm or so desiccating, are the *yamo* blowing down the steep valleys in Japan, the *zonda* down Andean valleys to the plains of Argentina, and the *tramontane* down valleys of central Europe. There are yet others of gale force and cold by reason of their origin over snow fields, such as those rushing down the steep-sided fiords of Greenland and Norway. All these *katabatic*, or fall winds, have profound effects not only upon local inhabitants but upon the careless aeronaut who does not take them into account.

There are two other types of local wind, the *whirlwind*—really a miniature storm center or tornado commonly found in warm and dry regions—and, when carrying dust, the *dust devil*. They are caused mainly by local convection from strong heating of the ground. At times they are quite violent, the gyratory movements of air with the strong updrafts attaining velocities almost equal to those in a small tornado; and they can cause serious damage to planes or cars that are carelessly driven through them. These winds will sometimes travel many miles in flat regions; their duration is said to be as many hours as they are thousands of feet in height. Their updraft may be strong enough to carry objects as large as cornstalks or packing cases hundreds of feet aloft. The load of dust in a well-developed dust devil is prodigious.

Fig. 10–25. Buildings covered by drifting soil during successive "black rollers." (*Courtesy U.S. Weather Bureau.*)

11

Atmospheric Optics

Atmospheric optics refers in general to anything seen as a result of light's passing through the atmosphere. Meteorologists, however, restrict the term's use to optical phenomena that seem to result from weather conditions or changes; to that extent, atmospheric optics is a legitimate field of weather science. Furthermore, certain optical phenomena observed in the atmosphere *do* reflect such weather conditions as differences in temperature at different elevations and the state of the moisture in the air—whether it is in the form of ice crystals or water droplets. These optical phenomena, therefore, are sometimes thought to be harbingers of fair or foul weather. It is better to think of them as manifestations of actual weather conditions rather than as prophets of weather to come. Whatever their relationship to weather, they are, and always have been, of great interest to mankind, and they deserve some mention here.

Phenomena Due to Refraction of Light

Refraction is the bending of light rays as they pass from a medium of one density into a medium of a different density. A stick thrust into clear water appears to be bent at the surface of the water. It is this quality of bending, or refracting, that is responsible for a number of interesting phenomena. One of the best-known examples is the appearance of the sun before it has actually risen above the eastern horizon, and another its visibility after it has actually disappeared below the western horizon. This is because the refraction of light at the horizon is 35 minutes of arc, or a little more than one-half of an angular degree. Figure 11–1 shows the effect of refraction. *AB* is the curved line which the sun's rays take in transit to the observer (*A*), but *AC* is the line of observation as it appears to the onlooker. Refraction has the effect of raising objects above their real location and thus lengthening the day by a few minutes in middle latitudes and by more than 24 hours beyond the polar circles.

Looming

Looming is essentially a special kind of refraction that occurs when there is a shallow but marked inversion of temperature, with only a few feet of cooler surface air below warmer air. Under this condition the rays of light passing upward from an object on the ground or on the ocean surface (or even below the horizon) are bent enough by refraction to make the reflection of the object appear inverted and at a higher elevation. This phenomenon is common over large bodies of water, near the seashore, and over cold surfaces, especially in arctic regions. Looming, sometimes called a *superior mirage*, is probably responsible for the expression "looms in the distance."

Mirage

Mirage (sometimes called *inferior mirage* to distinguish it from superior mirage) is essentially the opposite of looming. It occurs when a shallow layer of heated air is next to the ground with a decidedly cooler layer above it. It is not surprising, therefore, that it occurs most frequently on hot summer days and over level deserts; it is also known over bodies of water near the shore and is very marked on a paved highway in summer. Such conditions suggest the beginning of convection, but by the time convection itself begins, the mirage frequently disappears. The most common mirages appear as somewhat wavy, watery areas in the near distance. They are frequently observed in Death Valley which, as a consequence, was the site where many humans perished in gold-rush days because the "water" ahead proved to be an illusion; what appeared to be the surface of water was simply the image of the sky. Images of superficial objects, such as trees and buildings, that appear sometimes are always inverted and lower in height than the objects' actual elevation—again the result of refraction and reflection of light.

Twinkling

On cold and clear winter nights, especially near the horizon, the *twinkling* of stars, or scintillation, occurs. It is caused by three things: (1) a constant changing of the apparent positions of the stars due to refraction through many atmospheric layers of differing densities; (2) a change in the brightness of the stars due to the lens effect of the atmosphere that causes rays of some densities to concentrate at some points and be diverted at others, resulting in something similar to that which would occur with the simultaneous use of many convex and concave mirrors; (3) a change in color due to optical interference. *Interference* is the term used to indicate that the various rays of light received from the stars at a given moment may be of different lengths, a fact which causes certain rays to be out of phase and which in turn causes the destruction of certain colors (wavelengths) from moment to moment; therefore, those rays which are received will vary rapidly.

Refraction, looming, mirage, and twinkling would not occur if there were no atmosphere and if the gases in the atmosphere were not of different temperatures and especially of different densities.

Other Optical Phenomena

There are four other phenomena that occur less often, but usually more spectacularly,

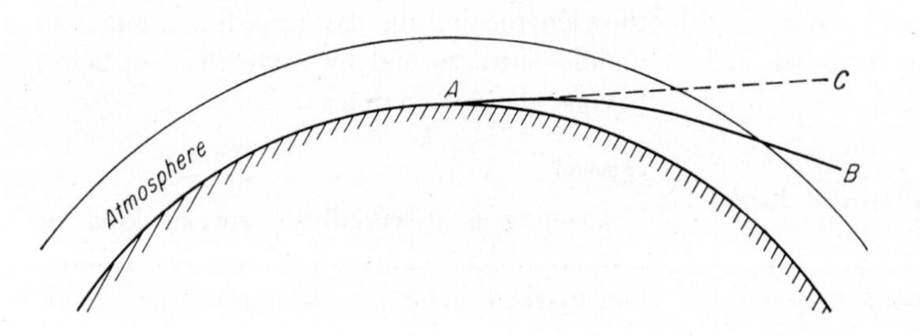

Fig. 11–1. Diagram to show the refractive effect of the earth's atmosphere.

Fig. 11–2. Solar halo. (Photograph by G. A. Clarke, U.S. Weather Bureau.)

than those caused by refraction of light as it passes through the air. They owe their appearance to foreign particles—chiefly moisture and dust that are in the air part of the time—as well as to the reflection, refraction, and diffraction of light.

Halos and Sun Dogs

Halos, often referred to as "rings" around the moon or sun, are formed rather frequently when the sky is completely or partially covered with a thin veil of cirro-stratus or, less commonly, alto-stratus clouds. As the light from either the sun or the moon passes through the ice and snow crystals of which those particular clouds are composed, it is refracted and also reflected. The angle of the observer produces a complete or partial circle around the luminary; the circle has a radius of about 22 degrees (Figure 11–2). A larger circle about 46 degrees in radius is sometimes present. At rare intervals a very complicated system of rings and partial rings may occur. Southern Michigan had such a display in the spring of 1950; it lasted in varying form all day. At that time the white ring, parts of three halos, three tangential arcs, small columns of light, and numerous *parhelia* were clearly evident. Halos may be white, or they may have the colors of the spectrum. Usually the most interesting aspect of these phenomena is the *mock sun*, or *sun dog*, technically referred to as *parhelion*. The sun dog makes its best appearance late on cold

Fig. 11–3. Primary and secondary rainbows. (*Courtesy U.S. Weather Bureau.*)

winter afternoons. It is only a small portion of a halo which appears on the right or left side of the sun, or sometimes on both sides. Its colors are generally red and yellow. Solar halos are probably as common as lunar halos, but the brightness of the sun limits solar halo observations to some extent.

Coronas

Coronas are small concentric rings that appear close to the sun or moon. The radii of coronal circles are from 1 to 10 degrees; at their best, they are somewhat colored, with the red on the outer edge grading into whitish or blue inside. Coronas are formed by *diffraction* as the light passes by the edges of water droplets of thin high clouds. Any color is caused by interference. In recent years the term *corona* has been extended to include a more common appearance, namely, the fuzzy-edged glow which occurs with fairly thick alto-stratus clouds that completely blot out the exact edge of the heavenly body. Not even a portion of a ring appears in this case.

Rainbows

Although the *rainbow* has no pot of gold at its end, it is doubtless the most beautiful of the atmospheric optical phenomena (Figure 11–3). It occurs almost exclusively when convection is evident, accompanied by showers or thundershowers. Whether it is a morning or afternoon bow, the primary bow can occur only when the sun is less than 42 degrees above the horizon. Furthermore, the weather must be partially clear so that the sun can shine on a shower which is on the opposite side of the observer; that is, the sun, the observer, and the center of the rainbow must be in a straight line. When the sun is on the horizon, the bow may extend 180 degrees from side to side and 42 degrees in height; at any higher elevation of the sun (up to 42 degrees), the bow will be correspondingly smaller. The bow is caused by the refraction and reflection of sunlight by the falling raindrops. The colors of the spectrum range from red on the outside to violet on the inside in the case of the primary or major bow; they are reversed in the secondary bow, which occurs fairly often at a higher elevation. Rainbows, often brilliant ones, may be seen in the spray of fountains, waterfalls, and even ocean waves. Less frequently observed but of equal interest is the *moonbow*, developed under the same conditions as the rainbow.

Crepuscular Rays

Crepuscular rays appear when the air has a high, or fairly high, dust or moisture content and when the sky is only partly filled with broken but fairly thick clouds of some cumulus or stratus type (Figure 11–4). The rays, which appear as fan-shaped light streaks extending downward from the sun, are the rays of the sun shining through the rifts, or openings, in the clouds and illuminating the particles of moisture and dust in the air; the dark areas in between the streaks are simply the cloud shadows. When the cloud conditions are the same and the sun is low in the sky, or just below the horizon, the rays extend upward and flare out from the sun; these are called *anticrepuscular rays*.

Sky Colors

Among the phenomena which are the result of particles always present in the air are the colors of the sky. Blue sky predominates, under cloudless conditions, when the sun is

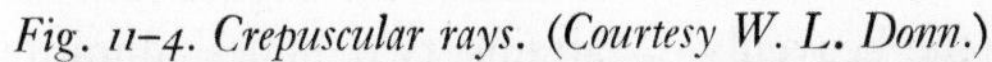

Fig. 11–4. Crepuscular rays. (Courtesy W. L. Donn.)

high in the sky. Near the horizon at such times, the appearance is gray, and close to the sun the blue becomes very faint or even whitish. The purer the air, the clearer will be the sky and the more intense the blue, like that common in deserts. *Selective scattering* of the blue rays by the multitudinous particles is responsible for the blue. What happens is that, as white light comes toward the earth, it is affected by the dust and moisture particles, both small and large, so that the long rays are allowed to pass through to the eye; one sees red and yellow as he looks directly at the sun, but the blue rays and to some extent the green (both short wavelengths compared with red and yellow) are scattered at right angles to the line of travel. Thus, to the sides of the sun, the blue rays cover almost all of the sky. If it were not for the foreign particles in the air, the sky would be black and the heavenly bodies would shine brightly on the black background, day and night.

When the sun is near the horizon, both morning and evening, the atmosphere through which the sunlight travels is much greater than when the sun is higher in the sky. At the low elevations of the sun, selective scattering is particularly efficient, with the blue rays appearing high in the sky and only the red, orange, and yellow (occasionally green) getting through the thick air and its moist and dusty freight. During sunrise and sunset, a small difference in the amount of moisture or dust may cause considerable variation in the low-sky colors. The purple color so often seen at sunrise and sunset is the result of the reflection of red rays against the dark side of the cloud.

Twilight and dawn are caused by the diffraction, scattering, and to some extent the reflection of sunlight by particles in the upper air when the sun is below the horizon. The duration of twilight or dawn varies with latitude and therefore with the angularity of the rising or setting sun. On the equator, where the sun always sets almost perpendicularly to the horizon, it lasts about 70 minutes. At higher latitudes it increases; it attains two hours' duration in the northern United States and several months in polar areas. *Astronomical twilight* refers to the length of time the sun is not more than 18 degrees below the horizon; *civil twilight* refers to the period of time the sun is between the horizon and 6 degrees below it; generally, it is too dark to do much work when the sun is more than 6 degrees below the horizon.

12

Climate

As indicated in Chapter 1, the conditions of the atmosphere at any given time make up the weather. All the weather which has occurred at any place makes up the *climate* of that place. Thus climate is much more than "average" weather, a definition which is often given. To describe the climate of a region adequately is obviously a difficult matter. It is not enough to mention the average or *mean* temperature and precipitation for the months and for the year, because the mean is rarely experienced.

This fact will be more readily appreciated if we examine, for example, the method of obtaining for any place the mean temperature for January over a period of twenty years, a relatively short period as far as climatic statistics are concerned. The usual method is to record both the maximum and the minimum temperature for each day of the month. At the end of the month, all the maxima are added together and divided by 31, the number of days in the month, which gives the mean maximum January temperature for that year. The minima are treated in the same manner. Then the mean maximum and mean minimum are added together and divided by two, the result being the mean temperature for the month. Then, to get the mean for 20 years, all the January means are added together and divided by 20, which gives the so-called "average January temperature," a value that furnishes only a broad concept of January temperature conditions for that place. The means or averages are valuable, however, particularly for comparative purposes. But they do not show extremes or give any indication of the frequency of large variations in temperature. The mean January temperature of 54.5° for Los Angeles does not disclose the fact that frosts sometimes occur there in that month, with serious loss to agriculturists in the area. The mean annual precipitation of 25 inches for Peiping does not tell the story of occasional droughts in that locality, with the consequent suffering of thousands through the failure of crops. The average September wind velocity of 8.2 miles per hour for Miami fails to indicate that the region is sometimes visited by hurricanes with wind velocities of over 100 miles per hour, sufficient to inflict appalling devastation of property and possibly of human life.

Again, the mean annual temperatures of New York City, Chicago, and Omaha are about the same. Mean figures belie the fact, however, that the seasonal extremes of Omaha are greater than those of Chicago, while Chicago has a greater range than New York. Omaha has the hottest summer and the coldest winter of the three places; New York is the most temperate with somewhat the coolest summer and warmest winter of the three; Chicago is in between.

Furthermore, the fact that New England infrequently experiences a hurricane did not deter one from wreaking great destruction there in 1938; another followed in the early fall of 1944, only to be doubled in 1954 when both Carol and Edna (each year's hurricanes are named alphabetically for women) ripped across Long Island and into the northeastern states; and the heavy rains and floods accompanying the weakened hurricane of August, 1955, are not soon to be forgotten. In June, 1953, despite the fact that the term *tornado* was never used in a prognostic sense for that area by the U.S. Weather Bureau in Boston, the most costly of all such storms up to that time tore into Worcester, Massachusetts, with tremendous damage. The day before (June 8), a considerable area of Flint, Michigan, where major tornadoes are not expected often, was devastated. While such storms may be unusual, even commonly unexpected, they are, from time to time, a part of the weather and also of the climate of the places affected; total climate must include them in order to be correctly portrayed. These and many other examples from different parts of the world could be cited to illustrate the fact that *climate is the aggregate of the weather*.

Elements and Controls of Climate

Since climate is *all* the weather, it follows that the elements of climate are the same as those of weather, although, as already pointed out, pressure in itself is not an important factor in climate. The elements may, therefore, be broadly grouped as *sunshine*, *temperature*, *moisture*, and *pressure and winds*. The term *moisture* embraces all forms of condensation and precipitation as well as humidity.

The *controls of climate* are the factors which act upon or influence the elements of weather and climate and which, combined with them, are responsible for climates. In Ward's *The Climates of the United States* are listed seven major climatic controls: *latitude*, *altitude*, *land and water*, *mountains as barriers*, *ocean currents*, *prevailing winds*, and *cyclonic storms*. To these might be added, as phases of land control, local topography, surface covering—such as vegetation or the lack of it—wetness, dryness, color, and snow cover. Although vegetation is a direct response to climate, it is believed to have an appreciable influence on temperature and precipitation. The significance of each of these controls will be readily understood in the study of later sections.

The prominence of the individual control varies with location and to some extent with the time of year. Where water is the dominant control, the climate is said to be *marine;* where land is dominant, the climate is said to be *continental*. *Mountain* and *plateau* climates are distinctive chiefly because of the prominence of the control of altitude and because of the increase in intensity of insolation and terrestrial radiation.

Solar Climate

The sun, or we may say latitude, is unquestionably the principal control of climate. Yet if it were the only control, all places on the same parallel would have the same kind of climate. The climate which any place would have by virtue of the solar radiation alone, i.e., if the atmosphere were absent, is termed *solar climate*. The moon has solar climate only. Solar climate is, therefore, the same for all points having the same latitude.

Computation of Solar Radiation outside the Atmosphere

The radiation coming from the sun is almost a constant quantity, so nearly constant, in fact, that it has been given the name *solar constant*. This radiation has been measured many times, and it is always found to be very close to 1.94 gram calories, a gram calorie being the amount of heat required to raise one gram of water (which is the same as one cubic centimeter of water) one degree centigrade. There are certain standard conditions to which any measurement of solar radiation must be reduced in order to give it the name of solar constant: the radiation is considered as incident on a square centimeter of surface exposed perpendicularly to the sun's rays for a period of 1 minute, at the outside of the earth's atmosphere, and at the earth's average distance from the sun. It may be thought of in this manner: if a gram of water in the shape of a cube could be taken to the outside of the atmosphere at the time of year when the earth is at its average distance from the sun, about April 1 or October 1, and if the cube of water could be so exposed that the sun's rays would strike a face of the cube perpendicularly, the radiation, or energy, received on that face would be sufficient to raise the gram of water 1.94 degrees centigrade in 1 minute. In a period of 12 hours, the total radiation under these conditions would be nearly 1,400 gram calories, or sufficient heat to raise 1,400 grams of water 1 degree centigrade.

Table 12–1 shows the relative amounts of insolation received at the outside of the earth's atmosphere on a horizontal surface during a 24-hour period, taking the insolation at the equator on March 21 as standard.

Table 12–1. Relative Values of Solar Radiation outside the Atmosphere

Latitude	Mar. 21	June 21	Sept. 22	Dec. 21
North pole	0.000	1.201	0.000	0.000
60°N	0.500	1.093	0.494	0.056
40°N	0.766	1.107	0.756	0.357
20°N	0.940	1.044	0.927	0.676
Equator	1.000	0.882	0.987	0.941
South pole	0.000	0.000	0.000	1.283

Measurement of Solar Radiation at the Earth's Surface

The radiation received from the sun at the earth's surface is never so great as the solar constant. The obliquity of the sun's rays, the distance of the earth from the sun, and the character of the air through which the rays pass are factors which reduce the values of solar radiation. Table 12–2 gives, for the most part, the actual values of radiation received on a horizontal plane at the earth's surface during a 24-hour period. They represent conditions over land more than those over water. A study of this table shows that at either equinox when the days are 12 hours long, there is no latitude where the insolation equals the solar constant for 12 hours. In fact,

Table 12–2. Daily Totals of Solar Radiation (Direct and Diffuse) in Gram Calories Received on a Square Centimeter of Horizontal Surface on a Clear Day

North latitude	Vernal equinox	Summer solstice	Autumnal equinox	Winter solstice
90	0*	896	0*	0
85	142	860	114	0
80	247	835	203	0
75	322	815	272	0
70	374	795	322	0
65	403	789	354	12
60	413	794	368	46
55	443	824	410	88
50	504	835	480	150
45	570	848	509	215
40	599	800	526	270
35	614	744	553	327
30	630	772	593	392
25	657	780	630	434
20	692	790	663	466
15	717	770	687	512
10	726	696	704	553
5	713	657	704	602
0	683	609	687	650

* These are not actually zero because the refractive effect of the atmosphere brings the sun just above the horizon on these dates. Light of the sky must also yield appreciable radiation.

SOURCE: Adapted from Kimball.

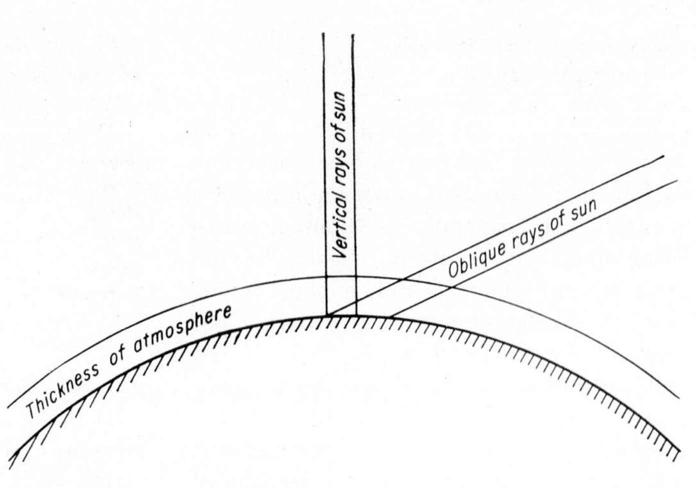

Fig. 12–1. Diagram to show the effect of the obliquity of the sun's rays. When the sun's rays are nearly vertical, less surface is covered by a given sunbeam (or "bundle" of rays) than when the sun is near the horizon. When the sun is low in the sky, the rays must pass through more atmosphere than when the sun is high; consequently, there is a great loss of solar energy through absorption by the atmosphere when the sun is low.

the highest value for any latitude at either equinox is only 726 gram calories, which is only about one-half the solar constant for 12 hours. A further study of the table reveals that the values of insolation vary greatly with the time of year and with latitude. The reasons for these variations need to be considered.

Causes of Variations

Distance from the Sun. It is evident that distance from the source of heat makes a difference in the amount of heat received. As pointed out in a preceding chapter, the earth is nearest the sun about January 3, the distance being approximately 91,500,000 miles; the sun is then said to be in *perihelion.* It is farthest from the sun about July 3 when its distance is approximately 94,500,000 miles; the sun is then said to be in *aphelion.* It might be expected, therefore, that January would be warmer than July in the northern hemisphere. Other factors prevent this state of affairs, although it may safely be said that January is warmer than it would be if the earth and sun were farthest apart in that month and that July is not so hot as it would be if the earth and sun were nearest each other in that month.

Altitude of the Sun. More significant than distance from the sun is the altitude of the sun above the horizon. The sun's rays are never vertical on the poleward sides of the tropic circles, and even within the tropics only one ray of the sun is ever vertical at one time. Thus we may say the sun's rays are oblique outside the tropics and rarely otherwise within the tropics. For example, at the time of the summer solstice on the 40th parallel north latitude, the sun even at noon is not at the zenith but 73½ degrees above the southern horizon; at the equinoxes it is 50 degrees above; and at the winter solstice it is only 26½ degrees above the southern horizon. This means, therefore, that at 40 degrees north latitude a given sunbeam must cover more horizontal surface in December than in March (Figure 12–1) and more in March than in June. Table 12–3 gives the altitude of the sun at noon at the equinoxes and solstices for latitudes of the northern hemisphere.[1]

Thickness of the Atmosphere. Furthermore, when the sun is low in the sky, its rays must pass through more atmosphere than when it is high (Figure 12–1). By reference to Table 12–4, showing the relative lengths of path of the sun's rays through the atmosphere at noon,

[1] To find the altitude of the sun at noon in degrees above the southern horizon, at either equinox, subtract the latitude from 90; to find the altitude of the sun at noon on the summer solstice add 23½ degrees to that figure; for the winter solstice, subtract 23½ degrees.

Table 12–3. Altitude of the Sun at Noon in Degrees of Arc above the Horizon, Assuming No Atmosphere

North latitude	Equinoxes	Summer solstice	Winter solstice
90	0	23.5	−23.5
85	5	28.5	−18.5
80	10	33.5	−13.5
75	15	38.5	−8.5
70	20	43.5	−3.5
65	25	48.5	1.5
60	30	53.5	6.5
55	35	58.5	11.5
50	40	63.5	16.5
45	45	68.5	21.5
40	50	73.5	26.5
35	55	78.5	31.5
30	60	83.5	36.5
25	65	88.5	41.5
20	70	86.5	46.5
15	75	81.5	51.5
10	80	76.5	56.5
5	85	71.5	61.5
0	90	66.5	66.5

Table 12–4. Relative Lengths of Path of the Sun's Rays through the Atmosphere at Noon

North latitude	Equinoxes	Summer solstice	Winter solstice
90	44.7	2.50	0
85	10.8	2.08	0
80	5.7	1.82	0
75	3.9	1.61	0
70	2.92	1.41	0
65	2.35	1.34	25.0
60	2.00	1.24	8.4
55	1.75	1.16	5.1
50	1.56	1.11	3.59
45	1.41	1.07	2.71
40	1.31	1.04	2.22
35	1.21	1.03	1.92
30	1.15	1.01	1.69
25	1.10	1.00	1.51
20	1.06	1.01	1.38
15	1.04	1.02	1.28
10	1.02	1.03	1.19
5	1.01	1.05	1.13
0	1.00	1.08	1.08

SOURCE: Adapted from Zenker.

one sees that at the time of the equinoxes, when the sun is directly overhead at noon at the equator, the vertical ray passes through just "one atmosphere," technically called "one air mass." At the same time of the year and at noon at 40 degrees north latitude, the sun's rays must pass through 1.31 times as much atmosphere, or through 1.31 atmospheres. When the sun is on the horizon, the rays must pass through nearly 45 atmospheres. As has already been shown, the atmosphere absorbs, scatters, and reflects insolation passing through it; consequently, the more atmosphere, the greater the absorption, scattering, and reflection. The student should recall that it is largely this reflection of sunlight, probably by the dust, perhaps by water vapor in the upper air, and to some extent by the molecules of atmosphere themselves after the sun has dropped below the horizon, that causes *twilight*.

Duration of Sunlight. If the days were all 12 hours long the values of solar radiation in the various latitudes could be easily compared. The changes in duration of sunlight together with the changes in seasons cause the insolation values to vary widely. Study again Table 12–2 and note how different the figures are for the various seasons. To understand these changes in solar radiation it is necessary to consider to what degree the hours of sunshine change with latitude and with time of year. Those changes may be easily determined by use of the "duration of sunlight chart," Figure 12–2, and by studying the tables. It should be noted here that the atmosphere has a *refractive* effect on the sun's rays, making the sun appear to be above the horizon when it is geometrically below. The chart takes this effect into account. Thus the duration of sunlight in the northern hemisphere is found to be a few minutes longer than it would be if there were no atmosphere to cause refraction, or bending of the sun's rays. It should also be noted that the higher the latitude the more significant is this refractive effect because the sun rises and sets more obliquely in high latitudes than in low. Equally significant is the greatly lengthened period of twilight in high latitudes.

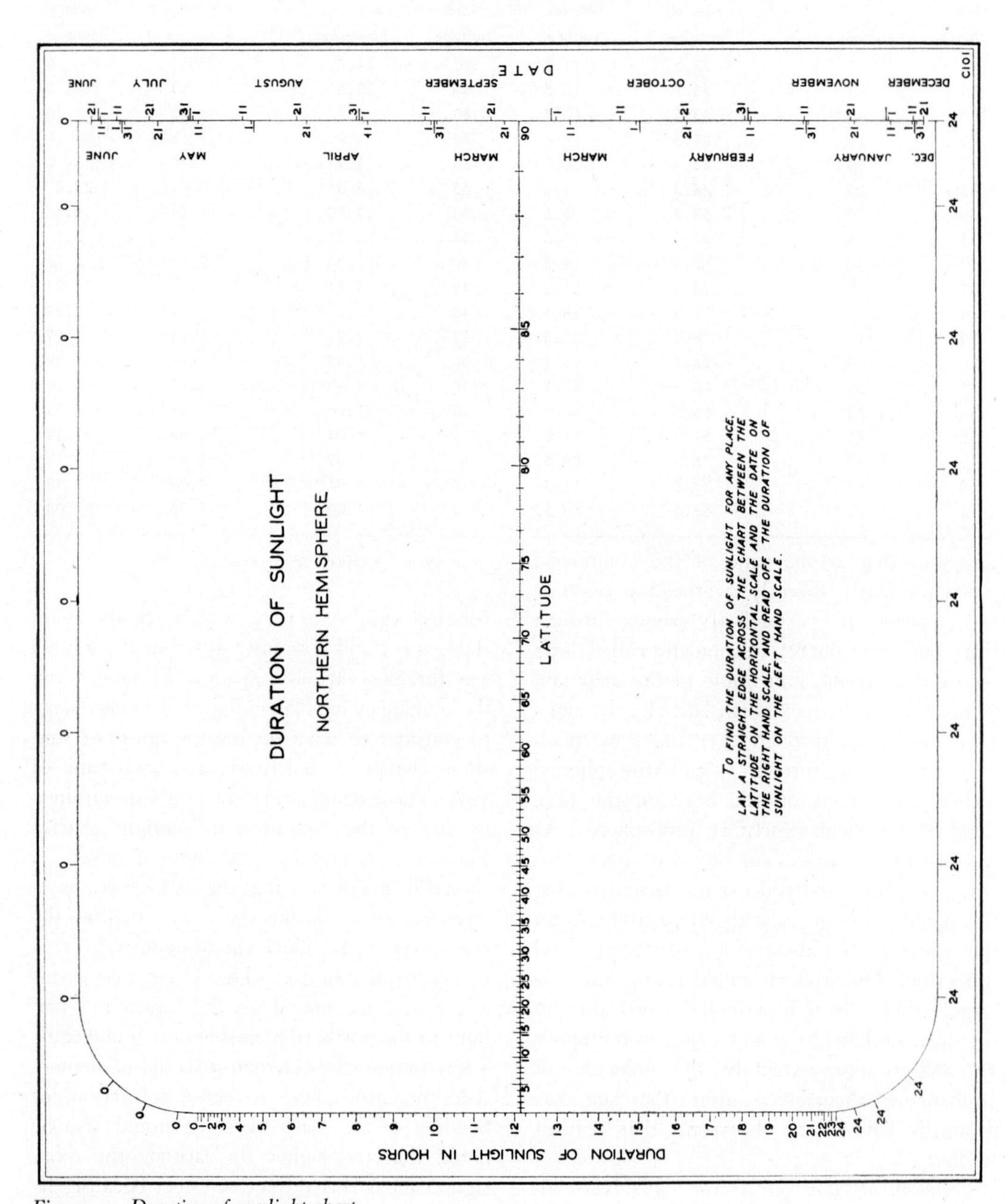

Fig. 12–2. *Duration of sunlight chart.*

Table 12–5. Duration of Sunlight in Hours and Minutes

North latitude	Dec. 21	Jan. 21	Feb. 21	Mar. 21	Apr. 21	May 21	June 21	July 21	Aug. 21	Sept. 21	Oct. 21	Nov. 21
0	12-7	12-7	12-7	12-7	12-7	12-7	12-7	12-7	12-7	12-7	12-7	12-7
10	11-32	11-38	11-52	12-7	12-24	12-37	12-43	12-37	12-24	12-8	11-52	11-38
20	10-55	11-7	11-36	12-8	12-43	13-9	13-20	13-10	12-43	12-9	11-36	11-7
30	10-13	10-32	11-19	12-9	13-3	13-47	14-5	13-48	13-4	12-11	11-18	10-31
40	9-19	9-48	10-57	12-11	13-30	14-34	15-1	14-36	13-32	12-13	10-56	9-48
50	8-4	8-47	10-29	12-13	14-7	15-58	16-23	15-44	14-9	12-17	10-26	8-47
60	5-52	7-7	9-44	12-17	15-6	17-35	18-52	17-42	15-10	12-23	9-41	7-6
65	3-34	5-38	9-10	12-21	15-52	19-26	22-3	19-36	15-57	12-28	9-6	5-37
70		2-28	8-18	12-25	17-7	24-0	24-0	24-0	17-14	12-35	8-13	2-24
75			6-43	12-34	19-39	24-0	24-0	24-0	19-53	12-47	6-33	
80			1-53	12-50	24-0	24-0	24-0	24-0	24-0	13-11	1-15	
85				13-19	24-0	24-0	24-0	24-0	24-0	14-36		
90				24-0	24-0	24-0	24-0	24-0	24-0	24-0		

Table 12–6. Dates of Sunrises and Sunsets

North latitude	Date at which the sun first appears above the horizon	Date at which the sun first remains continuously above the horizon	Date at which the sun first drops below the horizon	Date at which the sun first remains continuously below the horizon
70	Jan. 17	May 17	July 28	Nov. 25
75	Feb. 7	Apr. 28	Aug. 15	Nov. 6
80	Feb. 21	Apr. 14	Aug. 29	Oct. 22
85	Mar. 6	Mar. 31	Sept. 13	Oct. 8
90	Mar. 18	Mar. 18	Sept. 25	Sept. 25

Table 12–7. Duration of Days of Sunshine, of Absence of Sunshine, and of Alternating Sunshine and Absence of Sunshine

North latitude	Period of total absence of sunshine	Period of lengthening days and shortening nights	Period of total sunshine	Period of shortening days and lengthening nights
70	53	120	72	120
75	91	82	109	82
80	122	53	137	53
85	149	25	166	25
90	174	0	191	0

The three Tables 12–5 to 12–7 show some interesting features concerning the time when the sun is above the horizon. These tables not only indicate the importance of refraction, but they also show clearly the effect of the obliquity of the sun's course as it rises and sets. Both these factors account for the sunlight's duration being considerably longer at a particular latitude than the duration of darkness; this effect is brought out notably for latitudes beyond the arctic circle. Furthermore, the change in duration of sunlight is directly correlated with the course of the sun through the sky from sunrise to sunset at various times of the year; in the northern-hemisphere summer, the sun rises north of east, reaches a relatively high point in the sky at noon, and sets north of west, which means that the time required for the sun to take this course is considerably longer than in the northern-

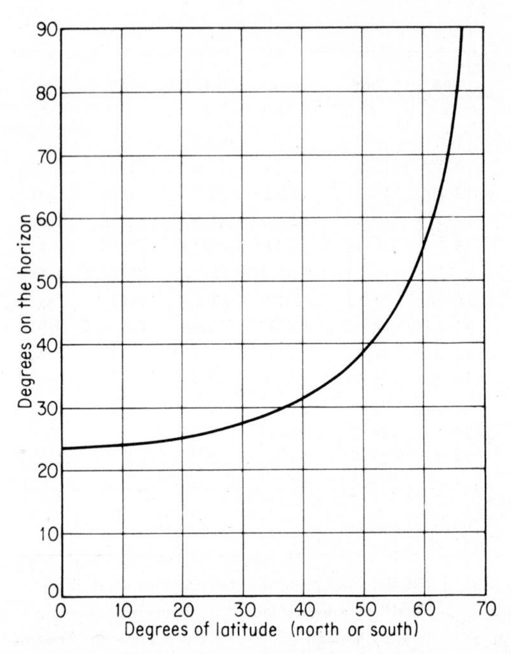

Fig. 12–3. Graph to show the number of degrees north of east at which the sun rises and the number of degrees north of west at which the sun sets on the summer solstice (about June 21). On the winter solstice (about December 21), the values are the same except for the fact that they represent the number of degrees south of east or south of west at which the sun rises and sets.

hemisphere winter when the sun rises south of east, is relatively low in the southern sky at noon, and sets south of west. The graph, Figure 12–3, shows the position of the sun at sunrise and sunset for all latitudes on the solstices.

In low latitudes, the values of solar radiation do not change greatly from season to season because the sun is always well above the horizon at noon and there is not much change in the duration of sunlight. Consequently, man does not have to change his seasonal habits in order to adapt himself to the resulting small changes in seasonal climate. In high latitudes, however, man's activities are inextricably tied up with changes in the intensity and duration of sunlight, which necessitates radical adjustments to the succession of the seasons. To the dweller in arctic regions, this annual succession of light and darkness, of intense solar radiation and then total absence of it, is of greater import than topography or soil or location or mineral resources—it is the one factor which determines most profoundly his habits and activities.

Summary

Table 12–8 shows in a simple summary the relative amounts of solar radiation for three latitudinal positions and also the effect of the earth's atmosphere upon that radiation.

Table 12–8. Insolation during a 24-hour Period

	Conditions	Latitude	Vernal equinox	Summer solstice	Autumnal equinox	Winter solstice
A	Outside the atmosphere	Equator	1000	882	987	941
		40°N	766	1107	756	357
		N.P.	0	1201	0	0
B	At the earth's surface	Equator	612	513	604	547
		40°N	407	655	402	123
		N.P.	0	489	0	0
B/A	Coefficient of transmission	Equator	61	58	61	58
		40°N	53	59	53	34
		N.P.	0	41	0	0

13

Physical Climate

Physical climate is simply solar climate modified by the atmosphere and by the various controls already referred to. The effect of latitude has already been discussed in the preceding chapter. We now consider the broad, general effects of altitude and land versus water. In succeeding chapters a more detailed study will be made of climatic elements as they are affected by the controls, concluding with a consideration of climatic regions and types of climate.

Mountain and Plateau Climates

Because of their altitude, mountains induce climates which differ markedly from those of lowlands; at various altitudes and on different sides of the same mountain, conditions may be in marked contrast. The altitude, however, as well as the barrier effect of mountains, has much the same kind of influence in all latitudes but not to the same degree. For example, high tropical mountains, even on the equator, have variations in temperature and vegetation that frequently run the gamut of climates, from tropical at their bases to polar on their snow-clad summits. In parts of Latin America, the zones of temperature and vegetation (and to some extent precipitation) are called *tierra caliente* (lowest and hottest), *tierra templada* (intermediate), and *tierra fría* (highest and coldest); Figures 13–1 to 13–3 reveal cultural aspects of life in these zones. Even greater distinctions are made in other places. Moreover, in middle latitudes where mountains are high enough, considerable variations are evident at different elevations, under different exposures to the sun or wind, and where there are differences in availability of moisture. Because of these great variations within short distances, mountain and plateau climates cannot be considered a type in the same sense that tropical deserts are considered a type. For this reason some geographers, botanists, and ecologists lump high mountains together, undifferentiated, saying they have not *a* mountain climate but *mountain climates*.

We have already seen that pressure always decreases as altitude increases, with important

Fig. 13–1. Tierra caliente. Sugar cane is well adapted to the constantly warm climate of this low-altitude, low-latitude section of the Cauca Valley in Colombia; abundant rain for much of the year with a definite drier period from December to March is the rule. (*Courtesy Standard Oil Company of New Jersey.*)

consequences to man; we have also seen that insolation, temperature, water vapor, and precipitation are definitely related to altitude. The student should take time here to review these relationships.

In general, the climates on mountains and plateaus are characterized by a decrease in temperature, pressure, and absolute humidity. They are also characterized by an increase in insolation and in relative humidity in the lower levels by day. As already indicated in Chapter 5, insolation increases with an increase in altitude at a lower rate than radiation, the increase of insolation in 2 miles being about 20 per cent while the increase in radiation for the same altitude is approximately 40 per cent. This fact accounts, in part, for the lower temperatures on mountains. The condition is due largely to selective absorption by the water vapor in the air. The decrease of temperature is greatest by day in summer and least by night in winter. In fact, inversions of temperature are frequent in winter, especially in latitudes outside the tropics. This seasonal variation in the rate of decrease in temperature is the result of strong heating at lower levels in the summer and strong cooling in the winter.

Mountains are also characterized by smaller ranges of temperature than are lowlands, but plateaus have higher temperature ranges than mountains; and the months of maximum and minimum temperatures tend to be delayed more on mountains. The maximum and minimum free-air temperatures are delayed, and this delay is felt on mountains because their limited surface can have only a restricted effect on the great mass of passing air. Where a snow cover persists late, the maximum temperature is further delayed. Diurnal (day to night) ranges of temperature are also lessened, and the daily maximum temperature tends to be delayed longer than on lowlands, particularly if the air is dry and the sky is clear. High valleys that are well enclosed have temperature conditions resembling those of continental lowlands rather than those of the surrounding mountains. In fact, the high valleys may have even more extreme conditions than lowlands.

At an altitude of 6,000 to 7,000 feet in the free air, and of about 10,000 feet on mountains, the absolute humidity is about one-half that at sea level; but in calm weather, relative humidity does not change much on the average with increase in height. Precipitation usually increases with increasing altitude, the zone of maximum precipitation being about 4,000 feet in the tropics and somewhat higher in extratropical latitudes; above this zone there is a decrease in water-vapor content; some mountains in arctic regions rise beyond the zone of appreciable precipitation. On mountains and plateaus, summer is usually the season of

Fig. 13–2. Tierra templada. Because of the moderate altitude of approximately 5,000 feet, temperatures are sufficiently reduced to permit the growing of coffee and plátano *(a variety of banana) in the Central Cordillera of Colombia, State of Caldas. Here both temperature ranges and rainfall distribution take on the characteristics of an equatorial regime, but without the high temperatures. (Courtesy Standard Oil Company of New Jersey.)*

Fig. 13–3. Tierra fría. Even the summers in Cerro de Pasco, Peru, at an elevation of over 14,000 feet, are cooler by 40 degrees, or more, than corresponding latitudes at sea level. Obviously, agriculture is strictly limited. (Courtesy Standard Oil Company of New Jersey.)

Fig. 13–4. Plateau climates, such as found here on this 15,000-foot plateau of central Peru, are likely to be deficient in rainfall and to have large daily ranges of temperature. Reduced pressure limits human activity, and agriculture is not a dependable occupation. The llama is especially tolerant of low atmospheric pressure. (Courtesy Standard Oil Company of New Jersey.)

maximum relative humidity, cloudiness, and rainfall because of the strong convection in that season. Most mountains, especially in trade-wind latitudes, have a rainy and a dry side unless they parallel the rain-bearing winds. Mountains have more precipitation than plateaus (Figure 13–4), and mountainous sections of plateaus usually stand out prominently as islands of heavier rain or snow. This is notably true in the western United States, in Yemen, and in Iran. Such elevations are often clothed with thick vegetation while the surrounding lower lands are arid. The water from these higher lands may be sufficient for irrigation, thereby permitting permanent settlements near the base of the mountains.

Winds tend to be more local in mountains and plateaus than in lowlands. Such winds are the mountain and valley breezes, chinooks, foehns, and others, some of which have been mentioned in earlier chapters; probably the most intensive study of these types of winds has been made in Switzerland (Figure 13–5). Certain high mountain ranges act as definite

Fig. 13–5. This moderately high valley in central Switzerland has many mountain climate characteristics: foehns, and mountain and valley breezes, particularly. Pastoral industries necessarily predominate. (Courtesy Standard Oil Company of New Jersey.)

Fig. 13–6. Marine climate is characteristic of most of Vancouver Island. The uniformly mild temperatures in both summer and winter, and the abundant rainfall (heaviest in winter), favor enormous trees like the Douglas fir. (Courtesy Standard Oil Company of New Jersey.)

Fig. 13–7. Continental climates do not necessarily mean heavy snow, but here in Vermont where the summer mean temperatures are under 70° and where the mean annual ranges of temperature are over 50 degrees, much of the winter precipitation has to be in the form of snow. (*Courtesy Standard Oil Company of New Jersey.*)

barriers to relatively shallow surface winds, thereby causing marked contrasts in climatic characteristics between the opposite sides. The Himalayas shut out from India the intense winter cold of Tibet and almost entirely prevent the north side's receiving moisture from the Indian Ocean; at the same time they cause abundant rainfall in India. The Rocky Mountains frequently prevent the penetration to the Pacific of cold waves from Canada; they also keep the moderating influences of the Pacific out of the interior. The Andes of South America are so lofty that the southeast trades do not blow over them. In Peru, the east sides are wet while the coast is desert.

From a physiological standpoint and because of climatic conditions, mountains of moderate altitude in middle and low latitudes are desirable as the permanent abode of man and as his playground because of their fairly equable temperatures, great dryness, and increased ultraviolet radiation.

Marine Climates

Marine climates exist wherever water is the dominant control. They are, therefore, found over the open ocean, on small islands surrounded by great expanses of water, and along coasts where the winds are continuously onshore. Marine climates are characterized by moderate temperatures with small diurnal and annual ranges, by high humidities, both absolute and relative, and by much cloudiness and rainfall (Figure 13–6). The reasons for these conditions have already been mentioned in Chapter 4.

Briefly summarized, the reasons for marine climate are: (1) Some of the insolation which falls on a water surface is reflected; (2) the rest penetrates to various depths and is almost wholly absorbed; (3) much of the heat absorbed is used in evaporation of the surface water; (4) some heat is lost by radiation from the surface; (5) much of the heat acquired by the surface layer is transferred downward by the process of convection; and (6) because of the great mobility and high specific heat of water, the temperature of a water body does not respond greatly to the heat absorbed. The great evaporation over water bodies makes the absolute and relative humidity of the air over them high. This latter condition favors great cloudiness that in turn tends to shade the water surface and, therefore, to reduce the insolation available for warming the water.

The slowness with which water bodies are warmed and cooled causes a retardation in the time of maximum and minimum temperatures in a marine climate. Although June is the month of greatest heating in extratropical latitudes in the northern hemisphere, August and frequently September are the months of maximum temperature in regions of marine climate. February and March are usually the coldest months. October is invariably warmer than April. In the tropics, where there is little difference in the insolation received from one time of year to another, the air temperatures over the water vary only slightly during the year, and the daily ranges may be as low as 2 or 3 degrees.

In a marine climate, the early morning hours are generally the cloudiest and rainiest because then relative humidity has attained its diurnal maximum. Autumn and winter are the seasons of greatest cloudiness and rainfall because then the annual cooling is in progress, thereby favoring high relative humidity. In both cases, the high humidity means relatively light air with considerable latent heat and, therefore, easy convection.

Continental Climates

Continental climate (Figure 13–7) is in general the antithesis of marine climate. The low specific heat of a land body permits rapid heating by day in summer and rapid cooling by night in winter. Where land is the dominant control, the climate is marked by wide ranges of temperature, both diurnal and annual. The limited amount of water available for evaporation favors low relative humidity, except in winter, although the high temperatures of summer permit, in spite of the low relative humidity, as high a water-vapor content as the sources can supply. The course of relative humidity is, therefore, opposite to the course of absolute humidity.

The quick response of a land surface to solar heating involves less retardation of maximum temperatures than a marine climate. In the northern hemisphere, July is the warmest month, and because of rapid cooling in winter, January is the coldest month. April is warmer than October, except in high latitudes where the snow cover retards the advance of spring.

The higher temperatures of summer and the greater water-vapor content (absolute humidity) of the air during that season favor more convection than in winter and therefore greater precipitation, but the average percentage of cloudiness may vary little with the seasons since relative humidity tends to be

Fig. 13–8. Amiens in northern France is a good example of a littoral climate: small mean annual ranges (July with a mean temperature of 63° and January with 36°) and adequate precipitation evenly distributed throughout the year. Such a climate favors a diversified agriculture. (Courtesy Standard Oil Company of New Jersey.)

Fig. 13–9. Deserts often represent extreme continental conditions. Here in Andrews County, Texas, the climate is not normally desert, but a long series of dry years and strong, desiccating winds have transformed this area into a desert—at least temporarily. (Courtesy Standard Oil Company of New Jersey.)

Fig. 13–10. The northwest monsoon brings abundant rainfall to western Java (7 degrees south latitude), while the southeast monsoon of winter favors a short, less rainy season. This type of rainfall, together with the constantly high mean temperatures, is ideal for rice and bananas. (Courtesy Standard Oil Company of New Jersey.)

greater in winter. The late afternoon hours are usually cloudiest and rainiest because then the process of convection is at its maximum, being favored by a combination of great surface heating and increasing relative humidity.

Littoral Climates

A climate which is neither continental nor marine but has some of the characteristics of both is often referred to as a *littoral* (coast) *climate*. Such a climate is best represented on east coasts in middle latitudes where the winds may blow part of the time from the land and part of the time from the sea, or near west coasts in middle latitudes where the moderating effects of the sea are still felt, as in France, Belgium, the Netherlands, and northern Germany (Figure 13–8). Temperature ranges, both daily and annual, are less extreme than in continental climates, and cloudiness and precipitation attain less distinct seasonal maxima or minima than in either marine or continental climates.

Desert Climates

Desert[1] climate is sometimes referred to as an extreme continental type because of the similarity in the high ranges of temperature and in the season of maximum and minimum temperature, humidity, cloudiness, and rainfall (Figure 13–9). Deserts may exist, however, where there are prevailing onshore winds, as along the coast of Peru and the west coasts of north and south Africa. In each case there are cold ocean currents and cold upwelling waters, so that winds of sea-breeze character blow from the zones of cold water to the greatly heated land. Under such conditions rainfall is generally impossible.

Two physical characteristics are especially associated with deserts: (1) A low relative humidity that results in a small amount of clouds and precipitation. The absolute humidity may be high, but unless there is sufficient cooling of the air, no precipitation ensues. (2) A great diurnal range of temperature. The usually clear skies that result from descending air permit intense solar radiation during the day, resulting in high temperatures; it also leads to rapid terrestrial radiation at night, usually causing temperatures to drop markedly. The Sahara attains temperatures of well over 100°[2] (even up to 135°) by day, only to drop to temperatures that are not far from freezing at night. Hence the light clothing worn by day contrasts with the relatively heavy clothing of the night, except among peoples who try to keep out the intense heat of the day by wearing a thick layer of clothes.

[1] A botanist's definition of *desert*, used in a broad sense by many geographers, is *a place which has little or no vegetation*. In such a category, such places as ice fields (glaciers) and mountainous areas above the tree lines and alpine meadows may be classed as deserts. Desert climate as used here, however, excludes these; it refers only to the relatively large (sometimes small) low- and middle-latitude deserts recognized in common parlance.

[2] Temperatures quoted are always those taken under standard conditions using a thermometer shelter, never those taken in direct sunlight.

Although they are not classed as deserts, both the Mediterranean and certain semiarid regions take on these same characteristics during the dry season when their weather resembles that of the deserts.

Monsoon Climates

In winter, continents favor high pressure and outflowing winds; in summer, they favor low pressure and inflowing winds. Over large continents where the annual ranges of temperature are large, these conditions are most pronounced. Oceans favor the opposite conditions—high pressure and outflowing winds in summer and low pressure and inflowing winds in winter. As already stated, prevailing winds that blow in opposite directions in summer and in winter are called *monsoons*. Such winds are most common in southern and eastern Asia (Figure 13–10), east central Africa, and in northern Australia.

A monsoon climate is marked in summer by high humidity—both absolute and relative—great cloudiness, and intense rainfall. In winter, humidity is lower, there is little or no precipitation, and skies are generally clear. In limited areas, as in parts of the Philippines, the coast of Indochina, and the east coast of southern India, monsoons may be onshore in the cool season rather than in summer; thus, it might be more accurate to say that the heaviest rain comes at the season when the monsoon is onshore.

Types of Climate

The foregoing are not all the climates of the world. They represent only highly specialized types, and they serve only as nuclei or centers. Between them are broad transitional zones that may have some of the characteristics of two or more of the major types. More detailed consideration of climatic types will be given as a conclusion to this study of climate in Chapters 18 to 22.

14

World Distribution of Temperature

Many vessels while at sea make systematic weather observations, reporting them either by radio or after arrival in port. Since the variability of weather elements over the oceans is less pronounced than over the continents, the smaller number of observations over the water is probably just as trustworthy as the greater number over land. Thus world maps of temperature, pressure, winds, and cloudiness show isometric lines over the ocean as well as over the continents. The trend of those lines may be relied upon in most cases as representing generalized conditions.

In constructing world maps of temperature, all the observations are usually reduced to sea level, the approximate lapse rate of 3.3 degrees per 1,000 feet being commonly used. To show actual temperatures over regions of considerable relief would result in a complexity too confusing for a general survey. It should be borne in mind, however, that such reduction may involve considerable error, especially for high altitudes in winter when inversions of temperature are frequent, as in the southwestern United States.

Ocean Currents

Ocean currents are so largely responsible for the irregularity of the isotherms over the sea and along coasts that it is necessary to note certain facts regarding them: name, number, extent, temperature, direction and amount of movement, and the factors which cause them (Figure 14–1). There are a number of currents and their extensions, called *drifts*, which merit special consideration: Atlantic Equatorial, Gulf Stream, North Atlantic Drift, Canaries, Labrador, Benguela, Brazilian, Pacific Equatorial, Kuro Siwo or Japanese, North Pacific Drift (Alaska Current), California, Humboldt or Peruvian, Falkland, East Australia, Okhotsk, West Australia, Mozambique, and West Wind Drift. It will be observed that the major currents in the open oceans move clockwise in the northern hemisphere and counterclockwise in the southern. The Indian Ocean is complicated by the fact that the northern part has a clockwise movement (although much of it is in the southern hemisphere), while the southern part is true to

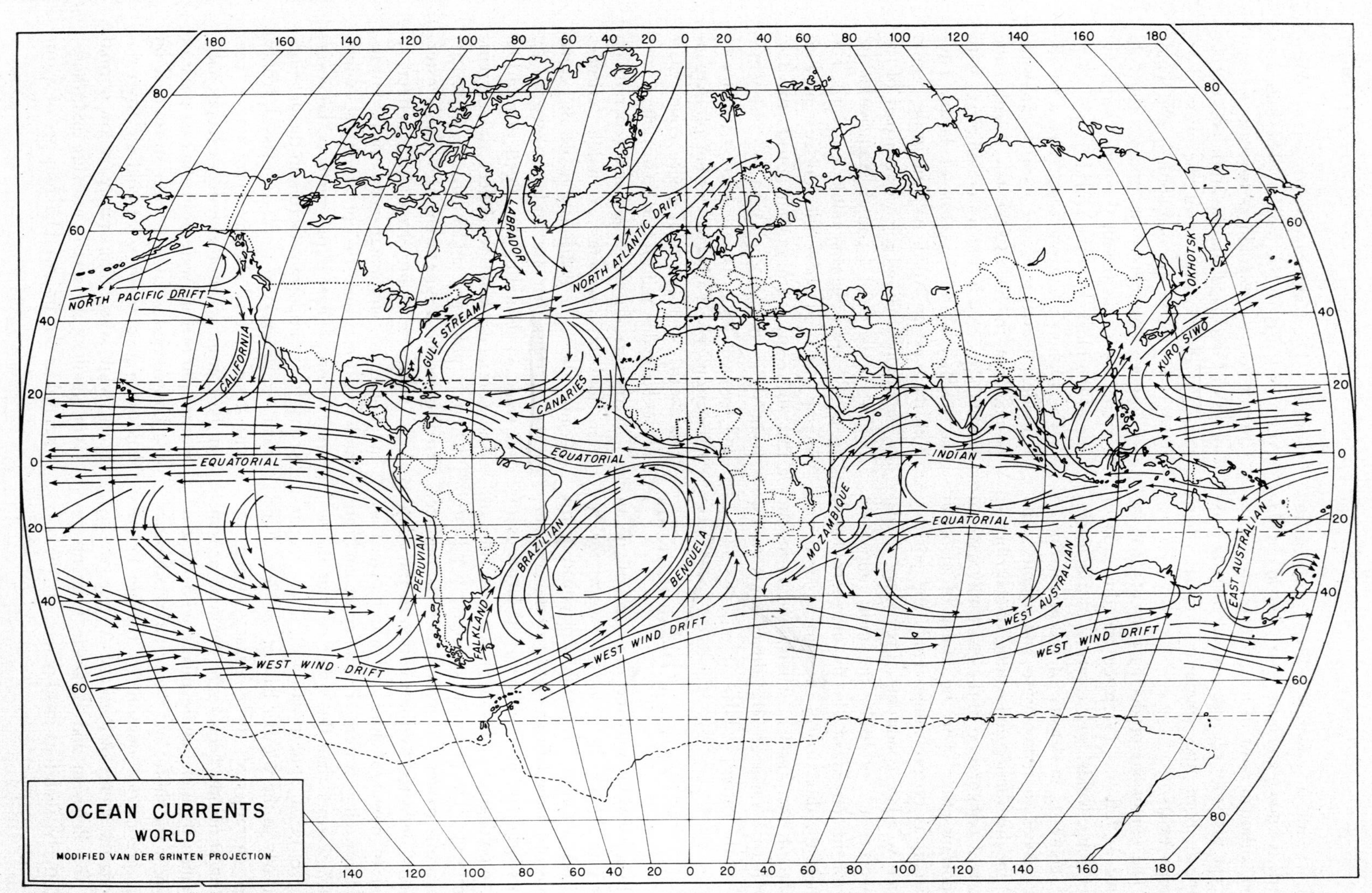

Fig. 14–1. *The average position and extent of the principal ocean currents. The length of the arrows gives some indication of the speed of the currents.*

form with a counterclockwise movement. Little is known about the north Indian circulation, although most of its currents are warm, except probably the ones paralleling the east African horn and southern Arabia where aridity exists.

The conditions producing currents may be listed as follows: (1) force and direction of prevailing winds, (2) rotation of the earth, (3) contrasts in density of the water due to differences in temperature and salinity, and (4) the distribution and volume of contributing rivers. Configuration of the coast may have an important effect upon the course of an ocean current, as do prevailing winds and possibly other weather conditions. Prevailing winds, however, are probably the most important factor in movement. In the northern hemisphere the direction of the currents is to the right of the surface winds, while in the southern hemisphere the direction is to the left of the surface winds.

Effects of Ocean Currents

In general, it may be said that all ocean currents flowing poleward are warm for the latitude while all currents flowing equatorward are cool for the latitude. Most of the cool currents originate as such between latitudes of 40 to 60 degrees and flow on the west sides of continents. The prevailing westerly winds are the prime movers. In most cases the surface waters, deflected by the rotation of the earth, tend to move offshore and are replaced by cold upwelling waters. The adjacent lands are much warmer than the water so that the sea breeze is of almost daily occurrence, particularly in the warmer half of the year. Fogs also characterize these coasts during the warmest months because the damp air at some distance from the shore mixes with the cold air near the shore, and the resulting advection fogs are carried onshore by the sea breeze. Along the shores of Peru and Angola, washed by the Peruvian and Benguela Currents respectively, such fogs yield practically the only moisture. The California, Canaries, and West Australia Currents are similar. The Labrador Current derives its coldness from the ice and snows of Greenland and the arctic and washes the east shore rather than the west. Fortunately for the people of Labrador, the winds are onshore only part of the time, yet this cold current assists in making Labrador's climate one of the bleakest in the world.

The Okhotsk Current is to Asia what the Labrador Current is to North America. Issuing from the Sea of Okhotsk, it bathes the Kurile Islands with cold water and mixes with the warmer Kuro Siwo east of Japan. Thus both the western North Atlantic in the region of Newfoundland and the western part of the North Pacific from Japan to Kamchatka are areas of converging and unlike currents. The mixing of the air above the cool current with that above the warm current lowers the dew point sufficiently in each case to cause dense and frequent fogs that are a great menace to navigation. The Falkland Current off Argentina is the southern hemisphere counterpart of the Labrador and Okhotsk Currents.

Warm currents are to be found generally on east sides of continents in middle latitudes because they are continuations of the equatorial currents that are pushed westward by the trade winds and by the effects of rotation. The continents tend to deflect the currents definitely poleward, and finally they are carried across the oceans by the prevailing westerlies. Warm currents have little effect in ameliorating continental conditions immediately to westward, except when the winds are onshore. These same currents, however, probably are important factors in the development and progress of warm-air masses, fronts, and storms which do have profound effects on the continents. For example, exceptionally warm Gulf Stream waters off the east coast of the United States favor greater storminess on land with great indrafts of cold air from the north, while water temperatures below normal favor smaller contrasts in pressure and, therefore, milder conditions. The influence of warm

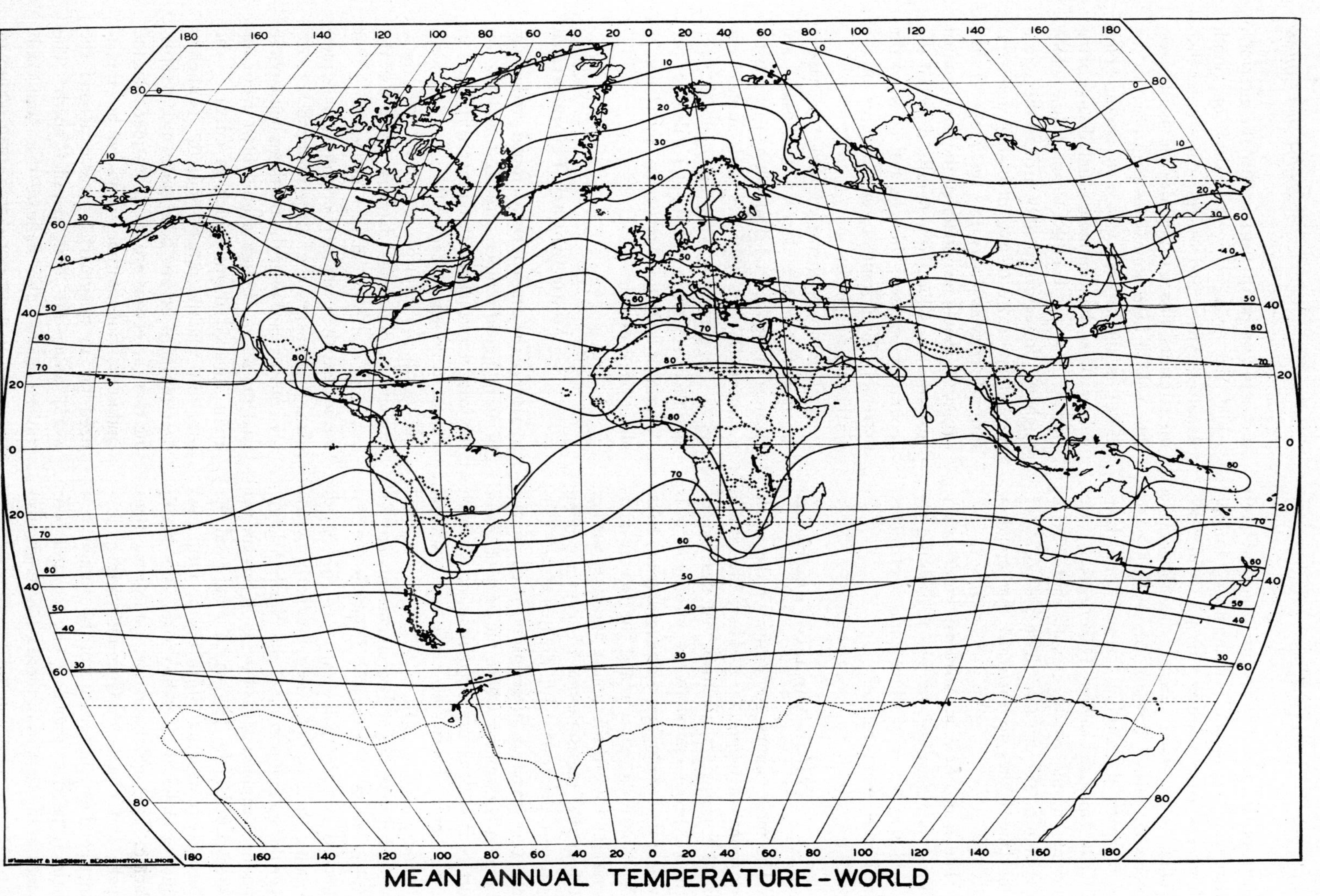

Fig. 14–2. Mean annual temperature at sea level.

currents is probably greater on the west margins of continents, as in northwest Europe, than on east margins because the prevailing west winds carry the moderating effects far inland. In the northern Pacific, the counterpart of the Gulf Stream is the Kuro Siwo or Japanese Current. The effects of warm currents in the southern hemisphere are of less significance because there are no important land bodies to eastward that could be affected.

The map of ocean currents, Figure 14–1, should be consulted freely in studying the world distribution of all the climatic elements, particularly temperature.

Mean Annual Temperature

The mean annual isotherms for the world are shown in Figure 14–2. There are certain broad features of this map that deserve attention. From our study of insolation it is evident that the highest temperatures are to be found near the equator and the lowest temperatures toward the poles; and it is further evident that the isotherms trend in a general east-west direction with the parallels. This east-west trend is more pronounced in the southern hemisphere where the area of water surface is so much greater than in the northern hemisphere. There are, however, several notable deflections, particularly in the northern hemisphere. Winds and ocean currents largely explain these deflections. There is a pronounced bending poleward over the North Atlantic because of more southerly winds over the eastern portion of the North Atlantic and the warming effect of the Gulf Stream and its eastward drift. The warm winds and the Kuro Siwo with its North Pacific Drift also cause a poleward bending over the North Pacific. The cold Labrador, California, Benguela, and Humboldt Currents cause a distinct bending equatorward of the isotherms in their respective regions. The equator-bent winds in the regions of these currents are also of some consequence in this regard. The west coasts of northern North America and northern Europe are warmer than the interiors, while the tropical west coasts of South America, Africa, and Australia are cooler than the interiors, disregarding altitude in all cases. The belt of high annual temperature (80° or more) in the tropics varies in width, being widest over the continents and disappearing entirely over the Pacific Ocean. A line which might be drawn completely around the world through the ridge or axis of highest temperature is referred to as the *mean annual heat equator*. It should be noted that the mean annual heat equator is located generally north of the geographic equator; this is a reflection of the extremely high summer temperatures of the northern hemisphere.

Temperature gradients, meaning the changes in temperature per degree of latitude, are greatest in high latitudes and along east coasts. Gradients in low latitudes and along west coasts are generally weak. In general, over land there is a change of about 1.5 degrees per 1 degree of latitude; over oceans, the change is about one-half that.

Mean January Temperature

Most of the characteristics found on the map of mean annual temperature are found also on the map of mean January temperature, only in a more pronounced degree (Figure 14–3). In the northern hemisphere, there are more isotherms shown for January than for the year and there are sharper deflections. This means that there are steeper isothermal gradients. At that time the isotherms bend noticeably equatorward over the northern hemisphere land masses as though a cold blast from the arctic invaded them, which in effect is the situation. On the other hand, the isotherms over northern oceans bend poleward, indicating that the oceans are much warmer than the land in the same latitudes. This can be emphasized by an imaginary traverse around the world along the 60th parallel north. The traverse would also indicate that land cools faster and to a greater degree than bodies of water.

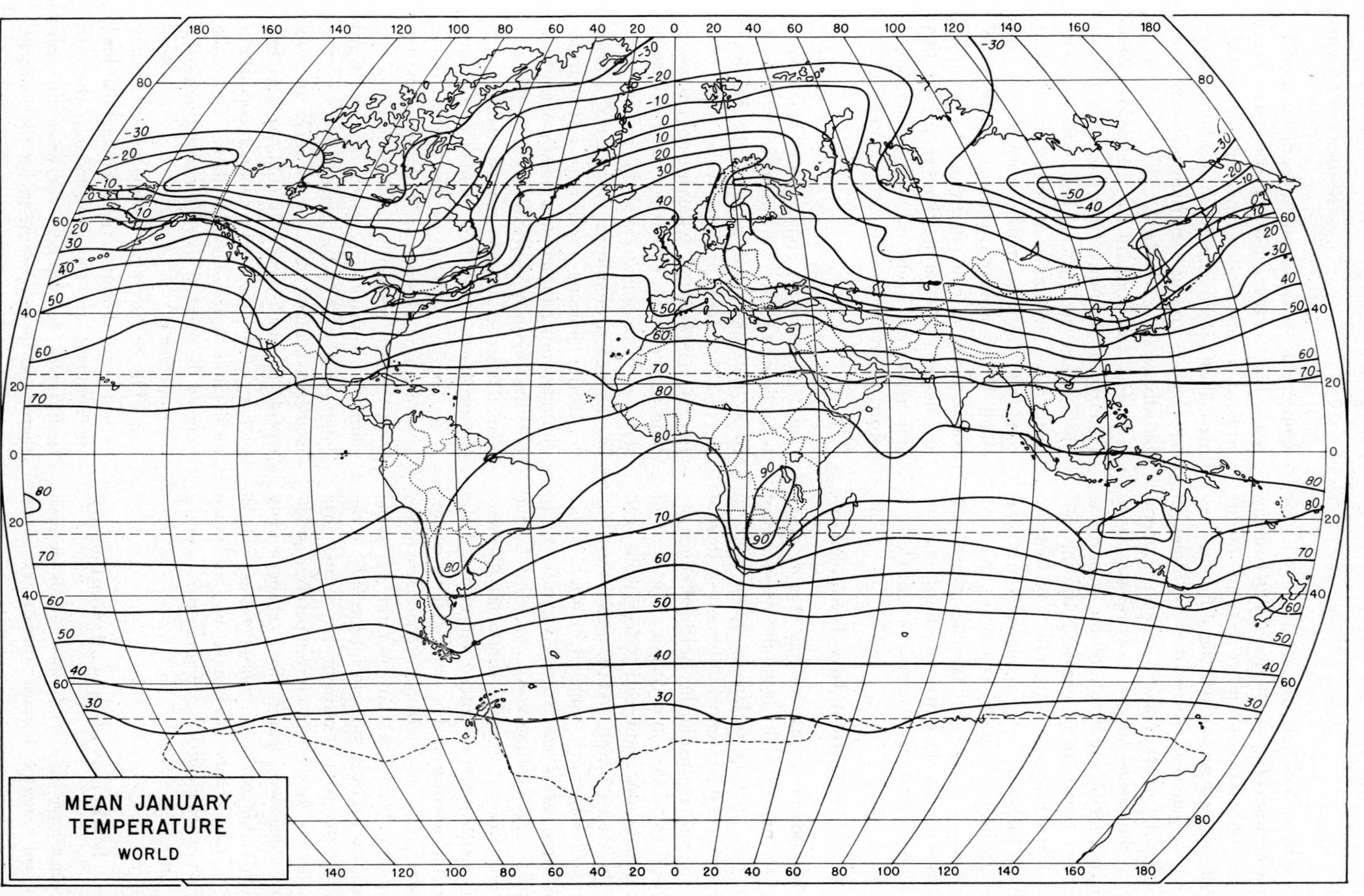

Fig. 14–3. Mean January temperature at sea level.

There are but slight changes during the year in the position of the isotherms in high latitudes of the southern hemisphere because of the small land masses. A striking difference between the January and annual maps is found in the location of the zone of highest temperatures, the *mean January heat equator* being almost everywhere south of the geographic equator. Small areas enclosed by the 90° isotherm are found in Australia and southern Africa. In North America and Asia, temperature gradients along east coasts are about 2.7 degrees per 1 degree of latitude; along west coasts they are less than one-third that, or about 0.8 degree per 1 degree of latitude.

Mean July Temperature

The isothermal map for July (Figure 14–4) departs more widely from the mean annual than does the map for January. In the southern hemisphere, the essential difference is that all the lines are pushed farther north, although the general trend does not change. The zone of greatest heating, however, is seen to be well north of the equator, and the area inclosed by the 80° isotherm is somewhat larger. There are large areas in North America, north Africa, and southwest Asia inclosed by the isotherm of 90°. On the continents the *mean July heat equator* is pushed far north of the geographic equator. The most striking change is seen in the northern hemisphere where the isotherms bend poleward over the lands and equatorward over the water, especially in the Pacific; that is, the continents are warmer than the oceans. In this season, temperature gradients along west coasts have changed only slightly, but over the continents, as well as along east coasts, they have become much weaker, being about 1.7 degrees per 1 degree of latitude.

Isotherms of Special Significance

There are five isotherms which deserve particular notice because they mark in an approximate way certain geographic conditions. The fact that the temperatures have been reduced to sea level means that the regions marked off on the maps by the isotherms have, in places, a greater extension poleward than actually exists on the land.

1. The 70° isotherm for the *coldest month* (January in the northern hemisphere, July in the southern hemisphere) marks the approximate poleward limits of unvarying tropical conditions. Annual temperature ranges are small. Where rainfall is sufficient, plants grow continuously throughout the year, for frosts are never experienced. The people of this zone gain their livelihood with a minimum of effort; clothing is of the lightest and simplest sort.

2. The 70° isotherm for the *year* is sometimes referred to as approximately the "poleward limit of palms"; it is more often referred to as the "poleward limit of the tropics." Beyond this zone palms may be found, but freezing temperatures, which are not unknown, limit their growth. Life is not quite so easy in the zone between this and the 70° isotherm for the coldest month; provision must be made for better housing conditions and warmer clothing during the coldest month when temperatures frequently drop to 50° or even lower. Plants, however, usually continue their growth throughout the year if the rainfall is adequate.

3. The 70° isotherm for the *warmest month* is sometimes considered the dividing line between warm temperate and cool temperate climates. It marks in general the poleward limits of corn. On the equator sides of this isotherm, summers are hot; winters are cool to cold but not excessively cold. At least five months of the year have summer conditions, mean temperatures of 60° or above. This is the zone of the greatest intellectual and industrial development.

4. The 50° isotherm for the *warmest month* is the poleward limit of trees. Beyond that line, the summers are too cool and too short to permit tree development. Between this isotherm and the 70° isotherm for the warmest

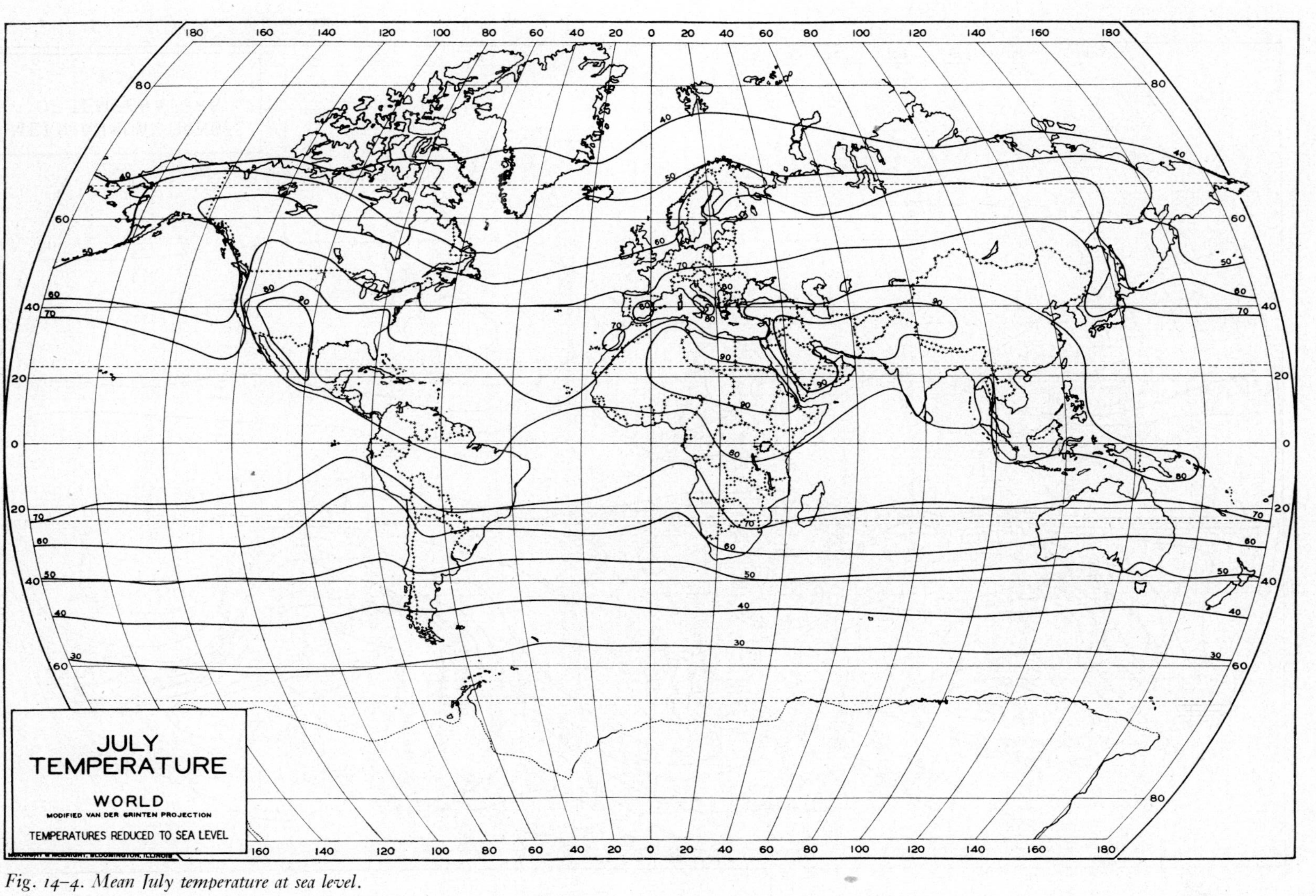

Fig. 14–4. Mean July temperature at sea level.

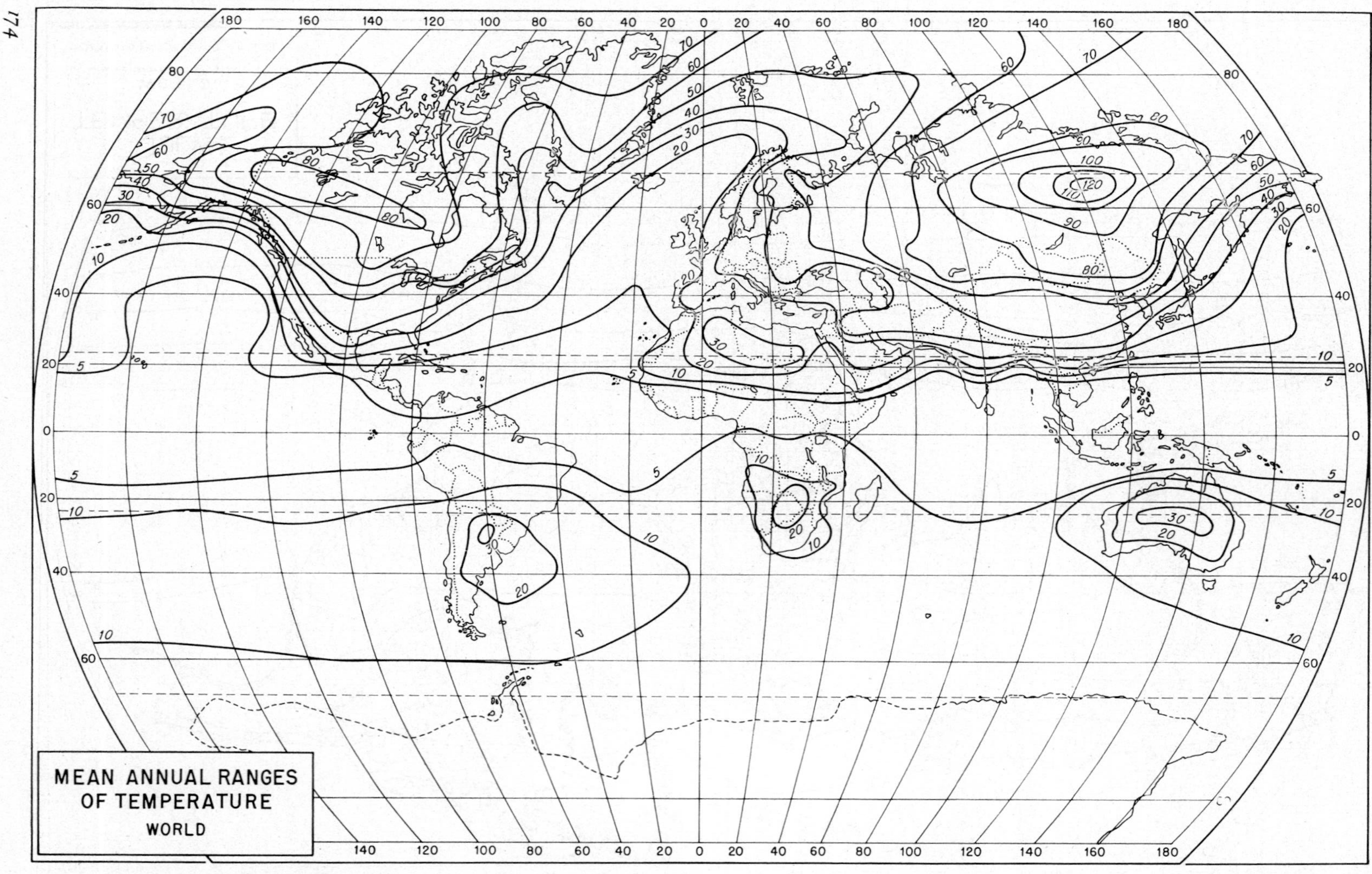

Fig. 14–5. Mean annual ranges of temperature.

month is the zone of cool temperate climates—short, cool summers and long, cold winters. Man must work hard during the long days of the short summer in order to provide food, warm clothing, and housing for the long, severe winter.

5. The 30° isotherm for the *warmest month* marks the limits of perpetual frost. On the poleward sides of this isotherm only ice and snow or barren rocks cover the landscape. Permanent human occupation poleward of this isotherm is impossible. Between this isotherm and the 50° isotherm for the warmest month is a treeless zone but one which supports during the short warm season abundant vegetation of quick-growing herbaceous plants. Hunting and fishing support the sparse nomadic population of this zone.

Mean Annual Ranges of Temperature

As would be expected from the large range in insolation, mean annual ranges of temperature are greater in high latitudes than in low and they are greater over land than over water, the amount of range depending upon the size of the land mass. Referring to the map, Figure 14–5, it is seen that Asia, the largest continent, has the greatest ranges. Africa, lying in low latitudes, shows relatively small ranges, but North America, because of its high latitudinal position, has ranges as high as 80 degrees. A remarkable feature shown by the map is the fact that, beyond the arctic circle in the North Atlantic, the mean annual range of temperature is under 20 degrees.

15

World Distribution of Pressure and Winds

Mean Annual Pressure

Study of the ideal planetary circulation (Figure 8–2) showed that certain definite pressure belts would be found if the earth were a homogeneous land surface. The existence of alternating land and water bodies, of ocean currents, of mountains, and of changing seasons causes certain modifications. The map of mean annual pressure (Figure 15–1) reveals the effect of these factors, except the seasonal changes. The pressure belts on this map are, however, in surprising accord with those of the ideal system. There is the equatorial belt of low pressure encircling the earth; the horse latitude belts of high pressure, a little less uniform, particularly in the northern hemisphere; and the subpolar belts of low pressure, interrupted by the continents in the northern hemisphere but clearly evident in the southern hemisphere. The absence of polar areas on this map projection prevents the showing of the polar high-pressure centers, although there is a suggestion of it across the top of the map. Striking features of this map are the regularity of the isobars and the distinctly lower pressures in the middle latitudes of the southern hemisphere. As on weather maps, the pressures on all of the world maps have been reduced to sea level.

January Pressure and Winds

The isobaric maps for January and July (Figures 15–2 and 15–3) depart more widely from the ideal distribution of pressure than does the map for the year. The axis of the equatorial low-pressure belt lies south of the equator in January. In the northern hemisphere, the horse latitude high-pressure belt is pushed far north over the continents. In Asia the high-pressure center takes a compromise position between the normal, which is near the 30th parallel, and the area of most intense cold near the arctic circle. The subpolar belts of low pressure are intense over the oceans but lacking altogether over the lands. In January in the southern hemisphere, the horse latitude belt assumes its normal position over the oceans but is weak over the lands, the great heating

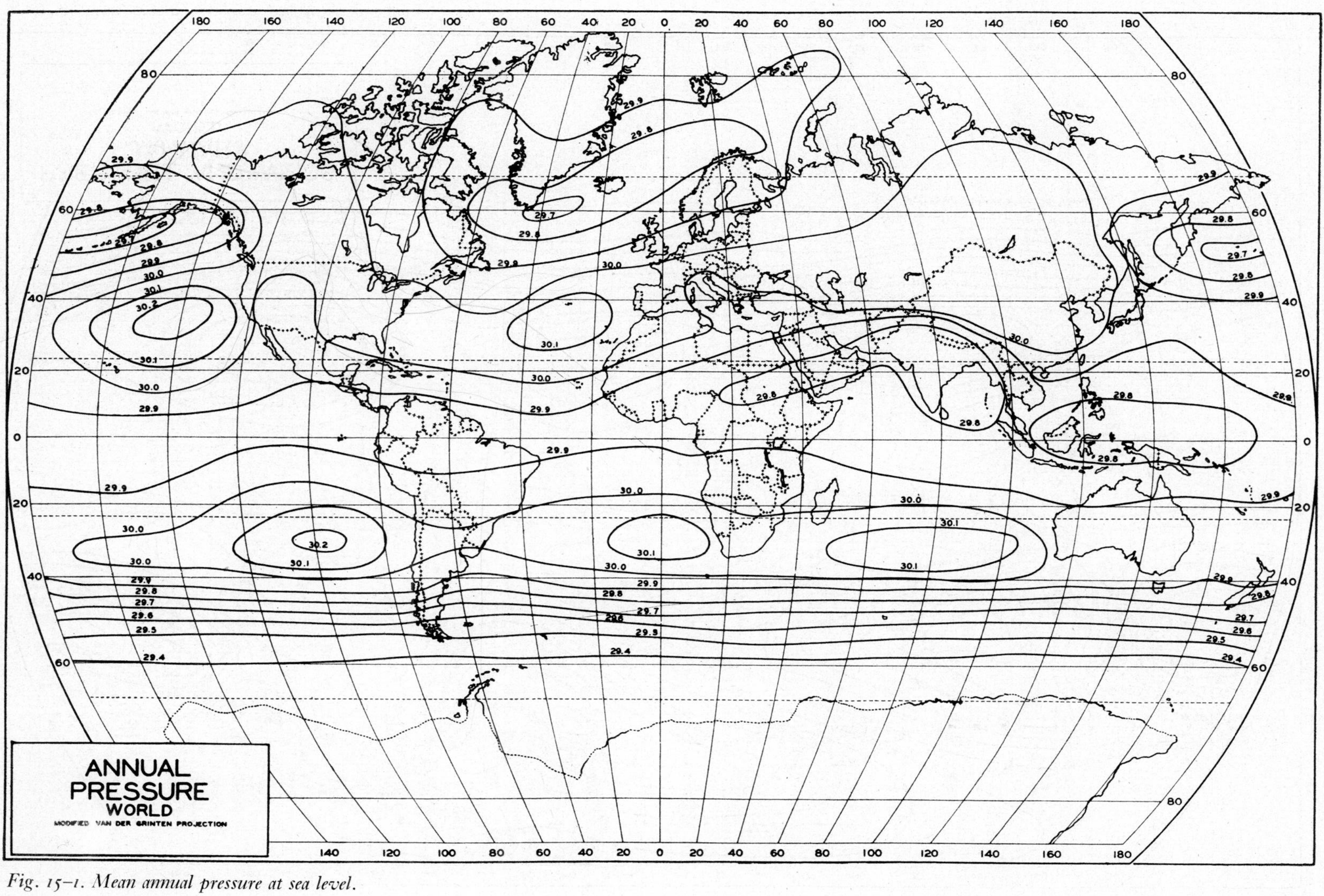

Fig. 15-1. Mean annual pressure at sea level.

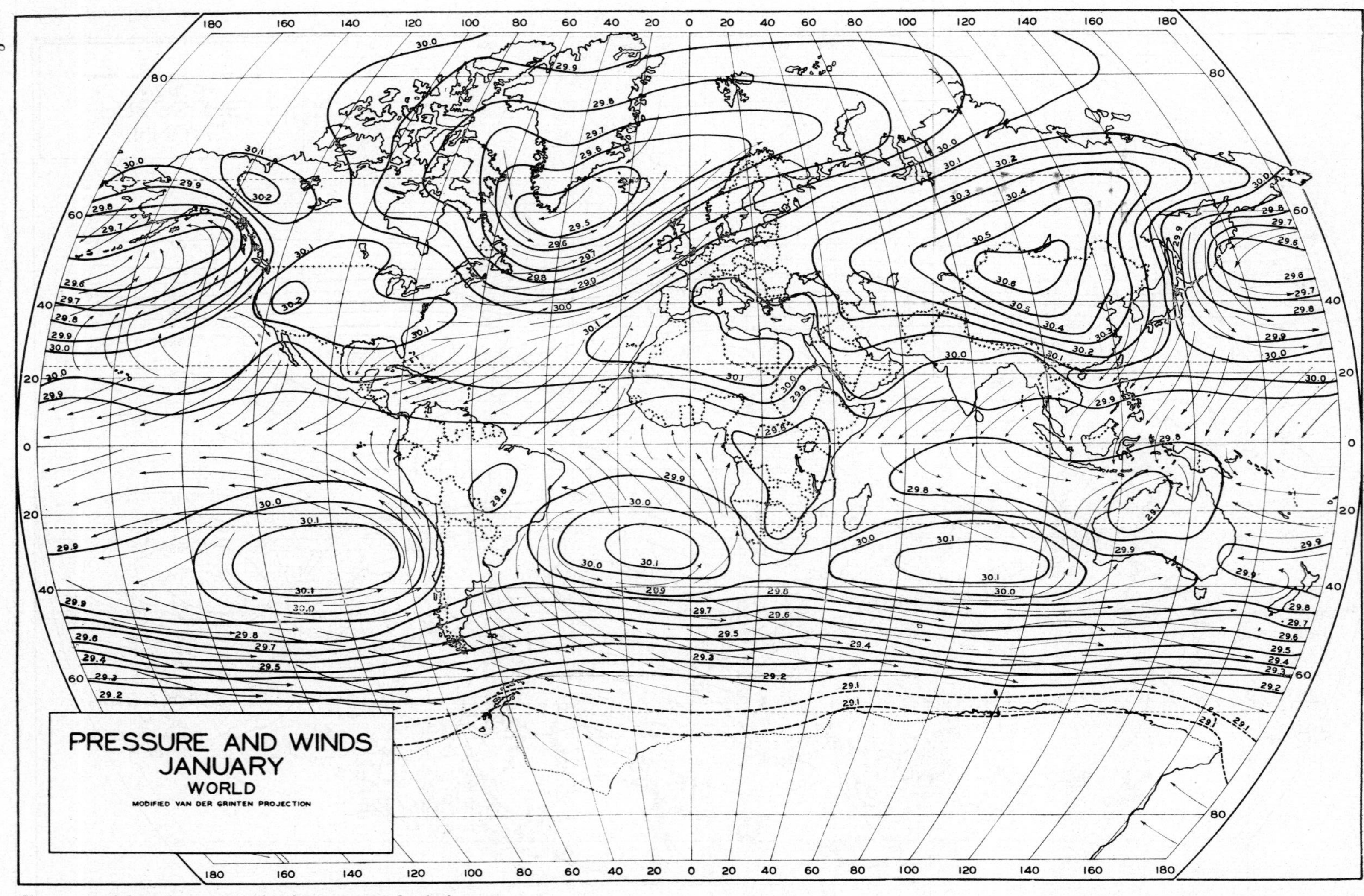

Fig. 15–2. Mean January sea-level pressure and winds.

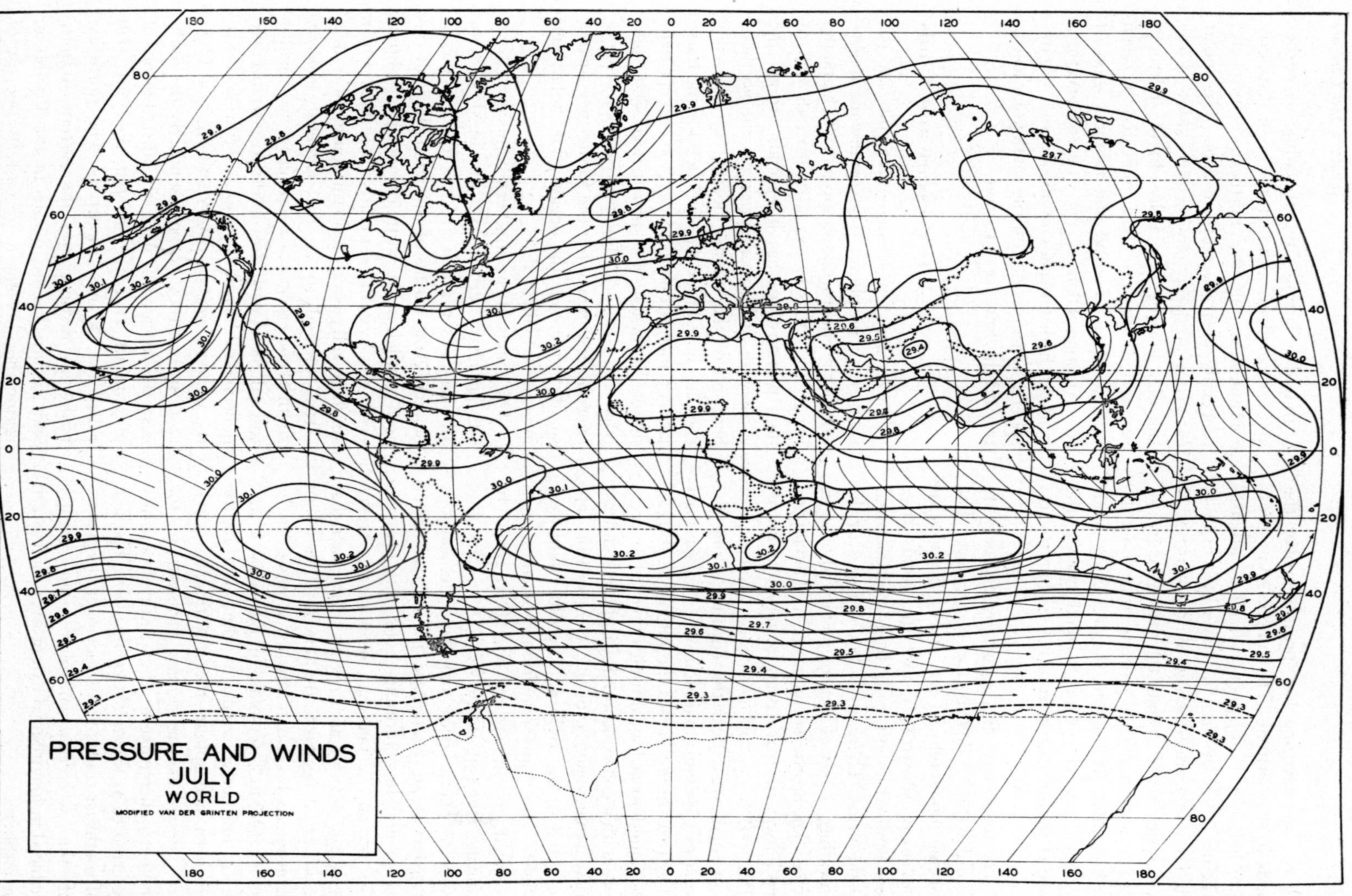

Fig. 15-3. Mean July sea-level pressure and winds.

by the sun favoring low pressures over the lands in that latitude. The subpolar low-pressure belt is very distinct as well as intense. The few observations which have been made in the far southern oceans and on the margins of Antarctica seem to indicate that pressure increases with increasing latitude from about the 65th parallel. At any rate, the prevailing winds are southeasterly, which would not be the case if the pressures were not higher toward the south pole.

The winds over the oceans are distinctly in accord with the distribution of pressure. The varying topography of the lands influences air movement so strongly that the winds have been omitted intentionally from the land areas of the maps. There are a few features of the prevailing winds of January which should be noted. The northeast trades are well developed and blow with marked uniformity in point of direction, force, and constancy, except over two small areas near the equator: along the coast of Colombia and along the Guinea coast of Africa. In both cases the intense heating of the humid air at low altitudes favors low pressure with a consequent indraft of the southeast trades that, upon crossing the equator, are deflected to the right and become southwest winds. This same condition, but more pronounced, is shown on the July map. In January, high pressure over Asia and low pressure over Africa and Australia cause a transfer of air from Asia to the two latter continents. In conformity with this pressure distribution, we find that the northeast trades cross the equator and, owing to the fact that winds are deflected to the left in the southern hemisphere, become northwest winds over the Indian Ocean just south of the equator and more northerly winds to the east. In the northern hemisphere these northerly winds are known as the *winter monsoon;* in the southern hemisphere they are frequently referred to as the *northwest monsoon*. The belt of southeast trades is distinct over the waters of the western hemisphere but greatly reduced in longitudinal extent over the eastern hemisphere.

In middle latitudes the prevailing westerlies are very evident, although they are distinctly southwesterly in the northern hemisphere, especially near western coasts. The polar easterlies are seen on the pole sides of the subpolar low-pressure belts. In the northern hemisphere they are best indicated between Greenland and Iceland and in the region of the Aleutian Islands. The few observations made in high latitudes of the southern hemisphere indicate winds off the south polar icecap toward the low-pressure belt.

July Pressure and Winds

There are few changes for July in the southern hemisphere. The northward migration of the equatorial low-pressure belt and the lower sun intensify the horse latitude belt of high pressure so that it extends around the earth with little interruption. In the northern hemisphere the intense heating of the lands completely nullifies the tendency for the horse latitude high pressure to develop, except over the oceans where it is pronounced. The subpolar and equatorial low-pressure belts merge over the continents, but the former persist to a small degree over the oceans. The polar high pressure is indicated.

The winds of the southern hemisphere have all the characteristics of the ideal system, although observations are too few to indicate definitely the nature of the winds beyond the 70th parallel. In the northern hemisphere, however, the system is greatly modified. Over southern and eastern Asia the southeast trades, often referred to as the *southeast monsoon* south of the equator, sweep across the equator toward the Asiatic low pressure and become deflected to the right to cause southwest winds, the *summer monsoon*, over southern Asia. Over eastern Asia they become southerly or southeasterly in accord with the distribution of pressure in that region, but they are still called the summer monsoon. The northeast trades are distinct over the eastern Atlantic and eastern Pacific, while in the Caribbean region they become more easterly because of

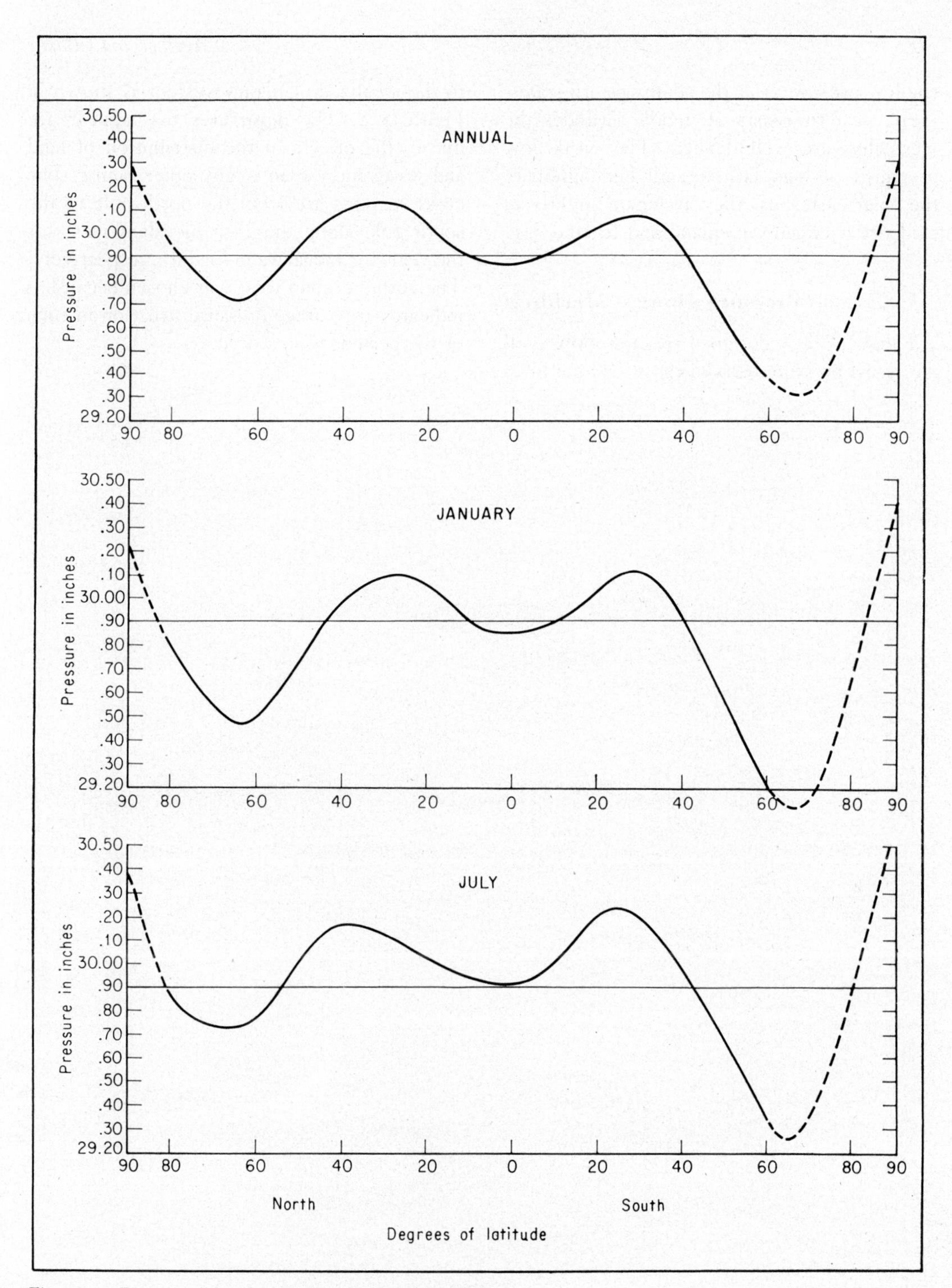

Fig. 15–4. Pressure along the 20th meridian, west longitude.

the low pressure over the southwestern states. Near western coasts of middle latitudes the westerlies are well defined. The weak low pressures of high latitudes all but annihilate the polar easterlies; they are again slightly in evidence between Greenland and Iceland.

Pressure along a Meridian

Figure 15–4 is designed to show how well the actual pressure belts of the world conform to the idealized planetary system as shown in Figure 8–2. The departures that appear are due to the effects of the distribution of land and water more than to any other factor. The cross sections are from the north pole to the south pole along the 20th meridian west for the year, for January and for July, respectively. The 20th meridian west was chosen because it indicates an average pressure situation in each of the periods represented.

16

World Distribution of Precipitation

Mean Annual Precipitation

The forms and causes of precipitation have already been mentioned. The student should review those facts. The world distribution of mean annual precipitation as shown by the map (Figure 16–1) bears a very close relationship to the distribution of pressure and winds. The equatorial low-pressure belt, with abundant moisture in its inflowing winds, has heavy rainfall, the greatest occurring near the coasts, or where mountains favor ascent of the winds, as along the east slopes of the Andes and Central America. A relief map will help greatly in the study of the effects of elevation upon precipitation.

The trades are normally drying winds because they arise from slowly settling dry air and because they blow from cooler to warmer latitudes. Consequently, the lands over which they prevail are dry or desert, unless the orographic influence is strong or they have passed over hundreds of miles of warm water before reaching land. The Sahara, Kalahari, Arabian, and Australian deserts are largely in trade-wind latitudes. Much of the time high pressures also prevail. Relief and convection aid in making windward coasts moist. Cold ocean currents, with a tendency to offshore movement because of the prevailing winds and deflective effect of the earth's rotation, favor the upwelling of cold water along Lower California, Peru, and northern Chile, Senegal, Southwest Africa, and Angola, thereby accentuating normal aridity of those regions. In the monsoon regions of trade-wind latitudes, relief causes a marked increase in rainfall, as in India and Ethiopia. Even in high Yemen the monsoons bring some summer rain to that thirsty land.

As would be expected, the horse latitude belts are regions of low rainfall, but because of their seasonal shifting, they are not distinct on the rainfall map. The northern Sahara, Arabian, Thar, Turkestan, Takla Makan, and Gobi are all deserts which lie within the horse latitude belt at some time of the year. In Australia, South America, and northern

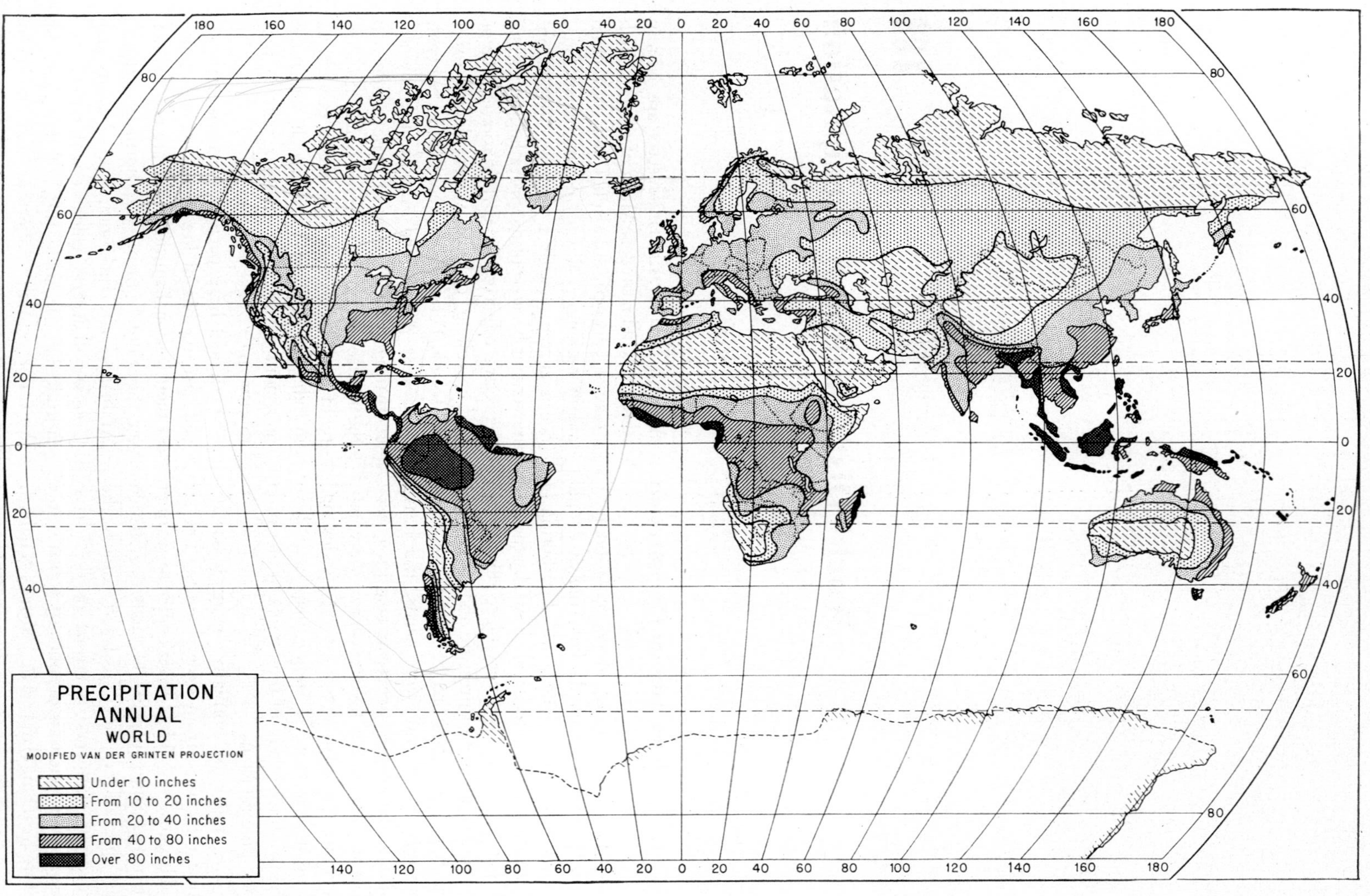

Fig. 16–1. Mean annual precipitation.

Mexico arid strips are found in this high-pressure belt. Continental interiors, if shut off from sea influences by mountains or distance, must necessarily be arid or semiarid.

West coasts in the belt of prevailing westerlies are invariably rainy, the cyclonic control is always strong, and altitude frequently has a marked influence. East coasts in those latitudes have less precipitation than west coasts, and the amount decreases with progress inland. The low absolute humidities of high latitudes mean a decided falling off in precipitation even though both the altitude and cyclonic controls may be marked.

It will be seen that substantial parts of the earth have less than 20 inches of precipitation annually; considerable areas have less than 10 inches. Relatively small areas, on the other hand, have over 80 inches yearly, which for most regions is considered an excessive amount.

Precipitation, November–April

Fully as important as the total yearly precipitation for any locality is the seasonal distribution. Indeed, agriculture may thrive in a region having a yearly precipitation of only 18 inches, as in parts of the Prairie Provinces of Canada, because most of it comes in the form of rain during the growing season. Eighteen inches of rainfall evenly distributed throughout the year would, in most instances, be insufficient for agriculture without irrigation.

The world map of precipitation for the winter half year of the northern hemisphere (Figure 16–2) is in many respects similar to the mean annual map, and like the latter, it conforms to the January wind belts. The equatorial low pressure, lying in general south of the equator, is a region of heavy rainfall. The northeast trades are everywhere dry except in a few places, as in Central America, the West Indies, and the Philippines, where orographic control is marked. This is the rainy half year for the east margins of the continents in middle latitudes of the southern hemisphere, while for west margins it is the dry, less rainy, season—note southern Australia, Cape of Good Hope, and central Chile.

In the northern hemisphere, west coasts are definitely rainy, while the interiors of the continents are definitely dry. Cyclonic storms, however, yield moderately heavy precipitation in the eastern United States. The winter monsoons of southern and eastern Asia afford little rain or snow to those regions, except where the winds are onshore. Almost everywhere north of the arctic circle, Norway excepted, the total precipitation for the winter half year is under 5 inches.

Precipitation, May–October

The map of precipitation for the other half year, May–October (Figure 16–3), shows conditions in the northern and southern hemispheres to be reversed. The equatorial belt of rains is north of the equator. Most of Australia and southern Africa are dry, and the middle portions of South America are definitely drier. The monsoon regions of Asia receive abundant rainfall in this season, as do the Caribbean countries. Large sections of Siberia, most of Europe—excluding the Mediterranean lands—and more than one-half of Canada and the United States receive more than 10 inches. Although cyclonic storms are weak during the warm half year, the higher temperatures and more abundant water vapor favor convectional showers that yield much of the precipitation of northern middle latitudes.

Variability of Precipitation

While mean annual precipitation maps constitute useful generalizations by showing average distributions, the precipitation variability map (Figure 16–4) takes into consideration departures from such means. It can be seen from the map that the humid areas of the world have a more reliable precipitation from

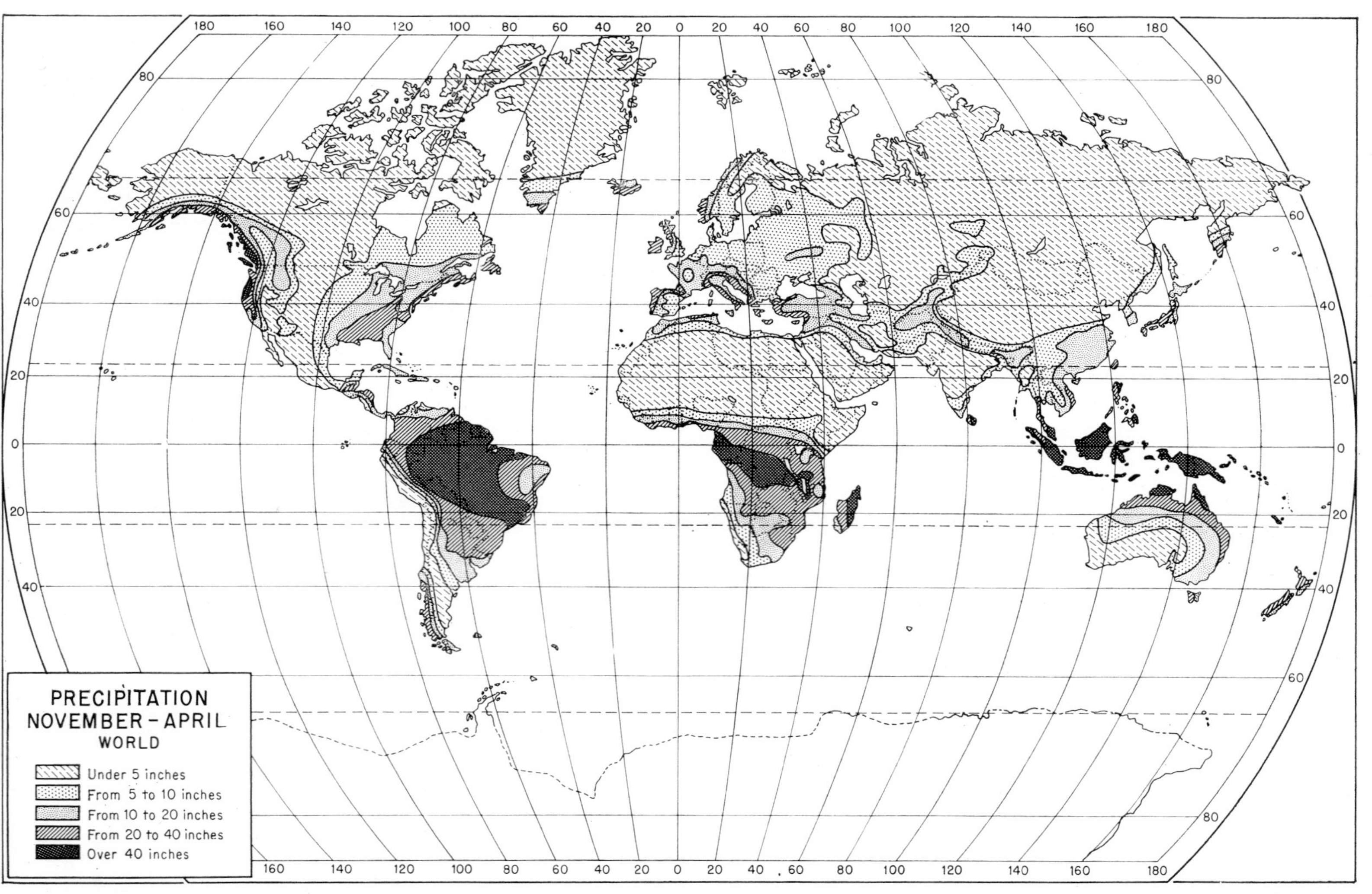

Fig. 16–2. Mean precipitation, November–April.

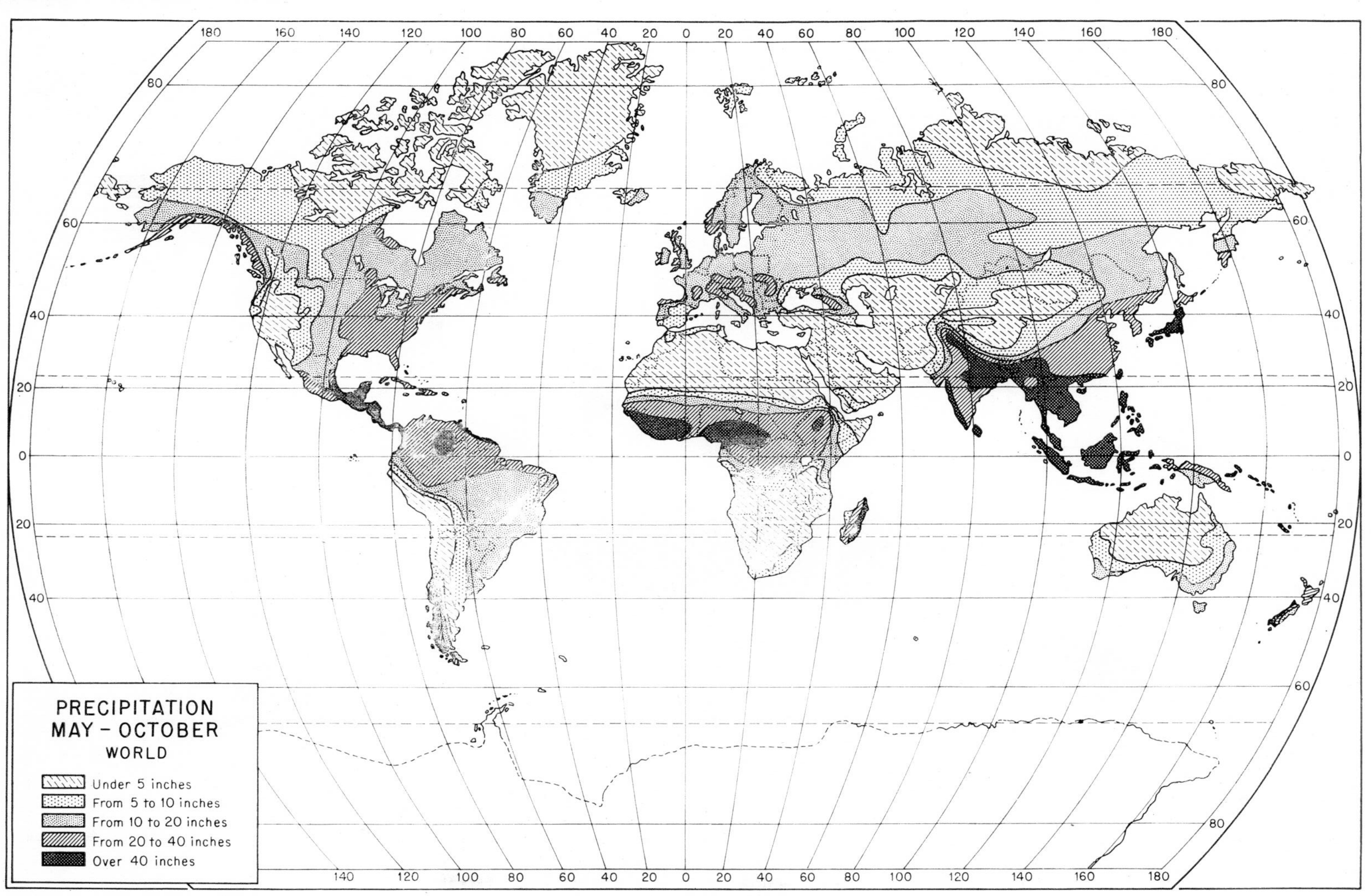

Fig. 16–3. Mean precipitation, May–October.

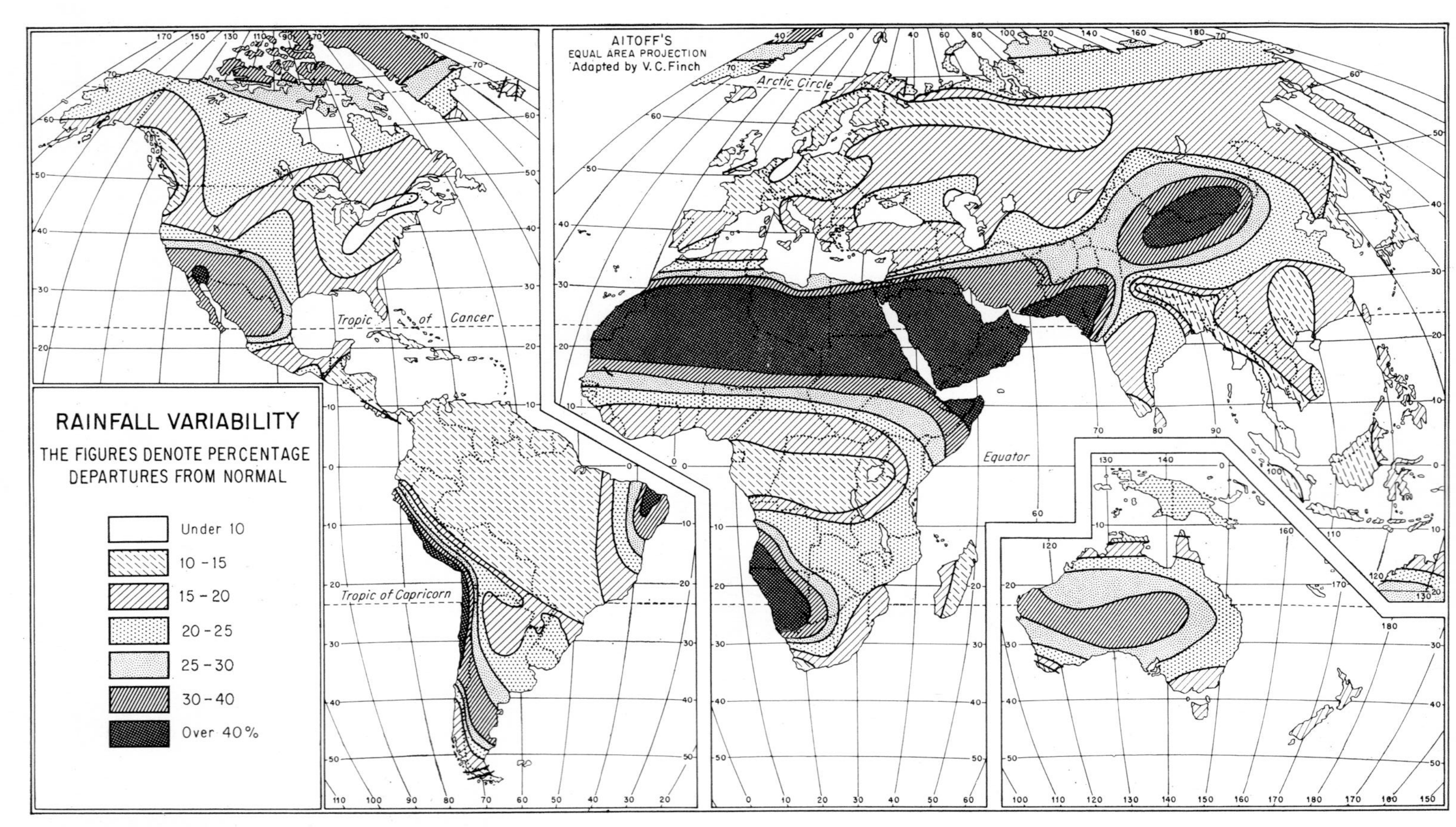

Fig. 16–4. Variability of precipitation. (After Biel.)

year to year; that is, they have less variability than do the dry areas. Most of the dry lands have annual variabilities between 30 and 40 per cent; deserts usually vary more than 40 per cent. There are many cases that illustrate these variations. The dust-bowl areas of the United States are well-known examples. The fluctuations from year to year are sufficient to place those areas sometimes in subhumid, sometimes in semiarid, arid, or even humid conditions. Frequently these sections of the United States have cycles of relatively rainy years followed by years of drought or relative drought. Some pronounced variations in precipitation in connection with steppe and desert climates are considered in Chapter 19.

Humid areas are also subject to some annual variations. Even Cherrapunji, India, one of the rainiest places on earth, has notable annual differences. The total annual precipitation in humid areas, however, is usually sufficient, despite fluctuations, to assure the usual agricultural pursuits. A small yearly variation in dry lands, on the other hand, may be the margin between crop failure and success. Concerning the variability of precipitation, this general rule may be stated: the less the precipitation, the greater the variability.

17

Classifying and Charting Climates

Classifying Climates

Because climate is made up of so many elements, some of which are highly variable from day to day and from season to season, it is impossible to determine upon any system of classification of climates that will be both simple and accurate. Consider the matter of temperature alone. If we should use the mean annual temperature as a basis of defining climatic regions, we should have no adequate idea of the range of temperature nor of the extremes and variability of temperature—all more significant than the single item of mean annual temperature. It is conceivable that a moderately simple and accurate map of temperature regions of the world could be made, based upon extremes of temperature as well as upon averages and ranges. It is conceivable, too, that an equally simple and accurate regional map could be made embodying the important phases of precipitation, sunshine and cloudiness, wind direction and velocity, relative humidity, or of any other phase of climate.

To obtain a complete classification which would embrace all the elements mentioned, it would be necessary to superimpose each of the maps upon a new map, or to trace on another map the lines demarcating the regions of temperature, precipitation, winds, and so on. Some idea of the complexity of the composite map may be gained by selecting four small paper squares of equal dimension, drawing at random not more than two straight lines on each square, and then transferring the lines of three of the squares to the fourth square. The number of geometric figures or regions on the fourth square might be as great as 256 and probably would not be less than 25. Similarly, the number of climatic regions would be large, though not proportionally so, since the regions for different elements have some relation to one another. Such a classification of climates could be simplified by eliminating many of the smaller areas. There would still remain, however, too many regions for a small map; to simplify the map further would partially destroy its accuracy. Thus great difficulty is always encountered in attempting to make a classification of climates which is both simple and accurate.

There are numerous accepted classifications of climates both in this country and abroad. Each classification has qualitative or quantitative criteria, or both, for the delimitation of its climates. Some classifications are complex; others are less detailed. Whatever the basis of division for the various climates recognized, and whether by name, symbol, or letter, the goal is essentially the same, namely: (1) to simplify the systematic classification of a tremendous congeries of minute areas throughout the world, each of which varies somewhat from all others in certain details; (2) to coordinate those areas which have similar climatic characteristics (although not necessarily the same) and to set limits upon the characteristics which are considered close enough to qualify those areas for a particular classification; (3) to give by analytical description the chief characteristics of unity within each climatic region thereby recognized and to compare and contrast climatic regions to some degree dissimilar; and (4) to show, by a map, the distribution of the climatic regions of the world, or portions thereof.

Because the criteria of delimitations are different in the various classifications, some of them accomplish one thing better than others do. The Köppen-Geiger classification is the one which seems to be most universally accepted. While in its original form it was quite detailed and intricate and seemed rather complicated for the beginning student of climate, the new edition has been modified and considerably simplified. It is this recently modified map which appears as Plate II, following page 216; and the discussions of climates which follow are based upon the Köppen system.

In consideration of the needs of those who may not have time to cover the fairly complete descriptions of climatic types found in Chapters 18 to 22, or who may want to use a classification less detailed than the Köppen system, a brief description of the classification of climates developed by the senior author appears in Appendix C. In addition to a general survey of the bases used for delineating the climatic regions, there is a summary of the chief characteristics of each of the 16 types shown on his map, Plate III, following page 216.

The Köppen System of Climates

The Köppen system of classification is largely quantitative. It is based principally upon the amount of precipitation, on both an annual and a distributional basis, and upon temperature, both monthly and annual. These elements, considered separately and together, constitute the framework for the divisions of climate. The criteria concerning precipitation and temperature that are used in establishing the climates were based originally upon natural vegetation; certain types of natural vegetation seemed to be limited by certain considerations of temperature and precipitation. Very definite numerical values have been established for these two elements, and the climatic regions based upon those values are used whether or not they conform to the original plant cover.

There are many objections to the Köppen system, just as there are objections to any other system of climatic classification. One criticism is that the various numerical criteria are arbitrary and that there is apparently no justification for the selection of some of them. For example, a large part of Illinois, New Jersey, Florida, and east Texas fall within the same climatic region because the temperature of the warmest month is above 71.6° and the temperature of the coldest month lies between 64.4 and 26.6°. The two latter temperatures seem to have little meaning when applied to a region where types of agriculture are so diverse as in these states. It would seem more logical to use a coldest month low limit of 40° instead of 26.6° because the lowest average January temperature in the so-called "cotton belt" is approximately 40°; this would exclude from the region the southern portions of

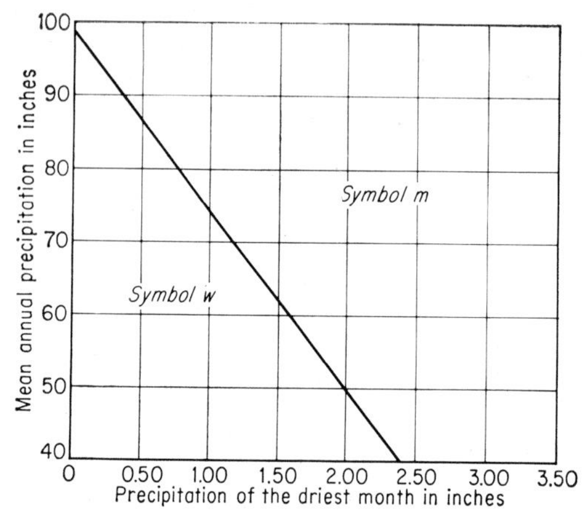

Fig. 17–1. A graph to determine whether a climate shall have the designation of m or w. If the point falls to the right of the diagonal line, the climate is classed as m; if to the left, as w. This graph is based upon the formula: if the driest month has more than 4 per cent of the difference between the total annual precipitation of the station and 98.5 inches (250 millimeters), the climate is m; if under that amount, then it is w.

Michigan, Wisconsin, Minnesota, and South Dakota, as well as most of Ohio, Indiana, and Illinois, and might include them with the region immediately to the north; this certainly would not seem objectionable.

By using the same January temperature limits of 26.6 to 64.4° and by applying a precipitation efficiency relationship based upon mean annual temperature and total annual precipitation, the Köppen system places parts of Idaho and Washington in the same region with parts of southern California, which is objectionable from the standpoint of natural vegetation as well as of agriculture. Some criticize Köppen for basing his types upon vegetation or climatic data alone rather than also relating them to location with respect to physical causes—wind and pressure belts, frontal systems or ocean currents.

Obviously, all such objections have merit, and obviously, too, no system of classification can be at once simple, complete, genetic, related to natural vegetation and human activities, and satisfy all the other requirements of the people who use it. In support of the Köppen system is the argument that anyone can apply the formula by the use of readily available climatic data. To change the formula in order to make more satisfactory boundaries would necessarily increase, in many instances, the number of regions, which would defeat the present advantage of simplicity. Some writers have modified the Köppen classification in order to take care of some of their objections and to adapt it to their needs, but that practice makes the use of the Köppen-Geiger wall map ineffective. Accordingly, the authors of this book, much as they would like to make changes, are following the Köppen system without modification. Except for the deletion of certain regions or areas too small to show on a small-scale map and the correction of a few boundaries where the lines do not conform to the climatic data, Plate II, following page 216 is the same as the latest Köppen-Geiger wall map.[1]

The Köppen Formula

There are five primary or major categories of climate that are given capital-letter designations of *A*, *B*, *C*, *D*, and *E*. Climates in the *B* category are termed "dry," since the potential evaporation during an average year exceeds the precipitation; the other four categories are considered moist, since the precipitation during an average year exceeds the evaporation. The temperature characteristics of each of the primary categories are as follows:

A: tropical; mean temperature of the coldest month at least 64° (actually 64.4°, which is the equivalent of 18°C). The Köppen system uses the centigrade scale; this book, however, will use the Fahrenheit equivalent to the nearest whole degree.

B: no temperature limitation.

[1] Köppen-Geiger, "Klima der Erde," Justus Perthes, Darmstadt, Germany, 1954. American distributor: A. J. Nystrom & Co., Chicago.

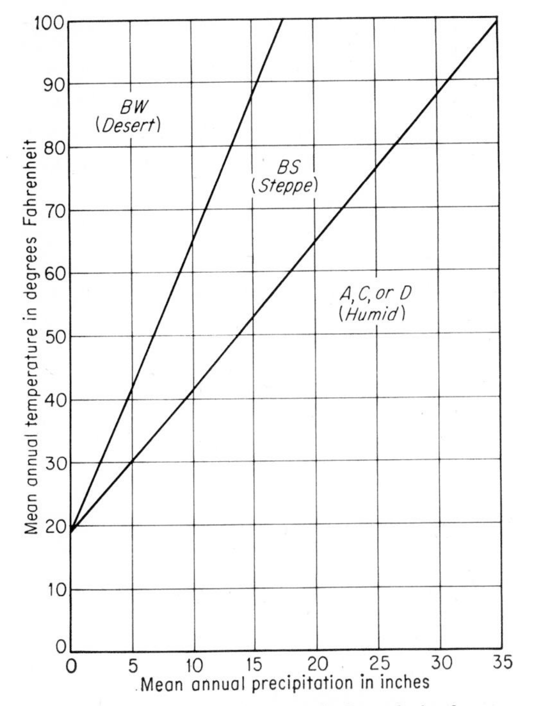

Fig. 17–2. A graph to indicate the boundaries between humid and dry climates when the annual precipitation is considered as evenly distributed. Notice that the mean annual precipitation that determines the boundary between humid and steppe climate is exactly twice that used to determine the boundary between steppe and desert climate. This relationship applies also to Figures 17–3 and 17–4.

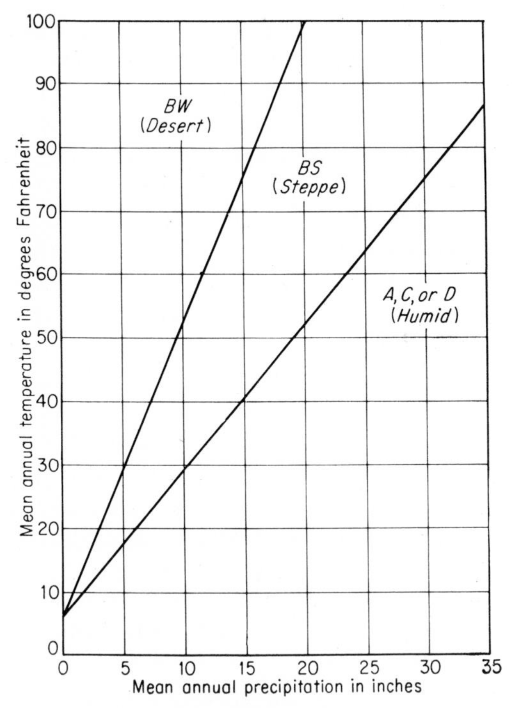

Fig. 17–3. A graph to indicate the boundaries between humid and dry climates when there is a summer concentration of precipitation.

C: mesothermal (middle or intermediate heat); one or more months with a mean temperature below 64° but no month with a mean temperature below 27° (−3°C); at least one month above 50° (10°C).

D: microthermal (small or little heat); the same as *C* except that the coldest month has a mean temperature below 27°.

E: arctic; mean temperature of the warmest month below 50°.

Each of the above major or primary categories, with the exception of *E*, is subdivided into two or more secondary categories by the addition of a second letter, capital letters with the *B* climates and small letters with the *A*, *C*, and *D* climates. Each of these secondary letters deals with some aspect of moisture; some letters differ slightly depending upon the major category with which they are used.

As used with the *A* climates:

f: moist (from the German word *feucht*, meaning "moist"); driest month has a mean precipitation of at least 2.4 inches (60 millimeters).

m: monsoon; has excessively heavy precipitation in some months which compensates for one or more months of less than 2.4 inches. The graph, Figure 17–1, indicates just how much annual and monthly precipitation is required.

w: winter dry period; driest month has less than 2.4 inches mean precipitation; the total yearly amount is less than that required for *m*.

As used with the *B* climates:

S: steppe; not only the amount of precipitation but its seasonal distribution determines the category as shown by the graphs in Figures 17–2 to 17–4.[1] In using these graphs it should be noted that precipitation is said to be concentrated in winter when more than 70 per cent of the total is received in the cooler six months (October –March); it is said to be concentrated in summer if more than 70 per cent is received in the warmer six months (April –September); if neither season receives more than 70 per cent of the total, the precipitation is considered to be evenly distributed. For the southern hemisphere, these seasons are reversed.

W: desert (from the German word *Wüste*, meaning "desert"); amount and distribution of precipitation as shown by the graphs (Figures 17–2 to 17–4).

As used with the *C* and *D* climates:

f: moist; precipitation well distributed throughout the year and not limited as in *w* and *s* below.

w: winter dry period; mean precipitation of driest month of the winter half year less than one-tenth that of the wettest month of the summer half year.

s: summer dry period; mean precipitation for the driest month of the summer half year less than 1.6 inches (40 millimeters) and less than one-third that of the wettest month of the winter half year.

The *B*, *C*, and *D* climates are still further subdivided into a third, or tertiary, category by the addition of a third small letter. Each of these tertiary letters deals with some further aspect of temperature.

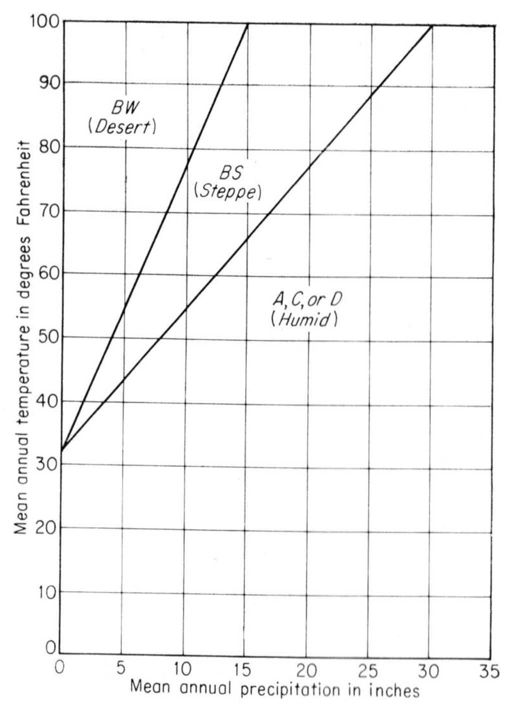

Fig. 17–4. A graph to indicate the boundaries between humid and dry climates when there is a winter concentration of precipitation.

As used with the *B* climates:

h: hot (from the German word *heiss*, meaning "hot"); mean annual temperature above 64°.

k: cool or cold (from the German word *kalt*, meaning "cold"); mean annual temperature below 64°.

As used with the *C* and *D* climates:

a: hot summer; mean temperature of the warmest month above 72° (22°C).

b: warm summer; mean temperature of the warmest month below 72°; at least four months with a mean temperature of 50° or more.

c: cool summer; mean temperature of the warmest month below 72°; from one to three months with a mean temperature of 50° or more.

As used with the *D* climates alone:

d: severe winter; mean temperature of the coldest month below −36° (−38°C).

[1] The lines for determining the boundary between humid and steppe climates are based upon the following three formulas:

$r = 0.44(t - 19.5)$ when the precipitation is evenly distributed throughout the year

$r = 0.44(t - 7)$ when there is a summer concentration of precipitation

$r = 0.44(t - 32)$ when there is a winter concentration of precipitation

in which r represents the mean annual precipitation in inches and t the mean annual temperature in degrees Fahrenheit. For the boundary between steppe and desert climates, use $\frac{1}{2}r$ in each case.

Summarized, the chief types of climate with their characteristics are as follows:

Af: All months hot and moist
Am: All months hot; some months excessively moist, which compensates for one to three dry months
Aw: All months hot; summer rains and winter droughts
BSh: Hot and semiarid (hot steppe)
BWh: Hot and arid (hot desert)
BSk: Cold winters; semiarid (cool or cold steppe)
BWk: Cold winters; arid (cool or cold desert)
Cwa: Hot, moist summers; mild, dry winters
Cwb: Warm, moist summers; mild, dry winters
Csa: Mild, moist winters; hot, dry summers
Csb: Mild, moist winters; warm, dry summers
Cfa: Hot summers, mild winters; all months moist
Cfb: Warm summers, mild winters; all months moist
Cfc: Cool summers, mild winters; all months moist
Dfa: Cold winters, hot summers; all months moist
Dwa: Cold, dry winters; hot, moist summers
Dfb: Cold winters, warm summers; all months moist
Dwb: Cold, dry winters; warm, moist summers
Dfc: Cold winters; short, cool summers; all months moist
Dwc: Cold, dry winters; short, cool, and moist summers
Dfd: Severe winters; short, cool summers; all months moist
Dwd: Severe, dry winters; short, cool, and moist summers
E: Polar or arctic climate: very short and cool growing season (tundra climate) or all months below freezing (icecap or perpetual frost climate)

Detailed study of the climatic map reveals that there is a definite pattern in the arrangement of the types of climate. There is a tendency for a climate to be repeated in about the same location in other continents, particularly if the continents are similar in size, in elevation, and in latitudinal extent. For example, all west margins of continents in lower middle latitudes have the characteristic and well-known "Mediterranean" type of climate—winter rains and summer droughts. Too, all places on west margins of continents in higher middle latitudes have climates resembling one another—mild, rainy winters and moderately warm, less rainy summers; and most places on east margins of continents in low latitudes that are exposed to trade winds throughout the year have essentially the same climate—all months warm to hot and rainy.

Climatic Data

Temperature and precipitation data for some 250 weather stations scattered over the world have been assembled and are found in the Appendix. The stations have been arranged alphabetically according to continents; this allows the student some discrimination in selecting the data representative of the type of climate he may be considering. He should bear in mind that there is usually a broad transition zone between the representatives of any two contiguous climatic regions; therefore, stations located on the margin of a particular climate may be transitional in their details of either temperature or precipitation, approximating the details of the margin of the other climate they are near. Many of the stations listed may have fairly broad variations in temperature or precipitation or both; nevertheless, they may be within the limits described for the particular climate under consideration.

Climatic Charts

A graphic representation of the temperature and precipitation for any particular station serves to fix in the student's mind the essential characteristics of the type of climate which the station represents. Probably the most com-

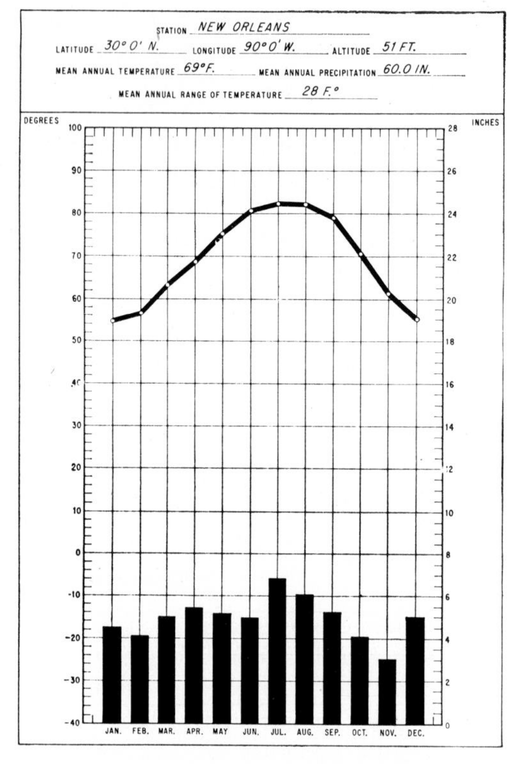

Fig. 17–5. An example of a line-and-bar graph.

mon and most useful type of chart is the line-and-bar graph, Figure 17–5, in which the line represents the course of temperature from month to month and the vertical bars represent the actual amount of precipitation in each month. The chief objection to this type of graph is that it lacks continuity and the student must be able to take the necessary mental jump to tie December to January. Other than representing the data so vividly, line-and-bar graphs are easy to make, and they are particularly useful in comparing one type of climate with another. In this book, each climate discussed is represented by a line-and-bar graph for a station whose data are typical for that climate.

Another type of climatic chart is the circular graph shown in Figure 17–6 in which precipitation is represented by bars marked off on the radii of the circle like the spokes of a wheel. Temperature is likewise ticked off on the radii, but the successive points are connected by a continuous line so that the area thus enclosed may be indicated by shading or stippling. This type of chart has the distinct

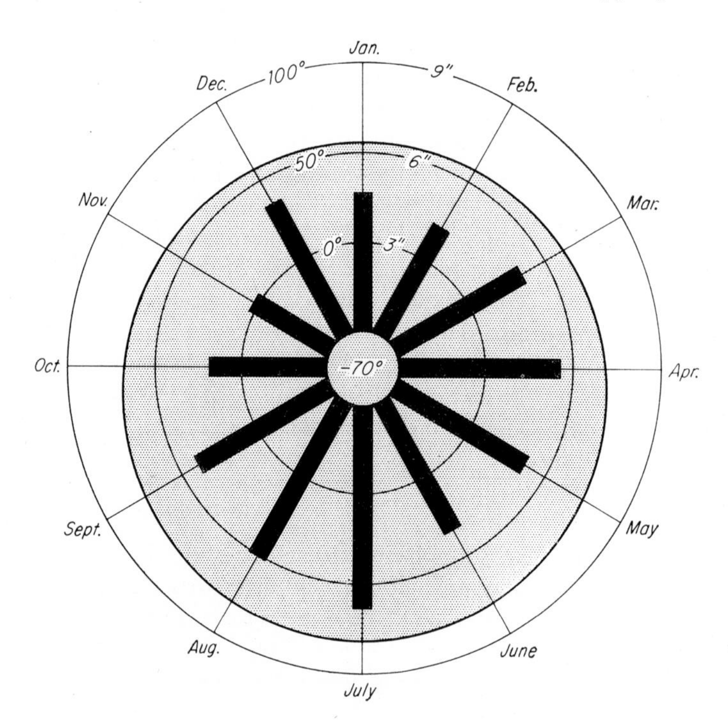

Fig. 17–6. An example of a circular graph (New Orleans).

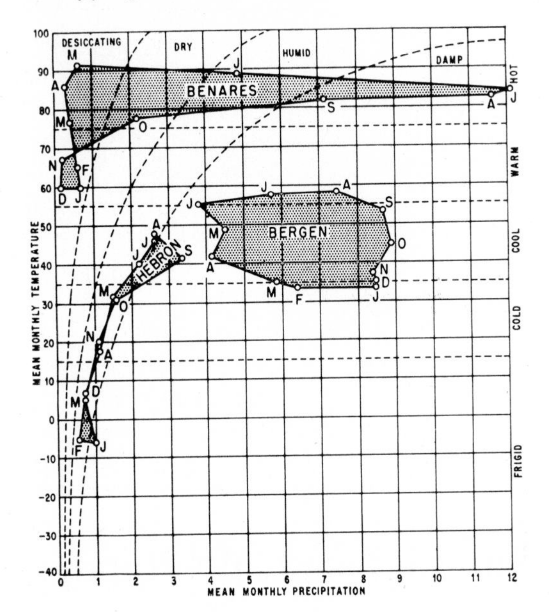

Fig. 17–7. Three examples of climographs.

advantage of continuity, but it is not easy to make, and the course of temperature is more difficult to visualize. The size of the shaded or stippled "eccentric" gives an indication of the degree of cold or warmth: if the area is large, the temperature is relatively high; if the area is small, the temperature is relatively low. For an equatorial or marine station, the shaded area is close to circular; for a high latitude, continental station, the area is decidedly eccentric.

For certain types of stations having either erratic monthly distribution of precipitation or a large annual range in temperature, or both, climographs are useful charts because they portray distinct patterns. Figure 17–7 shows the temperature and precipitation by months for three stations having completely different climatic characteristics. It should be noted that temperatures are plotted on the vertical scale and precipitation on the horizontal scale, and each point thus plotted represents the condition for one month. By connecting the points for successive months, an irregular polygon is formed; sometimes the lines may intersect, thereby forming two or more polygons. Such charts, where two variables are plotted, are called *isopleths*. Climographs have the advantage of continuity. Furthermore, by noting where the pattern for a particular station falls on the chart with respect to the dotted lines, a rather complete summation of the climate may be obtained. Thus Benares, India, is classed as warm and very dry from November to March, as hot and very dry from April to May, as hot and humid, or damp, from June to September, and as hot and dry in October. Obviously, the months from October to March would be the most pleasant, or at least the most comfortable, in Benares. Hebron on the Labrador Coast is always humid or damp, the winters very cold, the summers cool, and the transition seasons cold. Bergen is mostly cool and damp.

Blank charts of the types mentioned above are usually available from various publishing houses, particularly from map companies, or they may be made by the student.

18

The Humid Tropical Climates

The *A* climates are tropical, that is, always warm or hot, but they vary in the amount and distribution of precipitation. Near the equator, and in some locations where the trade winds are forced to ascend over mountains extending to latitudes as much as 25 degrees from the equator, precipitation is heavy every month, with a lush tropical rain forest resulting. In other parts of the tropics there is a wet season of varying length followed by a dry season in winter; those areas support less dense forests or various types of savannas. Most of the rainfall of the tropics is convectional or orographic, but tropical cyclones of variable intensity may swell the amount of precipitation. The annual rainfall is likely to be as much as 80 inches or more. Since no month is below 64°, there is no season or month when temperatures, on the average, drop below what we generally consider summer conditions; the monotonous repetition of high temperatures tends to promote lethargy. Almost one-fifth of the land area of the world comes under the *A*, or humid tropical, classification; somewhat more than one-half of this is *Aw*, with the remainder classed as *Af* and *Am*.

The Rainy Tropical (*Af*) Climate

Location

The *Af* climate—popularly called the "rainy tropical," "tropical rain forest," "wet tropical," or "wet equatorial climate"—extends as a discontinuous belt astride the equatorial portions of the earth. Certain equatorial areas, however, are either too high, and thus too cool, or too dry to be classed as *Af*. The width of the wet tropical or equatorial belt varies, but nowhere does the climate extend poleward 25 degrees, and most of it is within 10 degrees of the equator. The largest single land mass of *Af* climate is the interior portion of the Amazon River drainage basin. Parts of the Amazon itself, however, as well as many of its southern tributaries, are now known to have a dry season and are thus not classed as *Af*. But the rainy coast of Bahia, Brazil, the

Fig. 18–1. The continuously warm and moist weather of the Af climate in Malaya favors the intensive production of rubber. (Courtesy U.S. Rubber Company.)

limited area surrounding Santos, Brazil, and northwestern South America are smaller areas which are included. The last-named area extends northward along the eastern littoral of Central America as far north as the southern part of the state of Veracruz. The Lesser Antilles are also classed as wet equatorial. In Africa a narrow belt of French Equatorial Africa on and just north of the equator, and the northern half of the Belgian Congo are included. A small extension reaches to Lake Victoria. The eastern littoral of Madagascar is also *Af*.

On the mainland of Asia, only Malaya has sufficient rainfall throughout the year to be classed as *Af*. On the other hand, large parts of the Australasian islands do have sufficient year-round rain and thus qualify. Included are: Sumatra, all but the mountainous backbone of Borneo, Celebes, most of northern New Guinea, and many of the islands of Micronesia, Melanesia, and Polynesia.

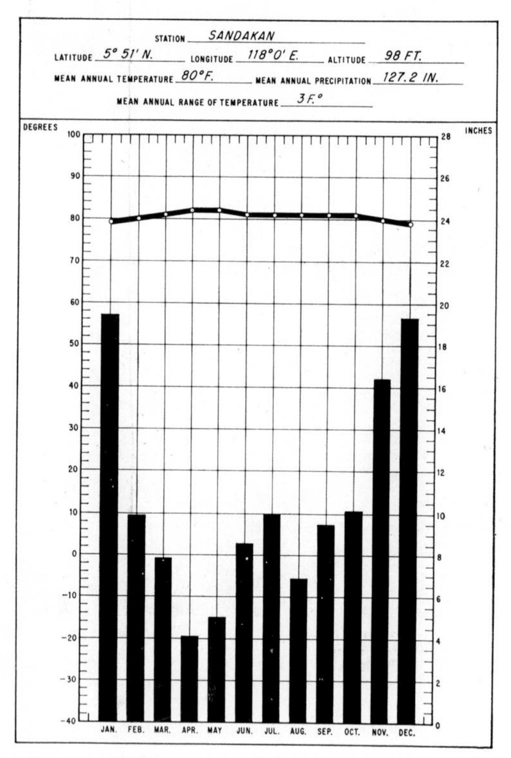

Fig. 18–2. Af climate.

Temperature

Regions having the *Af* climate have consistently high temperatures (all months average above 64°), as shown on the climatic chart for Sandakan, North Borneo, Figure 18–2. Mean annual temperatures are usually between 77 and 81°. Mean annual ranges of temperature are less than 5 degrees, sometimes only 1 or 2 degrees. Ocean Island in the Gilberts averages 81° every month, while Padang in Sumatra has a range of only 1 degree and Pontianak, Borneo, 2 degrees. Because diurnal ranges are greater than annual ranges, with night temperatures dropping characteristically into the low 70s, the night is frequently called the "winter of the tropics." Daily maxima are usually in the high 80s; they rarely rise as high as 90°, and thus temperatures are more moderate here than in many middle-latitude cities, especially in those located in the interior of North America or Eurasia. On the poleward edges of the climate especially, and sometimes elsewhere, a double maximum of temperature tends to occur, usually a month or so after the equinoxes; this, in turn, is reflected by a double period of maximum precipitation. Nouvelle Anvers, in the northern Belgian Congo, is an example.

Precipitation

It should be recalled that the equator and the areas extending a few degrees on each side of it constitute a belt of low pressure which encircles the earth. With trade winds converging and rising in this belt, called the *doldrums*, heavy convectional precipitation results. The doldrums as a whole comprise the world's greatest area of precipitation. It is the seasonal shifting of the doldrums, varying in width from 3 or 4 degrees to 10 degrees or more, that is responsible for the distribution of much of the precipitation in the tropics. Where the doldrums persist throughout the year, they bring the heavy year-round precipitation characteristic of the wet equatorial climate. Elsewhere within the tropics, poleward from the wet equatorial areas, and where the doldrums last less than a year, there is a dry period or, at best, insufficient precipitation to produce the lush vegetation of the constantly wet equatorial regions, except where noticeable monsoonal or orographic precipitation occurs. In the northern hemisphere summer, the doldrums shift northward; during the spring and fall equinoxes they are more nearly centered on the equator. During the southern hemisphere summer, they shift rather noticeably southward over the southern continents, bringing heavy rain which, however, decreases both in frequency and in quantity as it moves poleward.

Wet equatorial areas usually have at least 60 inches of precipitation annually. Many stations record total amounts of 80 or 100 inches; in the East Indies they mount higher, with averages of 126 inches and 178 inches at Pontianak and Padang, respectively.

No month is dry. The distribution, however, may be unequal. Although there are exceptions to the rule, stations located near the equator, or nearer the centers of the climatic belt, tend to have a more even distribution than those located on the poleward edges where the doldrums have somewhat less influence. Although it is true that the dry parts of the world have a greater annual percentage of variability in precipitation than the humid areas, even the doldrums sometimes have marked fluctuations from place to place and from year to year. Usually the variations are insufficient to cause economic distress but they are much greater relatively than the ranges of temperature.

Weather Synonymous with Climate

Where temperatures are always high and ranges are low, day follows day in an unending repetition of weather that varies little throughout the year. In addition, the same type or types of precipitation are expected at almost any time of the year. Many *Af* stations, however, have seasons of more rain and other seasons of less rain. With a constantly high relative humidity added to a usually high absolute humidity, it is not surprising that the sensible temperature is high much of the time; and cloudiness is common.

There are numerous descriptions of the daily rhythm in the equatorial rain forest which grows in response to perennially high moisture and temperature conditions. A typical morning dawns relatively clear, cool, and pleasant. Before long, by nine or ten o'clock, cumulus clouds begin to gather in the east and lazily float westward, obscuring the bright sun from time to time. By noon the temperature has risen 12 to 18 degrees or more, the humidity has become noticeable, and thunderheads may be seen. Sometime between noon and two or three o'clock, the heat, coupled with high relative humidity, becomes almost unbearable; almost all work ceases, and even animals refuse to move unnecessarily. It is incredibly still—the stillness before the storm. Soon the sun is darkened, lightning flashes,[1] and thunder rolls; almost immediately heavy rain falls. Frequently windy or even squally conditions accompany the downpour that may last an hour or two. As quickly as it began, the rain stops, the sky becomes clearer, the relative humidity goes down somewhat, the temperature drops, and life begins to move again. Of course, there may be another shower before nightfall, or there may be one during the early hours of the morning, at three or four or five o'clock; but most characteristically, the showers occur during the greatest heat of the day when convection is at its maximum. During the less rainy season, or seasons, there may be some days without precipitation, but most *Af* stations expect it at least two-thirds (frequently more) of the days of any year.

The rainy tropics are not entirely free from tropical cyclones; occasionally *Af* areas are visited by this destructive type of storm.

[1] Intense lightning is reported in interior locations, but apparently less of it is encountered in marine and littoral locations. The concensus is, however, that both lightning and thunder are characteristic of the wet equatorial regions.

Hurricanes are not so common in the equatorial areas as they are farther poleward on the edges of the tropics and in the subtropics, although they may spawn near the equator. More common than the destructive variety is a large, nonperiodic so-called "cyclone" of several days' duration that has a very weak gradient and apparently weak fronts, and about which little has been recorded. With the weak *Low*, however, comes a period of cloudiness, reduced temperatures, and more or less rain. As a whole, the wet equatorial regions are believed to have relatively few strong winds of destructive force, at least on land. Sea breezes, on the other hand, are the rule in marine and littoral locations and are extremely welcome; not only do they help to temper the midday conditions locally but they also help to make the coastal areas in general much more livable than the interiors. Sea breezes are usually not effective for more than 40 or 50 miles inland and sometimes less.

Responses to the *Af* Climate

The physical elements of any given area are inextricably interwoven, interdependent, and complicated. Climate, natural vegetation, and the regolith (parent rock on the earth's surface) determine, in large part, what soil or soils will develop locally. In turn, indigenous vegetation is largely a reflection of climate and also of the parent rock, type of soil, surface irregularities, or topography and slope, as well as drainage. While it seems best, almost necessary, to consider physical elements singly, it must be remembered that each of them modifies or affects the others constantly and is also affected by them. Therefore, the physical pattern of any area, indeed of any climatic type or region, is determined by the total of all its parts or elements. With cultural features superimposed upon the physical, it becomes evident that total geography is a complex interaction between man and his natural environment; each is partially changed or controlled by the other.

Natural Vegetation.[1] The combination of constantly high temperatures and no dry season results in the world's most luxuriant forest, which is known as the *tropical* or *equatorial rain forest*, or *selva*. It is a two- or three-tiered, broadleaf evergreen forest of huge proportions. The tallest trees tower like a canopy 130 to 150 feet above the surface of the ground; frequently the second tier of trees ranges from 80 or 90 feet to 120 feet; the lowest tier of trees is commonly 40 to 60 feet high. While the sun streams through the higher levels of trees to lower elevations, and here and there even penetrates to the ground, the trees, with their myriads of choking vines, lianas, and epiphytes, usually shut out so much of the sunlight that the ground is only dimly lighted at best. With little light available, there is usually no impenetrable growth of smaller plants and bushes on the forest floor; and progress on foot, though sometimes retarded, is not thwarted as it is in the *jungle*.

The trees hum with the activity of brilliantly colored birds of various descriptions, with insects, and with climbing animals, but large game animals are not numerous since they must forage principally in the more open monsoon forests and savanna lands. Nevertheless, the rhinoceros, elephant, anteater, wart hog, okapi, tapir, wild pig, and various antelopes occupy favored places, especially

[1] In studying the natural vegetation of each type of climate, the student should refer to the map "Natural Vegetation," Plate I, following page 216 in this book. The descriptions of natural vegetation in the text do not always conform exactly to the map because the latter, obviously, can show only the main types of vegetation. The student should relate the vegetation map to the Köppen-Geiger map of climatic regions and should try mentally to superimpose one map upon the other. The vegetation map was developed by A. W. Küchler and is taken from "Goode's World Atlas," 10th edition, 1957, edited by E. B. Espenshade, Jr., and published by Rand McNally and Company, Chicago. The map is used by permission of both the author and the publisher.

near the edge of the rain forest, and multitudes of fish, crocodiles, alligators, and hippopotamuses inhabit the waters. The ground is sometimes literally infested with crawling life—insects (bugs, ants, and termites), rodents, various reptiles, including snakes, and other animals.

Although the trees of the equatorial rain forest are large, their wood usually hard, and their stands normally thick, unfortunately from an economic standpoint the varieties are frequently rather widely scattered. Good and accessible stands of a single variety are considered rare.

Since there is neither a dry season nor a cold season, there is no wholesale shedding of leaves. Leaves do fall, imperceptibly, but the trees are never bare, except when an occasional strong wind strips some of the leaves.

Wherever much sunlight penetrates to the ground, real jungle appears along the edges of rivers, streams, and coastal areas, on steep slopes, where forest clearings of migratory agriculturists have been abandoned, and where certain fairly porous soils exist. Areal reconnaissance substantiates the fact that the true rain forest is interrupted frequently by smaller stands of impenetrable jungle that is a veritable tangle of low trees, bushes, shrubs, aerial roots, woody climbers, and strangling vines. The machete, which is similar to the midwestern corn knife, is an essential piece of equipment to the would-be explorer of the jungle.

Rain Forest Soils.[1] The *laterite* (from the Latin *later*, meaning "brick") is the residual soil considered most characteristic of the rainy tropics. Its usual reddish, bricklike color is the result of continuous leaching of the soil. The heavy rainfall with high temperatures causes certain mineral compounds to be carried out of the topsoil in solution, or as colloidal material (a process known as *illuviation*), and the finer solid materials to be removed mechanically by penetrating rain (a process known as *eluviation*); all these dissolved, colloidal, or fine-clay materials are carried to lower horizons and precipitated, or deposited, or else they are carried to lower elevations by penetrating rain or ground water. In an extreme case of laterization, it is mainly the insoluble hydroxides of aluminum and iron that remain on the surface, even the silica or sand being carried away; such soils are practically worthless, though they frequently achieve this condition through persistent cultivation, particularly in climates where there is at least a short dry season.

Furthermore, the hot humid conditions, aided by bacterial action, cause rapid oxidation and disintegration of the vegetative and nitrogenous materials. The residual soils are, therefore, poor in humus content, deficient in mineral foods, and thus low in productive capacity. Their porosity and granular nature render them easily tillable, but cultivation soon breaks down the soil structure, and fertilization is needed sooner than in middle latitudes. Abandonment follows within a few years of clearing the land, except where scientific plantation agriculture is practiced. As a whole, the laterites constitute one of the world's poorest groups of soils. Where complete laterization occurs, the surface becomes consolidated; it is then classed as a rock. In fact, some writers refer to laterite as a rock, not a soil, and to the soils as lateritic.

Occasionally alluvial soils of fairly high fertility are found in equatorial areas, but frequently they are located far inland—along the Congo or Amazon rivers or their tribu-

[1] Obviously space does not permit a detailed explanation of the origin and characteristics of soils. The student who wishes a brief description of the various soil types and something concerning their development may consult either, or both, of these references:

Vernor C. Finch and Glenn T. Trewartha, "Elements of Geography," 4th ed., pp. 435–464, McGraw-Hill Book Company, Inc., New York, 1957.

Guy Harold Smith, et al., "Conservation of Natural Resources," pp. 25–62, John Wiley & Sons, Inc., New York, 1950.

taries. So far, they have not been greatly used, for either plantations or even subsistence agriculture.

Occupations in the *Af* Climate

Although it is far beyond the scope of this book to attempt to portray, even in a small way, all aspects of man's activities in each type of climate that follows, it does seem desirable to mention some phases of human existence, especially those in which climate seems to have some bearing upon human endeavor. There is no disposition, however, to imply that climate is *the* factor that determines the type of activity, even in a primitive society; nor is it implied that climate is even a dominant factor in a particular activity. On the contrary, some activities mentioned may be carried on in spite of climate rather than because of it, and many are mentioned that have no apparent relationship to climate. Furthermore, there is no attempt to be complete in listing human activities or occupations; those mentioned are only suggestive of the many that are possible, whether they pertain to the culture of plants or animals, hunting, collecting, lumbering, manufacturing, trade, rendering of services, or any other pursuit. And there is no attempt to explain how climate *is* or *is not* significant in affecting the particular occupation or activity. To do so even in a cursory manner would require space which is not available in this book.

A population map reveals that in general the equatorial rain forest has a low density of population. Only five small areas (islands or coastal areas) are exceptions: the Hawaiian Islands, the rainy coast of Bahia, the Santos region, parts of the east Guinea coast, and southern Malaya, including Singapore. Besides the enervating climate, there are many factors which limit population. Scores of diseases exact their toll of lives, and animal life of many sorts deters settlement and exploitation. Poor soils, poor transportation, and vast distances fail to stimulate much agriculture, subsistence or commercial, and isolation is a deterrent. Yet nature provides enough food and shelter for the average inhabitant to get along without too much exertion. Civilization develops best only where there is a dry season or a cold season when man is forced to prepare for the nongrowing season, or else migrate or die; thus the *Af* portions of the world have made, so far, only a niggardly contribution. But there are exceptions to this, and there are possibilities of making better use of the wet equatorial regions.

Most of the inhabitants live a primitive life of hunting, fishing, collecting items from the forest, bartering, farming on a migratory basis (called "milpa," "ladang," or "fang" agriculture), or any or all of these occupations. It is the expansion of plantation agriculture, however, that probably holds the best promise for future use of the land. Much capital, chiefly American and British, has already been invested in the production of bananas in Central America; rubber and cinchona in Malaya, in parts of the East Indies, and to some extent in Amazonia; cacao along the rainy coast of Bahia; spices in the Celebes; and a variety of tropical products in Hawaii, the Lesser Antilles, Amazonia, and elsewhere. Under proper conditions, the collection, production, and sale of native forest products could also be increased. Such things as Brazil nuts, cinchona, manioc or cassava (tapioca), wild rubber, vanilla, papayas, and cabinet woods, such as mahogany, find ready markets outside the tropics. The rainy tropics may constitute one of the chief areas of potential growth in supplying products desired by populations that can afford them but cannot produce them.

The Tropical Monsoon (*Am*) Climate

Location

The *Am* (tropical monsoon) climate[1] makes

[1] In the Köppen system, the *Am* is a transitional type between the *Af* (constantly rainy) and the *Aw* (rather long, winter dry period). There are extensive areas in the eastern hemisphere subject to monsoons

its appearance principally where there are tropical littorals backed by highlands. The major exceptions are the lowland areas of the lower Amazon Valley and the Niger River Delta of the Guinea coast. As in the case of the Malabar Coast of India, coastal Burma, southern Ceylon, much of the Philippines, and elsewhere, the summer monsoon blows onshore steadily for several months, wafting warm, humid air (tropical maritime—*mT*) onto the land. Convection is at a maximum at that time, and, with orographic ascent added, heavy rainfall results. Where there is no highland backdrop, precipitation is chiefly convectional (lower Amazon and Niger). During the dry season, the wind is usually offshore, particularly in the Asiatic mainland locations. Such winds are generally dry and yield only small amounts of convective precipitation now and then.

Temperature

There is frequently a double maximum and a double minimum of temperature. The hottest season usually precedes the rainy season; the greater cloudiness of the rainy season is responsible for reducing the temperature several degrees. The mean monthly temperatures of the hot season are likely to be several degrees hotter than the warmest months of the *Af* climatic areas, and the cooler months of the wet season are often several degrees cooler. The mean annual range of temperature may be as great as 12 or 14 degrees at stations on the poleward edges of the climate: 14 degrees at Akyab, Burma, and 10 degrees at Rangoon; but it is much less at places nearer the equator, such as Calicut, India, 7 degrees and Colombo, Ceylon, only 3 degrees.

Precipitation

The *Am* climate differs from the *Af* climate in having a short dry season, although the mean annual precipitation is generally as high, or higher. The period of maximum precipitation in the northern hemisphere is usually from May to September or October, while the variable, less rainy season is a winter phenomenon lasting from two to four months. The total rainfall in the dry season is usually less than 10 inches; sometimes it amounts to only a few tenths monthly, as at Rangoon, Burma (Figure 18–4). In spite of the short dry season, however, the wet-season precipitation is great enough (40 to 80 inches or more) to keep the subsoil relatively moist and to produce, therefore, a fairly heavy monsoon evergreen forest with dense undergrowth as the indigenous vegetation. The trees are usually somewhat shorter and are less thickly spaced than in the *Af* locations, but nevertheless they constitute what is commonly classed as a "light rain forest."

Vegetation, Soils, and Occupations in the *Am* Climate

As pointed out earlier, it should be noted that the Köppen-Geiger map shows no boundary lines between the *Af* and the *Am* climates. This seems justifiable because there are many instances of rapid gradations between the two types (especially in the matter of rainfall), and the annual fluctuations of the boundary lines are often pronounced; thus any drawn boundary line would be highly compromised. In addition, the rainfall of the one, two, or possibly three dry months of the monsoon type—that is, monthly rainfall under 2.4 inches, but not necessarily rainless—is compensated for by the heavy rain of the other months. It is no wonder then that the vegetative aspects of the two types of climate are

---

which do not qualify as *Am* climate under the Köppen formula; on the other hand, Köppen classifies many areas as *Am* even though they do not experience the seasonal reversal of wind direction, a typical monsoon characteristic. Areas classed as *Am*, whether or not they have the reversal of wind, must have enough precipitation in the rainy season to make up for the deficiency in the dry season (see Fig. 17-1) in order to prevent the ground's losing its moisture to such a degree that the growth of dense vegetation is adversely affected. Tropical areas which do not have sufficient rain in the rainy season to compensate for the dry season are classed as *Aw*. The Köppen-Geiger map shows no boundary between the *Af* and the *Am* climates.

Fig. 18–3. The winter period of light rain in the Am climate permits clearing and cultivation of the land so that rice can be grown as here in Ceylon. (Courtesy British Information Services.)

very similar. In fact, the wetter portions of *Am* lands have tropical rain forests; it is only on the drier margins that the monsoon forests are less luxuriant and grade into semideciduous forests or savannas.

Furthermore, it has been shown that the *Am* climates are transitional between the perpetually rainy *Af* climates and the definitely dry winter, *Aw*, type. It may be argued that all climates of whatever type are transitional—that there are no sharp boundaries between any two types. But, unlike the vast stretches of the Russian and Siberian boreal forest lands or the expansive areas of the Sahara Desert where, in each case, one may travel many hundreds, or even thousands, of miles without experiencing any notable change in climatic characteristics, the monsoon climates change quite perceptibly, in point of rainfall at least, in traveling relatively short distances. This fact can be seen by a study of Table 18–1 which shows the pronounced changes in rainfall along a part of the Guinea coast. The sharp differences are not attributable to altitude because all stations lie below 400 feet.

Am climates, therefore, are closely related to the climates on their borders. Consequently, it has seemed unnecessary to discuss the natural vegetation and soils of the monsoon lands: the wetter margins have characteristics like, or similar to, those of the *Af* climates; the drier margins have characteristics like, or similar to, those of the *Aw* climates. Man's activities also assume many of the same transitional aspects: He produces rubber in the wetter portions and rice where there is a definite dry period; agriculture, other than commercial, is likely to be migratory on the

Fig. 18–4. Am climate.

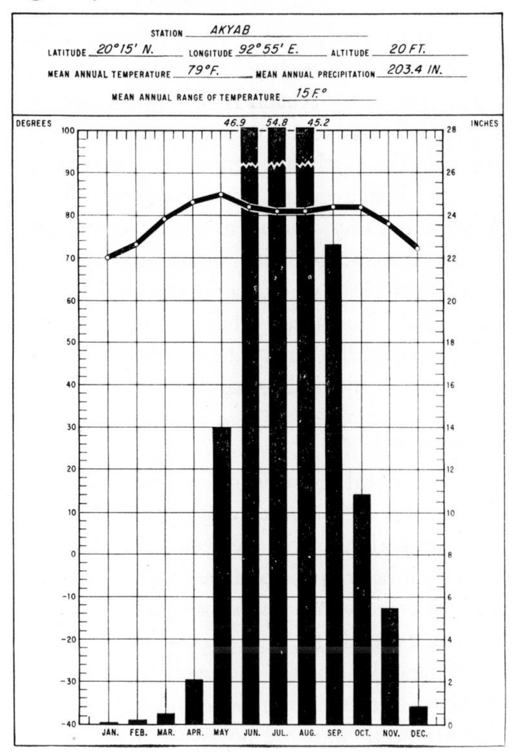

*Table 18–1. Monthly and Annual Rainfall along the Guinea Coast of Africa**

Station	Jan.	Feb.	Mar.	Apr.	May	June	July	Aug.	Sept.	Oct.	Nov.	Dec.	Year
Lagos (235)	1.1	2.0	3.7	6.3	10.1	19.2	10.2	2.4	5.3	8.6	2.4	0.9	72.2
Akasa (220)	2.6	6.5	10.0	8.6	17.0	18.6	10.1	9.3	19.3	24.7	10.6	6.5	143.8
Debundja (250)	8.0	10.9	17.1	17.3	24.8	59.7	64.4	57.7	65.2	45.2	26.6	15.1	412.2
Libreville (100)	8.9	8.5	13.3	12.8	7.8	0.3	0.1	0.7	3.9	14.0	14.8	9.7	94.8
Djole (450)	5.8	6.6	10.5	10.1	6.7	0.8	0.0	0.0	0.9	9.1	10.9	6.5	67.9
Banana	2.1	2.3	3.7	6.1	1.9	0.0	0.0	0.1	0.1	1.6	5.9	4.7	28.6

* The numbers in parentheses in the first vertical column indicate the distances between stations in air miles.

wet margins and settled on the dry margins; civilization is generally more primitive on the wet margins but becomes more highly developed on the dry margins. The student will be able to pick out many more instances of this sort after he has concluded his study of the moist tropical climates.

The Wet and Dry Tropical (*Aw*) Climate

Location

The *Aw* climate has a definitely dry season of variable length. It lasts from three to five, or even six, months and is followed by a definite rainy season; thus it is commonly referred to as the "wet and dry tropical climate." It is also called the "savanna climate," a name derived from the major vegetational associations. Typically, the climate is located adjacent to, but on the poleward sides of, the *Af* and *Am* climates into which it merges almost imperceptibly. In Africa it surrounds the *Af* area on the north (the Sudan), south, and east sides. In South America the *Aw* occupies a vast area south of the *Af* (called the *Campos*), a smaller area north of it (the *Llanos*)—including most of Venezuela and parts of Colombia—and parts of northern Brazil, British Guiana, and coastal Ecuador. Most of the Pacific side of Central America, as far north as the tropic of Cancer as well as portions of the eastern side, are also represented by *Aw*. Most of the West Indies and southern Florida as far north as Palm Beach are included. This is the farthest poleward extent of the climate (27 degrees). A considerable part of India, East Pakistan, and southeastern Asia also is classed as *Aw*, while limited areas of the Philippines and the southern parts of the East Indian Islands are included. Much of northern Australia is also wet and dry tropical.

Temperature

There is a greater mean annual range of temperature in the *Aw* than in the *Af* climate. This is partly because the wet and dry tropics extend farther poleward, reflecting a definite hot season and a cooler season. Annual ranges are likely to be over 10 degrees and occasionally as great as 15 or more degrees; Calcutta has a range of 21 degrees.

The period of greatest heat, the dry season, usually comes just before the rainy period. Temperatures commonly soar into the 90s, or even above 100°, during those stifling weeks (Figure 18–6). The high amount of cloudiness during the rainy season that follows prevents a part of the incoming solar radiation from reaching the earth and hence the temperatures at that time average several degrees lower. After the rainy period a cooler dry season ensues; it coincides with the months of lowest sun. With clear skies, rapid terrestrial radiation, and decreased relative humidity, the night temperatures may drop below 60°. This is the most enjoyable season.

Precipitation

The general location of *Aw* regions (from the equator to about 20 degrees) places them in a transitional area between the doldrums with their heavy precipitation, on the one hand, and the trade winds and subtropical high-pressure belts with their dry conditions, on the other hand. During the summer, the converging winds of the equatorial low and the doldrums that attend them move poleward with the sun. At that season, in effect, *Af* conditions prevail over *Aw* lands. In the winter, the opposite is true. At that season the trades, or subsiding air connected with subtropical high-pressure areas, move equatorward, spreading over *Aw* areas. The dry season is determined chiefly by latitudinal location; nearer the equator the dry season is shorter and precipitation is greater than on the poleward edge of the climate. Rainfall data as well as natural vegetation attest the fact that there is, in general, a gradual decrease of total precipitation poleward.

Characteristically, the wet and dry tropics have less precipitation than either the *Af* or

Fig. 18–5. Grass with scattered, flat-topped trees is characteristic of Aw climate where the winter season is definitely dry, as in tropical south Africa. (Courtesy American Geographical Society.)

Am; most weather stations record between 30 and 60 inches. Of greater importance than the actual amount, however, is its distribution. In the northern hemisphere, the wet season begins in April near the wetter margins (equatorial) but often not until May or June on the drier margins (poleward). The rainy season usually lasts about four months in the drier parts, but frequently six months, or longer, on the *Af* side.

The rainy period is very much like that of the *Af* regions. It is ushered in with violent thunderstorms and squalls. Eventually, however, long-continued and more general rains of the weak cyclonic type persist, interspersed with heavier downpours.

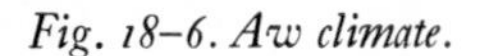

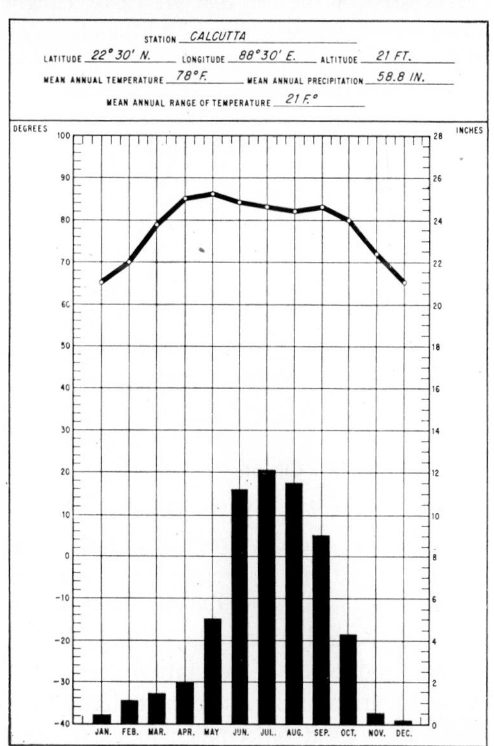

Fig. 18–6. Aw climate.

As the sun again retreats equatorward, dry conditions follow. In reality desert, or near-desert, weather then spreads over the *Aw* regions, bringing welcome relief from the oppressively high heat and humidity of the wet season. But it is likely to be a period of complete drought; on the *Af* margin, on the contrary, there may be no month that is without occasional precipitation.

With the cessation of rains, the vegetation changes in appearance rather rapidly, soon becoming brown and seared, withered, and seemingly dead. The dry savanna grasses are frequently burned at that time, supposedly in order to enrich the soil and to aid both mobility and hunting, since the dry season is the best for stalking big-game animals indigenous to savanna lands. Many of the trees, chiefly deciduous or semideciduous, shed their leaves, exhibiting well their characteristically flattened tops. Watercourses eventually fail to flow, and only here and there stagnant pools remain with floundering aquatic life. But the natives feast on the fish that are easily caught during the low-water stage. Although much of the vegetation is dormant, the dry season is not only more healthful but also more pleasant than the wet season.

It should be noted that there are rather great fluctuations in precipitation from year to year, even more than in the *Af* and *Am* climates; indeed, great annual fluctuations are characteristic. An average year may bring the right amount of precipitation, but a year or two may follow in which there is either too little or too much rainfall to maintain the usual economy.

Responses to the *Aw* Climate

Natural Vegetation. The transitional character of the wet and dry tropical climate is as well evidenced in its vegetation as in its precipitation distribution. On the equatorward side lie the tropical rain forests; on the poleward side dry-land types of vegetation take over. The wet margin of the *Aw* climate has a light deciduous tropical or semideciduous tropical forest with grasses and shrubby, or scrubby, undergrowth, except where the sun does not penetrate to the floor of the forest. Usually the trees are neither so tall nor so numerous as in the rain forest of the *Af* and *Am* areas and are thus more readily penetrated and more easily inhabited by elephants, fleet-footed gazelles, gnus, zebras, giraffes, lions, tigers, and leopards as well as many water-wallowing animals. All these animals range into the still more open grasslands farther poleward, too.

The most typical vegetation of the climate, however, is the savanna grass. In its purest state, a savanna grassland consists of tall, tough, and mostly inedible grasses that grow 10 or more feet high in the wet season. Rarely is a sod formed, each plant usually being distinct. More common than the pure grass savannas are the several combinations of trees and grasses. Various names are applied to them, according to the number and appearance of the trees and the height and spacing of the grasses. Three types are recognized by most geographers. The *high grass–low tree savanna* consists of grasses that attain maximum heights and are mixed with fairly closely spaced low trees. Often the trees are in clumps, presenting a parklike appearance. Farther poleward, the *acacia–tall grass savanna* is common. In it grasses are not so tall (3 to 5 feet) and flat-topped acacias constitute the chief variety of tree. Sometimes they appear in clumps, but often they are scattered. Nearer the dry edge of the climate, the *acacia–desert grass savanna* dominates. It has low, scrubby, thorny acacias with short dry grasses, both of which are adapted to withstand a dry season of eight or nine months. The name *thorn savanna* is sometimes applied to this type. Each of the savannas appears in response to the combination of climate, soil, slope, and other local determinants. Many savanna trees, such as the acacia, of which there are several hundred varieties, are flat-topped, except where there is sufficient rainfall to promote more mature growth. The flatness is suggestive of the fact, if it is not proof, that grasslands are indigenous to the climate and that trees are neither more numerous nor taller because precipitation is insufficient to keep the ground moist enough throughout the year to nourish full-fledged trees.

Soils of the Aw Climate. While little is known definitely about the savanna soils, it is believed they are closely akin to the laterites of the *Af* areas and are therefore of low productive capacity natively. The relatively new alluvial areas of the extensive flood plains and deltas, on the other hand, frequently provide not only excellent sites but also soils of good quality. These are often areas of major agricultural and commercial occupations.

Occupations in the Aw Climate

With respect to economic development, the wet and dry tropics present a widely varied picture. Some of the most backward tribes of both interior Africa and South America, plus the aborigines of northern Australia, are included in them; on the other hand, they contain Miami, Habana, Rio de Janeiro, Bombay, and Bangkok—some of the most progressive and interesting cities of the world. As a whole, both the total population and the population densities are greater than those of the *Af* and *Am* climates.

Hunting, fishing, and the collection of forest products mark the livelihood of tribal peoples in the less populated and less accessible areas of the interiors of Africa and South America. Such subsistence foraging, however, is not the typical mode of living for the majority. Farming of one sort or another, even though it is often precarious, is invariably the choice wherever the economy is relatively far advanced. The savannas constitute the real zone of tropical farming, embracing many of the world's best and largest tropical plantations, a large part of the world's subsistence rice areas, and extensive cattle lands of considerable value. While manufacturing is relatively unimportant in the wet and dry tropics, it does assume some importance in Thailand, Java, India, and in other places where handicrafts, processed foods, and even iron and steel are produced. With respect to trade, the savanna regions are better off than in manufacturing. They constitute an important source of many tropical plants and other raw products desired in the middle latitudes, so ports and trade facilities must be kept in good repair. Habana, Rio de Janeiro, Veracruz, Tampico, Bombay, Calcutta, Saigon, and Manila are world ports of considerable importance that are geared to care for the exports of their hinterlands and to act also as importers of manufactured goods from industrial nations.

19
The Dry Climates

General Characteristics

Dry climates are defined as those in which the moisture taken up and held by the atmosphere is greater than the moisture added to the land by precipitation. This definition is based upon rather intangible, or hypothetical, considerations since there is no satisfactory method of measuring the amount of moisture taken into the air, especially when we need to consider the loss of moisture through transpiration by plants and animals as well as the loss through direct evaporation from the ground. Even though evaporation, as from an open receptacle of water, is not difficult to measure, there are relatively few installations in dry lands which have facilities for that purpose. Furthermore, it has been definitely established that the rate of evaporation depends upon many factors, such as the temperature and relative humidity of the air, the volume and speed of the passing air, atmospheric pressure, and certain other minor factors. Obviously, to take all factors into account would complicate any method other than direct measurement in determining the amount of evaporation.

Since temperature and its effective ally relative humidity seem to be most significant in determining evaporation—the amount of moisture which the atmosphere can include being approximately doubled for every increase of 20 degrees (Table 3–2)—it seems feasible, though probably unscientific, to relate temperature and precipitation and to ignore other factors. This is exactly what the Köppen system has done in establishing the limits of the dry climates. In effect, the system states that there is a direct relationship between the mean annual temperature and the mean annual precipitation—that the higher the temperature, the greater the precipitation requirement to keep a climate from being dry. In addition, it states that less precipitation is required if it is concentrated in winter, that more is required if it has an even distribution throughout the year, and that most is required if there is a summer concentration (the exact meaning of winter and summer concentration has already been indicated in Chapter 17). This relationship is shown in the graphs, Figures 17–2 to 17–4. Although the boundary between the dry and humid climate in each case is indicated by a straight line, the relationship is actually logarithmic and, therefore, should be indicated by a parabola, or curved line. In other words, the precipitation requirement decreases much more rapidly with decrease in temperature than any of the graphs show. This is not too serious for our purposes, however, because the direct relationship as shown by the straight lines holds fairly true for short temperature spans, such as 40 to 80°, the span which takes in a large percentage of the stations concerned. It should be noted that, for the same mean annual temperature, the difference in mean annual precipitation between a summer-concentration type and an even-distribution type is the same as the difference between the even-distribution type and the winter-concentration type. For example, at 60° the precipitation boundaries for the three types are: summer concentration, 23.5 inches; even distribution, 18.0 inches; and winter concentration, 12.5 inches—both differences being 5.5 inches. This holds true for the boundary between the desert (*BW*) and the steppe (*BS*) as well as for the boundary between the steppe and humid climates (*A*, *C*, or *D* classification). Furthermore—and as pointed out earlier—for the same mean annual temperature, the precipitation requirement for the humid-steppe boundary is exactly double that for the steppe-desert boundary in each of the three types of precipitation concentration. It is seen, therefore, that regardless of the merits of the Köppen criteria, the formula is certainly systematic and relatively simple to apply. The student will find these graphs very helpful whenever he is called upon to determine whether a given station should be classed as humid, steppe, or desert.

Although the boundaries are almost wholly determined by temperature and precipitation, there are other conditions that govern the location and extent of the dry lands, such as local differences in exposure to winds, the extent and location of high-pressure areas, the frequency, or lack, of cyclonic activity, orographic influences, natural vegetation and agricultural development of the land, and even soils in some cases. All these not only contribute to the cause and origin of dry lands, but they also serve in establishing the boundaries where climatic data are few, or lacking altogether. Doubtless certain "exotic" streams like the Nile, Tigris, Euphrates, Indus, and the Colorado, although having their origin in humid lands, flow across dry lands and con-

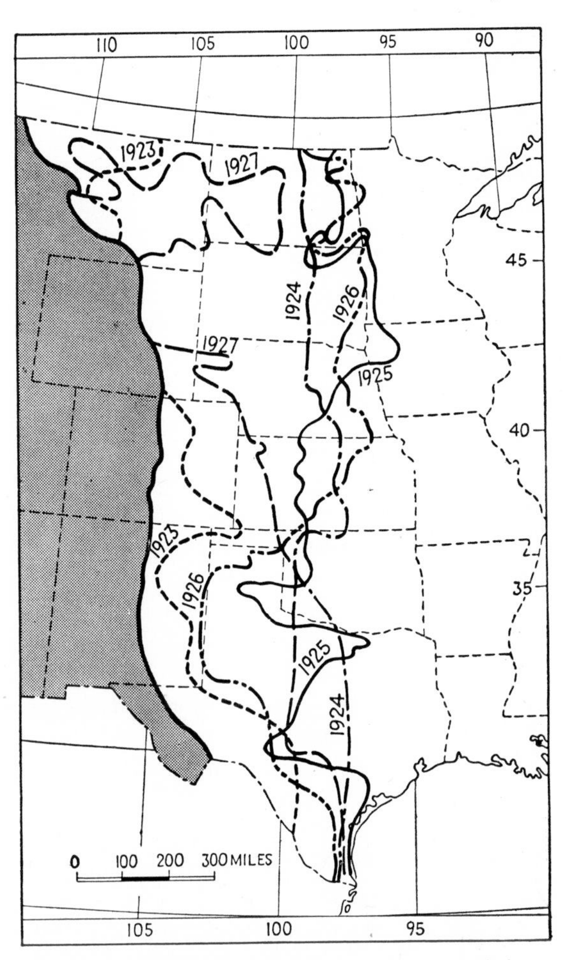

Fig. 19–1. Annual fluctuations during a five-year period of the boundary between humid and dry climates in the central United States. (After Kendall.)

tribute moisture to the air at least in a narrow band, thus modifying dry climates in a small way, directly or indirectly. Certainly, cold ocean currents have a definite influence, causing lower temperatures, increased aridity, and pronounced fogginess.

General Location

Impressive as are the large areas of the world that have *A* climates, even more remarkable are the vast areas of *B* (dry) climates. Köppen indicates that somewhat over one-fourth of the earth's land surface and a similar proportion of the oceans have dry climates. The most extensive dry lands extend eastward from the great bulge of Africa at Dakar across Saudi Arabia, Iraq, and Iran and far into India, somewhat past Delhi. The second extensive area begins along the lower course of the Dniester River and extends eastward across Soviet middle Asia, through the Dzungarian Gap, the Tarim Basin, and Gobi Desert. Were it not for the Elburz Mountains of northern Iran, these two belts would be joined. Much of the western United States and northern Mexico, small areas in northern South America, and isolated parts of Spain complete the areas of dry climates in the northern hemisphere. In the southern hemisphere, three-fourths of Australia, the southwestern tip of Madagascar, much of southwestern Africa, Patagonia and associated areas east of the Andes, and coastal Chile and Peru west of the Andes complete the list. Among the inhabited continents, Europe has the least of the world's arid and semiarid regions, followed by South America, North America, Asia, Africa, and Australia in proportionally increasing order.

Precipitation

The dry climates have seven characteristics: (1) meagerness of precipitation; (2) uncertain amount of precipitation, that is, great extremes in the seasonal and annual amounts and in the distribution of the precipitation received (Figure 16–4); (3) potential evaporation that is always greater, often many times greater, than the actual precipitation; (4) generally low relative humidity; (5) much sunshine; (6) considerable wind; and (7) struggle for, and conservation of, available water resources by plants and animals as well as humans.

Although all dry lands are deficient in precipitation, the wetter parts are distinguished as semiarid (*BS*), while the drier parts are arid (*BW*). As a rule, the semiarid areas surround the arid areas because they are situated so as to receive more precipitation—usually either of middle-latitude cyclonic origin or of doldrum association, as will be shown later.

Within deserts, rain comes at such infrequent intervals, sometimes months or even years apart that, the average amount is never great, although occasional cloudbursts do occur. Under the latter conditions much erosion results. Where temperatures are high, most of the precipitation is of the convectional type. While relative humidities are characteristically low, absolute humidities may be quite high, but clouds may yield no precipitation. Because of the excessive heat, falling rain is frequently evaporated before it reaches the ground.

It will be recalled that, as a rule, the drier the climate, the less reliable is the precipitation. Therefore, while both *BW* and *BS* areas witness great fluctuations from year to year, steppes have not only more rain but it occurs more often, is more certain, and by nature is usually more widespread than in deserts. Nevertheless, because of rather great fluctuations from year to year, even in steppe regions, agriculture is often precarious. Lured by a series of rainier-than-usual years, farmers are likely to bring under cultivation ground that may be quite incapable of producing a crop during an ensuing series of drier years. Dust bowls result. Good desert land with assured irrigation is often better for agricultural purposes than semiarid land that is subject to the caprices of nature.

Because of the great variability of precipitation in the dry lands, the line which divides the deserts (*BW*) from the steppes (*BS*), and again the steppes from the humid climates, is likely to shift radically from year to year. This fact is clearly brought out by reference to the map of the United States showing the fluctuations in the boundary between the dry and humid climates (Figure 19–1).

Temperature

The clear skies of dry lands lead to the receiving of much heat by day and to the rapid loss of it by night; thus the diurnal ranges are characteristically high for their latitudes. Annual ranges of temperature, while also usually great for their latitude, are usually not so great as the daily ranges.

Since latitude is the chief control of temperature, dry lands in the high latitudes have

Fig. 19–2. The hot steppes (BSh) are generally too dry for crops without irrigation; grazing probably constitutes the best use of such land. The scene is in Bandera County, Texas. (Courtesy Standard Oil Company of New Jersey.)

cool or cold winters, while those in or near the tropics have no such season, although they do have occasional low temperatures at night. Because of the marked seasonal differences, deserts are divided into low-latitude, or hot, deserts (*BWh*) and the middle-latitude, or cool, deserts (*BWk*); steppes are likewise divided into low-latitude, or hot, steppes (*BSh*) and middle-latitude, or cool, steppes (*BSk*). While these divisions are not entirely satisfactory, they serve in a general way to distinguish the hot from the cooler dry lands. The third letter, *h*, is used for both deserts and steppes when the mean annual temperature is above 64°; the letter *k* appears when the mean annual temperature is below 64°.

Deserts and steppes are windy places. This is partly the result of considerable convection during the day and of sparse vegetation that offers little frictional retardation of air movement. While convectional currents are primarily vertical in movement, they foster horizontal movements, too; in areas largely devoid of vegetation where many convectional circulations may occur within a relatively small area, frequent but often strong and variable winds mark the climates. The night is by contrast usually much quieter, often very still. This aids in nocturnal cooling. But the following day, especially in the deserts, brings the winds again, and the sky is often obscured by dust, sometimes to great heights.

The Low-Latitude Steppe (*BSh*) Climate

Location

In considering the low-latitude, or hot, steppe climates it is well to recall two things: (1) they characteristically surround the low-latitude desert climates, as in Australia, and (2) they are transitional zones of precipitation between the arid climates on one hand and the humid climates on the other, as in North Africa. The low-latitude semiarid lands appear chiefly in the latitudes from 15 to 35 degrees, although they may extend closer to the equator, as in the extreme northern part of South America, in the Sudan of Africa, and in southern India. The hot steppes, like the hot deserts, usually reach the west coasts.

Precipitation

Because of the fact that hot steppes flank hot deserts on either their poleward or equatorward sides, and frequently both, their rainfall distribution is similar to, and is determined by, the rainfall regime that exists on the rainy side. The *BSh* area on the north side of the Sahara has a winter maximum of precipitation because at that season the middle-latitude cyclones and fronts which bring rains to the Mediterranean lands also extend somewhat farther south, bringing rains to the northern fringe of the Sahara. Similarly, the steppes of Iraq, southern Australia, northwestern Mexico, and the southwestern United States have a cool-season maximum and a hot-season minimum.

For several months of the year, often as many as seven or eight, hot steppes with winter precipitation are controlled by subsiding high-pressure air masses (subtropical highs) that are not favorable to much condensation or precipitation; therefore dry conditions result. During the winter season, when cyclonic storms may occur, overcast days with precipitation are not unusual; neither are importations of cool air that follow the passage of cold fronts. Yet, even the winter season is one of prevalent sunshine and generally healthful weather conditions.

The Sudan, which flanks the Sahara on its southern side, has a summer maximum of precipitation and a dry winter season. The summer rainfall of the Sudan, like that of the *Aw* climatic belt next to it, is the result of the poleward shift of the doldrums that brings a short period of raininess with it. Since the *BSh* part of the Sudan is farther poleward from the center of the doldrums than the adjoining *Aw* area, both the length of the rainy season and the amount of precipitation are therefore

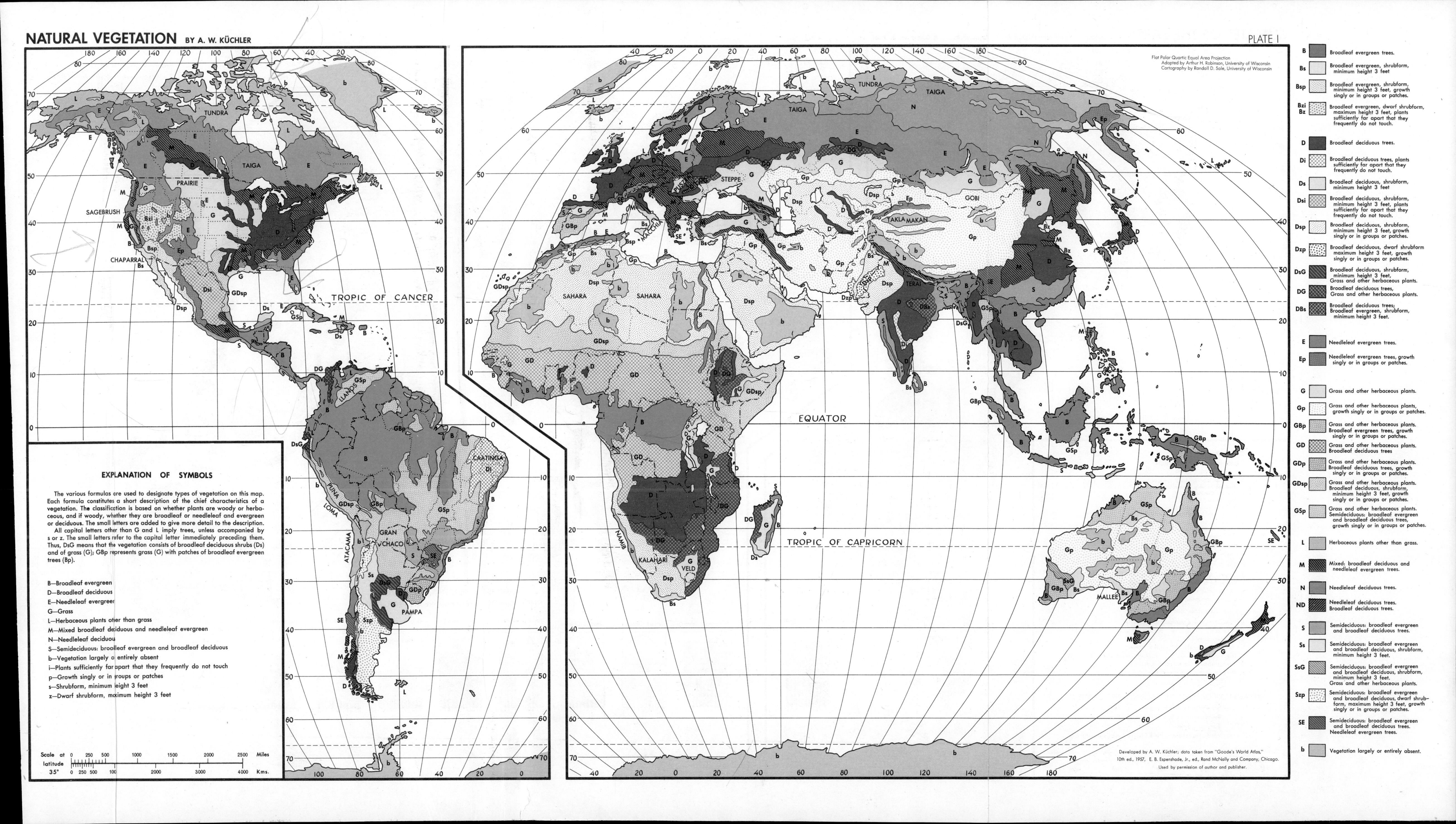
NATURAL VEGETATION BY A. W. KÜCHLER
PLATE I
Flat Polar Quartic Equal Area Projection
Adapted by Arthur H. Robinson, University of Wisconsin
Cartography by Randall D. Sale, University of Wisconsin
TUNDRA
TAIGA
PRAIRIE
SAGEBRUSH
CHAPARRAL
TROPIC OF CANCER
LLANOS
CAATINGA
PUNA
LOMA
ATACAMA
GRAN CHACO
PAMPA
STEPPE
GOBI
TAKLA MAKAN
SAHARA
TERAI
EQUATOR
NAMIB
KALAHARI
VELD
TROPIC OF CAPRICORN
MALLEE
EXPLANATION OF SYMBOLS
The various formulas are used to designate types of vegetation on this map. Each formula constitutes a short description of the chief characteristics of a vegetation. The classification is based on whether plants are woody or herbaceous, and if woody, whether they are broadleaf or needleleaf and evergreen or deciduous. The small letters are added to give more detail to the description.
All capital letters other than G and L imply trees, unless accompanied by s or z. The small letters refer to the capital letter immediately preceding them. Thus, DsG means that the vegetation consists of broadleaf deciduous shrubs (Ds) and of grass (G); GBp represents grass (G) with patches of broadleaf evergreen trees (Bp).
B—Broadleaf evergreen
D—Broadleaf deciduous
E—Needleleaf evergreen
G—Grass
L—Herbaceous plants other than grass
M—Mixed broadleaf deciduous and needleleaf evergreen
N—Needleleaf deciduous
S—Semideciduous: broadleaf evergreen and broadleaf deciduous
b—Vegetation largely or entirely absent
i—Plants sufficiently far apart that they frequently do not touch
p—Growth singly or in groups or patches
s—Shrubform, minimum height 3 feet
z—Dwarf shrubform, maximum height 3 feet
Scale at latitude 35°
Miles
Kms.
B Broadleaf evergreen trees.
Bs Broadleaf evergreen, shrubform, minimum height 3 feet
Bsp Broadleaf evergreen, shrubform, minimum height 3 feet, growth singly or in groups or patches.
Bzi Bz Broadleaf evergreen, dwarf shrubform, maximum height 3 feet, plants sufficiently far apart that they frequently do not touch.
D Broadleaf deciduous trees.
Di Broadleaf deciduous trees, plants sufficiently far apart that they frequently do not touch.
Ds Broadleaf deciduous, shrubform, minimum height 3 feet
Dsi Broadleaf deciduous, shrubform, minimum height 3 feet, plants sufficiently far apart that they frequently do not touch.
Dsp Broadleaf deciduous, shrubform, minimum height 3 feet, growth singly or in groups or patches.
Dzp Broadleaf deciduous, dwarf shrubform maximum height 3 feet, growth singly or in groups or patches.
DsG Broadleaf deciduous, shrubform, minimum height 3 feet, Grass and other herbaceous plants.
DG Broadleaf deciduous trees, Grass and other herbaceous plants.
DBs Broadleaf deciduous trees; Broadleaf evergreen, shrubform, minimum height 3 feet.
E Needleleaf evergreen trees.
Ep Needleleaf evergreen trees, growth singly or in groups or patches.
G Grass and other herbaceous plants.
Gp Grass and other herbaceous plants, growth singly or in groups or patches.
GBp Grass and other herbaceous plants. Broadleaf evergreen trees, growth singly or in groups or patches.
GD Grass and other herbaceous plants. Broadleaf deciduous trees
GDp Grass and other herbaceous plants. Broadleaf deciduous trees, growth singly or in groups or patches.
GDsp Grass and other herbaceous plants. Broadleaf deciduous, shrubform, minimum height 3 feet, growth singly or in groups or patches.
GSp Grass and other herbaceous plants. Semideciduous: broadleaf evergreen and broadleaf deciduous trees, growth singly or in groups or patches.
L Herbaceous plants other than grass.
M Mixed: broadleaf deciduous and needleleaf evergreen trees.
N Needleleaf deciduous trees.
ND Needleleaf deciduous trees. Broadleaf deciduous trees.
S Semideciduous: broadleaf evergreen and broadleaf deciduous trees.
Ss Semideciduous: broadleaf evergreen and broadleaf deciduous, shrubform, minimum height 3 feet.
SsG Semideciduous: broadleaf evergreen and broadleaf deciduous, shrubform, minimum height 3 feet. Grass and other herbaceous plants.
Szp Semideciduous: broadleaf evergreen and broadleaf deciduous, dwarf shrubform, maximum height 3 feet, growth singly or in groups or patches.
SE Semideciduous: broadleaf evergreen and broadleaf deciduous trees. Needleleaf evergreen trees.
b Vegetation largely or entirely absent.
Developed by A. W. Küchler; data taken from "Goode's World Atlas," 10th ed., 1957, E. B. Espenshade, Jr., ed., Rand McNally and Company, Chicago.
Used by permission of author and publisher.

CLIMATES OF THE EARTH

According to KÖPPEN-GEIGER MAP Revised 1953 by R. Geiger and W. Pohl

Flat Polar Quartic Equal Area Projection
Adapted by Arthur H. Robinson, University of Wisconsin
Cartography by Randall D. Sale, University of Wisconsin

Climatic Regions
(Explanation of climatic syllabi)

First Letter

- A, C, D Sufficient heat and precipitation for high-growth trees. Average monthly temperature over 64°.
- A Tropical Climates
- B Dry Climates.
- C Humid Mesothermal Climates. Coldest month between 64° and 27°.
- D Humid Microthermal Climates. Warmest month over 50°. Coldest month below 27°.
- E Polar climates. Warmest month below 50°.

Second Letter

- S Steppe Climate.
- W Desert Climate.
- f Sufficient precipitation each month.
- m Tropical forest climate despite dry period.
- s Dry period in summer of respective hemisphere.
- w Dry period in winter of respective hemisphere.
- [w] Same, overlapping into other hemisphere.
- s′, w′ Single precipitation period shifted to the fall.
- w″ Long drought in winter, short in summer.

Third Letter

- a Warmest month over 72°.
- b Warmest month below 72°, at least 4 months over 50°.
- c Less than 4 months over 50°. d Same, coldest month below —36°.
- h Dry-hot, mean annual temperature over 64°.
- k Dry-cold, mean annual temperature below 64°.

EXPLANATION OF LIMITS AND BOUNDARIES OF THE MAP

- ——— Boundaries of climatic regions
- —·—· Boundaries of a, b, c, d in between the sphere of A, C, D Climates (Lowland)
- ········ Within the sphere of the B Climate Boundaries of h & k

Scale at latitude 35°: 0 250 500 1000 1500 2000 2500 Miles; 0 250 500 1000 2000 3000 4000 Kms.

TROPIC OF CANCER

EQUATOR

TROPIC OF CAPRICORN

EXAMPLES FOR KÖPPEN'S CLIMATIC ZONES.
ANNUAL RANGE OF TEMPERATURE AND PRECIPITATION.

- Tropical Rainforest Climate — Singapore — Af
- Savanna Climate — Yola (Nigeria) — Aw
- Desert Climate — Aswan — BW
- Steppe Climate — Denver (Colorado) — BS
- Tropical Monsoon Climate — Allahabad — Cw
- Mediterranean Climate — Palermo — Cs
- Marine Humid Mesothermal Climate — Hamburg — Cf
- Short Summer Humid Microthermal Climate — Moscow — Df
- Transbaikalian Subarctic Climate — Yakutsk (Siberia) — Dw
- Tundra and Ice-Cap Climates — Upernivik (Greenland) — E

Original map published by Justus Perthes, Darmstadt, new edition, 1954, by R. Geiger, Munich.
Data used by permission of R. Geiger.

CLIMATIC REGIONS OF THE WORLD According to CLARENCE E. KOEPPE

Flat Polar Quartic Equal Area Projection
Adapted by Arthur H. Robinson, University of Wisconsin
Cartography by Randall D. Sale, University of Wisconsin

TROPIC OF CANCER

EQUATOR

TROPIC OF CAPRICORN

WET EQUATORIAL
All months 70° or more; annual ranges of temperature less than 5°. Annual rainfall at least 60"; all months moist.[1]

TRADE WIND LITTORAL
Mean annual temperature at least 70°; annual ranges generally under 20°. Rainfall more than 35"; no month rainless.[1]

WET AND DRY TROPICAL
Mean annual temperature at least 70°; annual ranges under 15° except in regions of strong monsoon influence. Rainfall more than 35"; distinct winter dry season.[1]

SEMIARID TROPICAL
Mean annual temperature at least 70°; annual ranges under 25° except in regions of strong monsoon influence. Rainfall between 10" and 35"; at least five winter months with less than 1".[1]

TROPICAL DESERT
Mean annual temperature at least 70°. Rainfall under 10" to 14", depending upon temperature.[1]

DRY SUMMER SUBTROPICAL
Coldest month above 45°. Annual precipitation not less than 10" with winter maximum, wettest month usually having at least three times the precipitation of the driest month.

HUMID SUBTROPICAL
Coldest month above 40°. Annual precipitation at least 35"; all months moist.

COOL MARINE
Coldest month between 30° and 45°. Annual precipitation at least 35" with winter maximum; no month rainless.

COOL LITTORAL
Coldest month under 40°; annual ranges generally under 40°. All months moist.

HUMID CONTINENTAL
Coldest month under 40°; at least three months above 60°; annual ranges more than 40°. Annual precipitation not less than 16" on poleward margins, nor less than 22" on equatorward margins; summer maximum.

SEMIARID CONTINENTAL
Temperature same as in Humid Continental. Precipitation between 6" and 16" on poleward margins, and between 12" and 22" on equatorward margins; summer or double maximum.

INTERMEDIATE DESERT
Temperature same as in Humid Continental. Precipitation under 6" on poleward margins, and under 12" on equatorward margins; summer or double maximum.

MARINE SUBPOLAR
One to four months under 30°; at least one month above 50°; not more than two months above 60°. Winter maximum of precipitation.

MODERATE SUBPOLAR
Temperature same as in Marine Subpolar. Summer maximum of precipitation.

EXTREME SUBPOLAR
At least five months under 30°; at least one month above 50°; not more than two months above 60°. Summer maximum of precipitation.

POLAR
All months under 50°.

Scale at latitude 35°: 0 250 500 1000 1500 2000 2500 Miles; 0 250 500 1000 2000 3000 4000 Kms.

[1] Includes areas where altitude may reduce the general temperature level as well as the annual totals of rainfall, although the annual courses of these elements are the same as in the lowlands. Apply normal lapse rate of 3° per 100 feet.

Data used by permission of A. J. Nystrom & Co., Chicago.

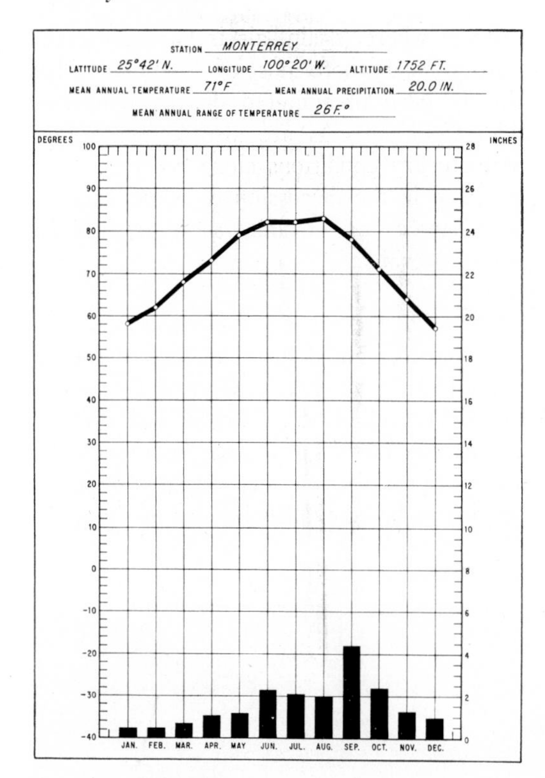

Fig. 19-3. BSh climate.

less than those of the *Aw* climatic area. While the rainy season is short, it partially compensates for this by the heaviness of its convectional showers. On the other hand, since the rain falls during the hot season, evaporation is high; therefore, as a whole, precipitation is less effective than in the case of the tropical steppes with a winter maximum of precipitation. By the same analysis, more precipitation is required to produce a steppe with a summer maximum of precipitation than one with a winter maximum. At Wyndham, Australia, two months are rainless and four months have less than 1 inch of precipitation each, and most of the precipitation is distributed unevenly throughout the remaining six months. Monterrey, Mexico, is an example of a low-latitude steppe with a summer maximum of precipitation (Figure 19-3).

Temperature

Whereas tropical deserts hold the records for absolute high temperatures and for absolute diurnal ranges, they are about the same as tropical steppes in mean annual temperatures, in mean annual ranges of temperature, and in highest mean monthly temperatures. These are detailed in the following section on the low-latitude deserts.

The Low-Latitude Desert (*BWh*) Climate

Location

Of the austere regions of the world perhaps none is so arresting as the low-latitude or tropical desert region. Few climatic or vegetational areas are as extensive as the Sahara and Arabian deserts, and few are viewed with as much interest and questioning. Except for the polar deserts of ice and snow, no other regions are quite so inimical to settlement. Besides the great expanse of hot desert across northern Africa, Arabia, Iraq, Iran (in part only, some higher elevations being classed as cool desert), and into India, smaller areas in the northern hemisphere appear in the southern part of the Great Basin in the United States, extending southward on both sides of the Gulf of California and also into part of the interior of northern Mexico. In the southern hemisphere, hot deserts occupy almost 40 per cent of interior and western Australia, occur in interior southwestern and coastal western Africa, and make a limited appearance in coastal Peru and northern Chile between the Andes and the Pacific.

The usual locations of the hot or low-latitude deserts are from about 15 to 30 degrees latitude in both hemispheres. They are located on the western, or leeward, sides[1]

[1] While Africa may appear to be an exception to the rule, since the desert extends completely across it, climatically Africa functions with Eurasia, with which it is so closely tied; thus the Sahara should be considered as an extension of arid southwestern Asia.

of the continents in the latitudes where either high-pressure areas or trade winds are operative. The high-pressure areas are, physically, regions of descending air, warmed by compression, that are picking up moisture—just the opposite of what is necessary to cause either cloudiness or precipitation. Where trade winds appear, the effect is the same since they blow from cooler to warmer latitudes, thus picking up moisture and warming concurrently. In addition to this, locations far in the interior of continents are either too far from moisture-bearing winds or else the winds simply do not blow in the right direction to import moisture to such areas. In many places deserts also occur on the lee sides of mountain systems where, again, air is descending and warming and where most of the moisture has already been precipitated on the windward side. Cool currents are also known to accentuate dry conditions along coasts, as was indicated in the discussion of ocean currents in Chapter 14. Under these conditions, aridity prevails.

Precipitation

Tropical or hot deserts are located too far equatorward to be benefited much by middle-latitude precipitation of frontal and cyclonic origin, too far poleward to be affected by doldrum precipitation, and too far toward the interior to be controlled by any of the moist

Fig. 19–4. This is an extreme type of tropical desert (BWh) in Nejd of Saudi Arabia; even though there may be some rainfall, it is difficult for plants to gain a foothold in such terrain. (Courtesy Standard Oil Company of New Jersey.)

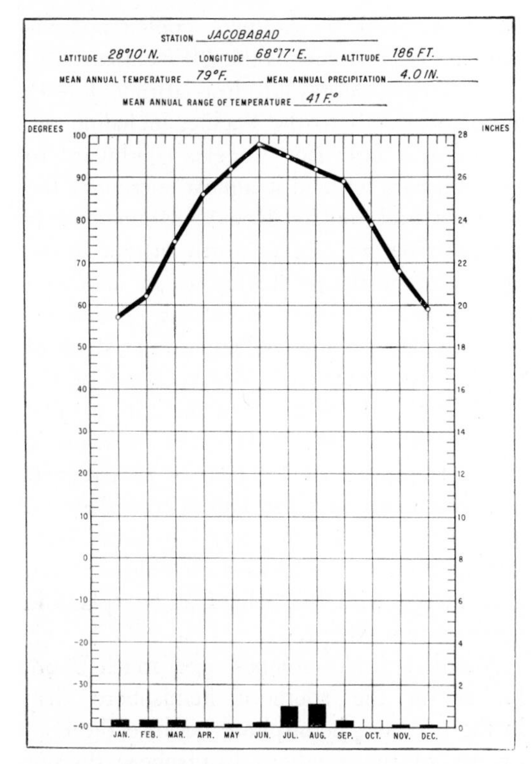

Fig. 19–5. BWh climate.

tropical air masses that originate on the eastern sides of continents.

The slight precipitation that falls at infrequent intervals is almost always both convectional and local in nature. Widespread precipitation is extremely rare in most of the hot deserts. As previously indicated, no set amount of precipitation distinguishes the hot deserts from the hot steppes, but the amount is usually from 10 to 15 inches. Most of the hot desert lands, therefore, receive less than this. The approximate mean annual precipitation in inches at Yuma is 3, at Baghdad 7, at Khartoum (Sudan) 6, and at Jacobabad 4 (Figure 19–5). Some desert stations record practically nothing annually, or the amount is frequently too small to be measurable. This is true of Aswan (Egypt), In Salah (Algeria), and Iquique (Chile). The northern desert of Chile rarely receives rain, although the absolute humidity, as evidenced occasionally by towering cumulus clouds, is often considerable. At Iquique, for example, there was no rain for four years, but in the fifth year one shower brought 0.6 of an inch. To illustrate how misleading precipitation figures may be, that meant that the mean annual precipitation for the five-year period was 0.12 inch, yet it was very unevenly distributed, as is true in most deserts. Unreliability is characteristic of the rainfall of the hot deserts; statistics give the variability—departure from the mean annual precipitation—as over 40 per cent.

In spite of the great amount of surface convection in low-latitude deserts, the subsiding air aloft is generally very stable, and therefore it usually prevents the lower air from rising high enough above the condensation level to produce thunderstorms. On occasion, however, severe thunderstorms do result, and almost as frequently the intense heat of the desert evaporates much of the rain before it reaches the ground. At times a middle-latitude frontal storm will invade the poleward edge of low-latitude deserts; under such visitations the desert is greatly benefited.

Sunshine is the lure of the desert; skies are predominantly clear, both day and night. While the percentage of possible sunshine received varies greatly, many resorts capitalize on the high amount compared with humid areas. The Sonora Desert of Mexico and the southwestern part of the United States, for example, receive annually almost 85 per cent of the sunshine possible; Yuma receives 88 per cent. With that much sunshine it is not surprising that Yuma has 55 times as much potential evaporation as it does precipitation in the hot season, nor is it surprising that the relative humidity frequently drops below 10 per cent.

Temperature

The highest official temperature ever recorded comes from the low-latitude desert. Azizia, Libya, enjoys the distinction of having the record of 136.4°. The highest in the United States, 134°, was recorded in Death

Valley. Temperatures in the sun, however, run much higher, and they are the temperatures to which man and beast are customarily subjected.

The hottest month usually averages between 82 and 95°, bringing with it even more scorching and desiccating conditions than the previous months. In the northern hemisphere, July is usually the hot month; occasionally June or September excels it. Daily maxima frequently reach figures between 100 and 110°. In some places the maximum temperature may be above 90° for three or four months. The nights offer only relative relief from the day temperatures, with the night minima usually between 70 and 80°; even so, the hot-season diurnal ranges may be 30 to 40 degrees.

The coolest month usually averages somewhere in the 50s, except near the equator where it may be as high as 70°, as at Khartoum. Daily maxima are characteristically in the 60s, 70s, or even the 80s, while minima dip down into the 40s and even the 30s. Freezing temperatures bring light frosts from time to time. This places the diurnal ranges in the neighborhood of 40 to 50, or more, degrees during the low-sun period, and they are noticeably greater than in the high-sun season.

Hot deserts have greater mean annual ranges of temperature than any other type of tropical climate; ranges of 25 to 35 degrees are not unusual. For example, the range at Khartoum is 21 degrees, at Yuma 36, and at Baghdad 46. It is noteworthy, however, that the cool-season diurnal ranges are usually greater than the mean annual ranges of temperatures.

The Middle-Latitude Steppe (*BSk*) Climate

Location

Almost without exception the *BWk* areas of Asia and North America are surrounded by *BSk*, the middle-latitude, or cool, steppe climate. The latter, however, also appears in other places in the northern hemisphere. In North America, for example, cool-steppe climate not only surrounds the *BWk* climate (except where it merges into low-latitude desert) but also extends to the Pacific, including the coastal area near Los Angeles southward to the 30th parallel, and it almost surrounds the southern and middle Rocky Mountains. It extends from 25 degrees north latitude in the Sierras of northern Mexico as an unbroken climatic belt east of the Rockies to 52 degrees north in Alberta and Saskatchewan. Much of the United States and Mexico west of the 100th meridian and north of the 20th parallel is either *BS* or *BW*. Middle-latitude semiarid lands stretch across Eurasia as an irregular but continuous belt for almost 100 degrees. They appear again in central Turkey, Iran, Afghanistan, and northwestern Pakistan, and even merge with the Sahara at one place in northwestern Africa.

Middle-latitude steppes are much more limited in the southern hemisphere than in the northern hemisphere because the continents of the southern hemisphere do not extend far south (Africa and Australia), or they become narrowed longitudinally (South America). Australia, oriented in similar fashion to Eurasia with a long east-west axis, has a narrow belt which comes down to the sea along the Great Australian Bight. A small section appears in South Africa. In South America, cool steppes make limited appearances on the west coast and in the high plateau of Bolivia, but the largest South American area surrounds the Patagonian Desert (*BWk*), extending, as in North America, from about latitude 25 to 50 degrees south.

Precipitation

Middle-latitude steppes have the same controls as the middle-latitude deserts. Being transitional between the deserts and humid lands, however, they are also affected by humid controls. Thus, while anticyclones still prevail in the winter, except on the subtropical margins near seas, they are inter-

Fig. 19–6. On lands of this sort near Cut Bank, Montana, the cool steppe (BSk) climate favors wheat production. (Courtesy Standard Oil Company of New Jersey.)

rupted more frequently by cyclones than is true in the deserts. In the summertime, more moisture is transported inland to the middle-latitude semiarid lands than to the arid ones because they are more accessible. In the United States, *gT* or *aT* air masses spread over the Great Plains from time to time in the summer, bringing either frontal or convectional precipitation.

The amount of precipitation varies from place to place and also rather greatly from year to year. In most of the cool steppes, the amount averages between 8 and 22 inches annually, which is notably less than the amount received in the tropical steppes. While Tehran (Iran) has only 9 inches and San Diego 11, Glasgow (Montana) has 13, Denver 14, Goodland 18 (Figure 19–7), and Kimberley (South Africa) 18. In each of the foregoing, summer is the period of maximum precipitation, except in San Diego and Tehran, which are located near Mediterranean lands where summers are dry. Summer precipitation is largely convectional and occasionally frontal, while winter precipitation is mainly frontal, much of it in the form of snow. Blizzards and severe cold waves are common cold-season phenomena in the Great Plains steppes of the United States and are also known by various names in Asia. When chinooks descend from the Rockies onto the Great Plains, they often raise the temperatures considerably, sometimes rapidly. This has the effect of melting the snow, sometimes depriving winter wheat of its protective covering and thus causing damage or loss of the crop if the ground freezes before

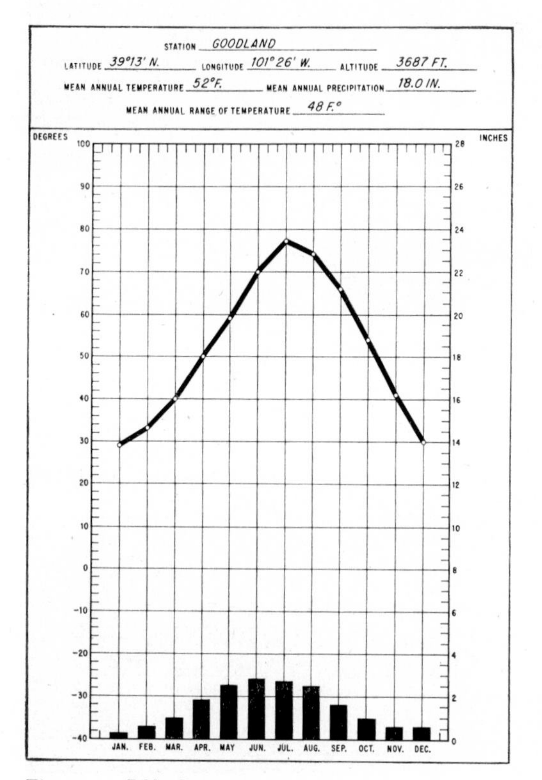

Fig. 19–7. BSk climate.

another snowfall covers it; at the same time it uncovers grasses on range lands and thereby cuts cattle losses which might result through starvation.

Temperature

Temperatures of representative *BSk* stations are similar to those classed as *BWk*. The cold month (usually January) at Williston (North Dakota) is 6°, at Denver 30°, and at Tehran 34°. Usually one to three, or even four, winter months average 32° or less. Absolute minima as low as −25 or −30° are not uncommon. Places that are located either nearer the equator or near the ocean are exceptions; in those cases, neither cold-month means nor absolute minima are so low.

The month of maximum temperatures in the northern hemisphere is usually July, except in marine locations where summer heating is retarded (San Diego, August). Most interior locations average around 70 or 75° at that time. Helena averages 68°, Goodland 77°, and Tehran 85°. Absolute maxima are: Lusk (Wyoming) and Denver 105°, Rocky Ford (Colorado) 106°, Spearfish (South Dakota) and Williston (North Dakota) 110°, Lamar (Colorado) and Miles City (Montana) 111°.

Mean annual ranges of temperature vary considerably with location, although most frequently they are between 40 and 60 degrees. Denver has a mean annual range of 42 degrees, Goodland 48, Tehran 49, and Pierre (South Dakota) 59. The range at San Diego is only 15 degrees and that of Kimberley 26, while that of Williston is 63 degrees.

It thus becomes apparent that middle-latitude steppes—and middle-latitude deserts, too—are far from tropical, except for maximum summer temperatures that may frequently resemble the lower latitudes. Higher latitudes, the usual interior locations far removed from marine influences, and lee locations account for the temperature regimes which indicate rather great seasonal extremes compared with either of the dry climates of the low latitudes.

The Middle-Latitude Desert (*BWk*) Climate

Location

The middle-latitude deserts are not so extensive as their counterparts in the low latitudes, with which they sometimes merge. Middle-latitude deserts are located rather far in the interior of Asia, where they are most extensive, and in North America, but they are primarily coastal and are of smaller extent in the southern hemisphere. They are thus either on the lee sides of mountains (as in the cases of Patagonia, the Takla Makan, Gobi, and Mongolian deserts of interior Asia, and the Great Basin of the United States); or far from oceans and moist masses of air (the situation of Soviet middle Asia east of the Caspian Sea). Unlike the low-latitude deserts, they are not

Fig. 19–8. Owing in part to a moderately even distribution of precipitation, the cool desert supports a varied type of drought-resisting vegetation. Joshua trees are fairly common on the BWk lands of the United States; these are on the Mormon Mesa in Lincoln County, Nevada. (Courtesy U.S. Forest Service.)

affected by trade winds but by high pressures with their settling air much of the time, especially in winter. Both the Peruvian Current and the Benguela Current are probably factors in their appearance on the western side of South America and in southwestern Africa.

Precipitation

Interior locations and semipermanent high-pressure systems preclude the formation of much precipitation in the cold season, except when and where cyclones or orographic ascent may bring modest amounts of either rain or snow. The period of maximum precipitation is the summer at which time the bare, or nearly bare, ground heats rapidly, thus producing much turbulence and convection and some thunderstorms. The summer season is also quite windy, especially during the daytime. Although during the summer the warm or hot interiors of both North America and Asia produce seasonal low centers and apparently produce at least partial monsoonal systems, whereby moisture-bearing winds are wafted onto the land masses, the *BWk* areas are too far inland to benefit much from such winds; the moisture they contain is largely lost en route.

Compared with the *BWh* climate, however, the middle-latitude desert climate usually has both a more even distribution of precipitation throughout the year (Figure 19–8), and a more reliable annual occurrence. The mean annual precipitation varies from place to place, depending upon the factors mentioned. While Swakopmund (Southwest Africa) has less than 1 inch annually, Santa Cruz (Argentina) has 6 and Mendoza (Argentina) 8. Fewer parts of the middle-latitude deserts have the excessively dry conditions that *BWh* areas have, during which no rain may fall for months or even years.[1] Also, while the moist margins of the *BWk* climate receive less precipitation than the moist borders of the *BWh*, the lower temperatures combined with a more even seasonal distribution of precipitation are often more effective for the growth of natural vegetation and crops than the hot deserts.

Temperature

Middle-latitude deserts (sometimes called "cool deserts" in contrast to the hot low-latitude deserts, but not to be confused with the botanist's ice deserts or tundra deserts) have a very definite cold season. The summer is hot, although rarely so hot as in low latitudes. Tecoma (Nevada) and Turfan (Sinkiang) have three months below freezing; Kashgar (Sinkiang) and Lovelock (Nevada) have two months. Other *BWk* stations located farther equatorward or on the coasts of Africa or South America have no months that average below freezing. The coldest month in Santa Cruz is 33° and at Swakopmund 55°. Both of these are coastal. The coldest month at Tecoma (Figure 19–9) is 23°, at Lovelock 30°, and at Turfan 13°.

Warmest-month temperatures average 61° at Santa Cruz, 70° at Tecoma, and 90° at Turfan. Turfan is located below sea level where the effects of descending air are partially responsible for the 90° that is higher than for other *BWk* areas. Also, air drainage in the winter is responsible for the low month average of 13° at Turfan.

[1] A few places, such as Swakopmund, Southwest Africa, have several rainless months. High pressure and the cool Benguela Current accentuate the dryness.

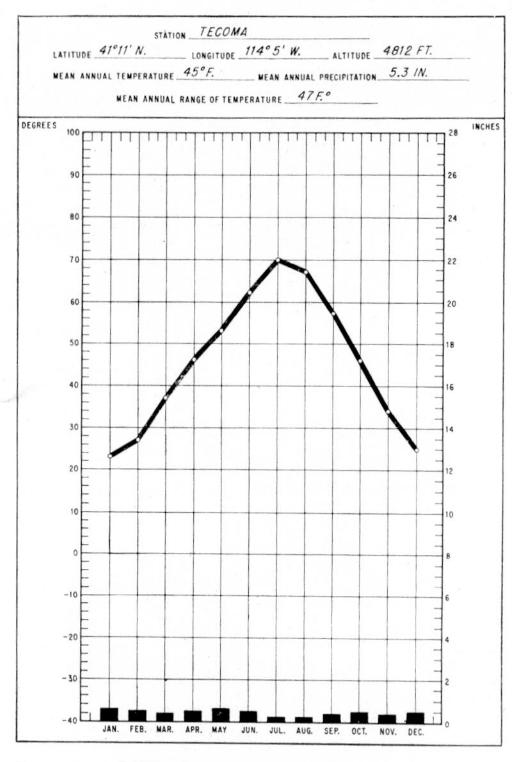

Fig. 19–9. BWk climate.

The mean annual range at Turfan, 77 degrees, is larger than in representative *BWk* stations. On the other hand, the mean annual ranges at Santa Cruz (28 degrees) and Swakopmund (9 degrees) are smaller than usual; again they reflect marine influence. More representative of cool deserts are the ranges at Lovelock (45 degrees) and Tecoma (47 degrees).

Diurnal ranges are high in the summer, when maximum temperatures may hit between 100 and 110°, or even higher, and minimum temperatures may occasionally drop to 40°, or lower. During the cold season, ranges are not so great, although they are still likely to be relatively high for their latitudes, except at places modified by marine conditions. Minimum winter temperatures at interior locations are frequently below freezing, sometimes for many days. Some weather stations that have mean monthly temperatures below freezing have temperatures approaching the zero mark. Turfan, with a January mean of 13°, must have considerable zero weather.

Responses to the Dry Climates

Natural Vegetation

Only in favored places do dry areas exhibit much vegetation. Valleys and other low spots, given a good and retentive soil plus accessibility to subsurface moisture, often support higher forms of vegetation than those that exist in more level areas. Permanent watercourses are commonly lined with trees, giving the appearance of oases in dry lands. Orographic areas, if high enough, may support scattered trees, pseudo forests, or true forests with trees rather sparsely spaced. Both deserts and steppes exhibit great varieties of annuals, shrubs, xerophytes, scrubby forms of humid land species, numerous varieties of woody plants, and occasionally scattered low trees.

Desert. To the biologist a desert is a place that supports little or no vegetation. Some deserts have no vegetation. They may be sandy wastes (*ergs*) with crescent-shaped dunes called *barcanes*, or they may be rocky and stony regions called *desert pavements*, or bare rock areas sometimes called *hamadas*. All of these unvegetated land forms appear in parts of the Sahara, parts of the Sonora and Mojave deserts of North America, and in other deserts of the world.

Some of the most barren desert areas are the result of the excessive accumulation of evaporated salts rather than a minimum of available moisture. Alkali flats, called *playas*, are representative of such salt-encrusted basins, where little, if any, vegetation exists. Some thick-leaved plants, however, are salt tolerant; they may thrive in such places.

Most deserts, however, have more or less flora. In some places low forms of rather thickly spaced annuals predominate. Within a short distance, taller, more sparsely spaced plants may take over. Again, large areas may be devoted to certain cacti;[1] the same species may not be common again for many miles. The desert plant variety is great, and whether the local vista includes different kinds of cacti, or other xerophytes, scrub trees, sagebrush, greasewood, annuals, or even scattered bunch grass in the most moist places, whenever rain falls the desert bursts into unbelievably beautiful florescence. Since most of the plants are either drought-resistant or drought-enduring, eventually they again become drab after the effect of the rain has passed and await the return of moisture and a fresh impetus to blossom forth. Whatever the local desert vegetation, it is largely a response to the amount of moisture locally available for plant growth.

Steppe Vegetation. The ability to produce at least prevalent grasses, and on the humid margins and in the wetter areas a continuous carpet of grasses, distinguishes the steppes from the deserts. While thorny bushes and scrub trees and countless annuals may be common, especially on the drier margins of the steppes, bunch grasses of various varieties predominate. Natively, trees appear only in favored places, as in canyons, along rivers, and in higher, more moist situations.

The more humid eastern part of the Great Plains of the United States and the black earth steppes of the Soviet Union supported indigenously an unbroken short-grass association, except in the localities where porous soils with their significantly reduced water retentive powers were located. Some of these natural grasslands remain as grazing grounds for cattle and sheep, but unfortunately large parts of them have been ploughed and sown with small grain crops, especially wheat.

[1] Illustrative of this fact is the concentration of certain cactus associations within rather limited areas in Arizona. Near Ajo is the Organ Pipe National Monument; east of Tucson is Saguaro National Monument. While these plant groups appear elsewhere, too, it is interesting that about 60 miles north of Phoenix the saguaro disappears altogether.

During the dry years, or cycles of years, which inevitably occur, some of the finest soil blows away; what is left is frequently useless. Under such circumstances, at least locally, desert conditions follow. In cool-steppe areas it takes at least 100 years, and usually many more, to produce an inch of topsoil even under the most fortuitous circumstances.

In North America, the steppes east of the Rocky Mountains are actually sometimes semiarid, sometimes arid, and sometimes humid, as far as annual precipitation is concerned (Figure 19–2). Because of this great fluctuation, steppe agriculturists are constantly faced with the necessity of conserving moisture from year to year by various dry-farming methods. Even then, ploughed land is subject to considerable deflation as well as to desiccation during dry spells. Little wonder then that conservationists, soil scientists, agricultural schools, and governmental representatives constantly advocate retaining the grasslands in their native state rather than ploughing and cropping them, and little wonder, too, that many consider good desert land, if irrigation water is assured, a much better agricultural risk than semiarid lands with their erratic precipitation.

Soils of Dry Climates

Some dry soils that have not been used are excellent both in structure and in texture and sometimes in mineral content; what they usually lack, other than moisture, is humus. While desert soils as a whole have small prospects of greatly expanded use, except under optimum conditions, steppe soils, on the contrary, are often excellent and are extensively grazed and frequently cropped. Most steppe soils are classed as *pedocals*.[1]

Desert Soils. Because of the slow weathering of rock in deserts, the resulting regolith, from which any soil must form, is usually not only thin but little decomposed. In uneven terrain, however—and most deserts are by no means flat—wind and even water erosion often carry to lower elevations whatever residual materials result from the mechanical (very little chemical) breakup. The finer, and usually better, soil materials are concentrated in valleys and lowlands. Valley bottoms, *alluvial fans*, alluvial-filled *bolsons*, and *piedmont alluvial plains*, if sufficient irrigation water can be obtained, comprise the most important arid lands for agricultural pursuits.

The dominant desert soils, really thin but also coarse and stony, rocky, or pebbly veneers with tiny pockets of finer materials, are gray in color, but others have casts of yellow, brown, pink, or red. Lack of decomposed vegetable matter precludes the black color which is common to the more humid soils. Most desert soils are also little leached, because of the dearth of water and chemical decomposition, and so are high in available mineral constituents.

Steppe Soils. The best steppe soils are the black *chernozems*,[2] which are rich in decayed grass residue and are only partially leached of their minerals. They are generally fine-textured and granular-structured so as to enhance proper friability upon ploughing, and they are generally classed as one of the world's best

---

[1] The term *pedocal* applies to a grassland soil developed under subhumid or semiarid conditions where a zone of lime carbonate accumulates at the depth of rainwater penetration, seldom more than a few feet. In humid climates where rainwater penetrates to the ground water table, there would be no zone of lime carbonate accumulation because the lime would be carried away in solution. Because of the limited leaching process in steppe climates, pedocals are inherently soils of high fertility. Pedocals vary in color from almost black through dark brown and light brown to almost gray, the color depending upon the amount of humus which in turn is dependent upon the luxuriance of the grass cover. *Pedo* refers to soil, *cal* to calcium (lime).

[2] *Chernozem* is a Russian word meaning "black earth." It is a pedocal, high in humus, and it develops on the more humid margins of the steppes.

soils, if not the best. On drier margins of the steppes, where a less continuous grass cover prevails, brown and chestnut brown soils are common. While ploughed steppes may be subject to denuding winds, and thus to the loss of part or all of their topsoil, in similar fashion they may also receive loess and other minute dust particles from deserts as partial compensation. The most important deterrent to greater use of middle-latitude semiarid soils is erratic and uncertain precipitation in both summer and winter. The soils of the hot steppes are believed to be somewhat less valuable than the soils of the cooler steppes.

Occupations in the Dry Lands

Historically, both deserts and steppes have been scenes of early civilizations based largely on grazing and on irrigation agriculture, where local conditions permitted. The scantiness of both precipitation and natural vegetation, coupled with the inability to gain water for irrigation, dictated low, or sparse, population densities. Only at focal trade centers, located frequently on navigable portions of rivers, on seas, or in oases, did large population concentrations eventually arise. The call of the desert, with its therapeutic effects on both the ill and the vacationer, is of more recent date. Exploitation of the dry lands for minerals is also a twentieth-century innovation.

Deserts

Great distances between oases and between nomadic encampments; dust-laden, often desiccating and violent winds, such as the simoom of the Levant, the brickfielder of Australia, the haboob of Sudan, the harmattan of the western Sahara, and the khamsin of Egypt; parched, thirsty, dusty, sparse vegetation; and above all, great heat by day: these are some of the things the desert wayfarer encounters, whether he is a geologist, agriculturist, nomad, or merchant. So far, man's conquest of the large deserts is negligible; for the most part he must adjust to them if he remains in them. While this is less true of middle-latitude deserts than of low-latitude deserts, coping with deserts still presents many problems not inherent in moist places.

Despite sparse natural vegetation, the deserts of the world are the grazing grounds of many nomadic herds of sheep, goats, cattle, and horses or ponies. In some desert places, such as Patagonia, grasses are sufficient to support sedentary, or partially sedentary, grazing in addition to very limited grain agriculture. Oases, with dates, wheat, barley, sometimes vegetables and fruits, cotton, and other products that are locally important, appear wherever sufficient water for irrigation is obtainable. Most desert oases are small. Others that depend on large sources of water, usually rivers such as the Nile, Colorado, Tigris, Euphrates, Amu Darya, Syr Darya, and Indus, are much larger and contribute substantially to the world's products.

Huge supplies of oil, both in reserve and in production, in southwestern Asia; important copper production in the United States, Peru, and the U.S.S.R.; nitrates in Chile; borax in Death Valley; gum arabic in Sudan; black opals, gold, and lead in Australia; mineral salts in both the Caspian area of the Soviet Union and in the western United States, are a partial list of minerals being produced in notable amounts in deserts. Mineral salts (nitrates and borax), due to excessive evaporation in dry atmospheric conditions, are clearly the result of aridity; some of the other minerals, such as petroleum, were formed under more humid conditions in past geologic ages.

Steppes

More favorable conditions for man exist in steppes than in deserts. Therefore the population density, while still light, is heavier than in deserts; cities are somewhat more common. While the more humid grasslands are usually better foraging grounds than the steppes, their carrying capacity, or carrying power,[1] being much higher, they are not so extensively used for range cattle, sheep, and goats. More semiarid country is devoted to stock raising, either nomadic or sedentary, than to any other gainful pursuit. Undoubtedly animal industries will continue to dominate the occupations of the steppes in the foreseeable future.

From an areal standpoint, crop agriculture, chiefly by dry-farming methods, is second to stock raising. It is practiced principally on either the more moist or cooler margins of the steppes and in oases, such as the Platte River oasis in western Nebraska and the series of oases at the base of the Rockies north of Denver. As previously indicated, farming in a climate that is subject to such great fluctuations in precipitation tends to have many problems, some of which curtail steppe farming completely at times, unless water is available for irrigation.

Mining has become important in the steppes. Oil, coal, copper, silver, and diamonds are among the chief minerals mined. Locally, manufacturing of varied kinds is important. In North America and the U.S.S.R., meat packing and the processing of minerals are engaged in. The most highly industrialized area of the Soviet Union, the Donets Basin, is completely within the steppes. Tourist attraction, however, is not so great as in the deserts.

[1] *Carrying power* and *carrying capacity* are terms used to indicate the number of cattle, sheep, horses, or other animals that a given area can support. The better the grazing land, the higher the carrying capacity.

20

The Humid Mesothermal Climates

Middle-latitude climates, in general, have both a summer and a winter season in contrast to the tropical climates. They are located between the constantly cold and the constantly hot climates. Except for the high mountainous areas (classed as *E*) and the dry climates, the remainder of the middle-latitude climates, both mesothermal and microthermal, are exempt from continuous cold, from continuous heat, and from continuously dry conditions and are classed as humid, although the *Cs* climate, because of the dry summer, is actually subhumid.

Mesothermal Climates

The mesothermal (meaning "middle heat") climates extend as far north as the 70th parallel near North Cape, Norway, and as far south as the Strait of Magellan; they are not confined entirely to the middle latitudes. Regardless of latitude, however, their locations are such that their temperatures are within the limits prescribed as mesothermal. Only the *Cfb* climate has a small enough mean annual range of temperature to be classified as truly *temperate* (meaning "moderate"); the rest of the *C* climates have summers either too warm or too hot to be classed as temperate. Winters are considered mild. It will be recalled from Chapter 17 that all *C* climates have one or more months with a mean temperature below 64°, at least one month above 50°, and the coldest month above 27°.

In the northern hemisphere, the poleward margins of *C* climates are usually bounded by *D* climates (although occasionally by *E* climates, as in Iceland, northern India, and along the southern coast of Alaska); in continental interiors, they are bounded sometimes by *D* climates, as in California, Turkey, Russian Turkestan, and eastern Europe; but more frequently by *B* climates, as on the Great Plains, in Russian Turkestan, and northern China (which is also true of the southern hemisphere, as in Argentina and Australia). Equatorward margins are largely bounded either by *A* climates, as in Brazil and southeast Asia, or by

B climates, as in both North and South Africa and in Southwest Asia. The *C* climates appear in three characteristic locations: (1) on the eastern sides of continents in lower middle latitudes, (2) on the western sides of continents in middle latitudes—narrow coastal strips in North and South America where mountains are close to the sea but far inland in Western Europe where mountains do not greatly interfere with the eastward penetration of marine influences—and (3) in the interior of continents in low latitudes and lower middle latitudes where either altitude or winter monsoons reduce the general temperature level (mostly *Cw* areas, as in Mexico, southern Africa, and Southeast Asia).

Sometimes, the *Cfa* climate is incorrectly considered *humid subtropical;* it extends too far poleward for such a distinction. Neither can the *Cfb* climate be termed *west-coast marine*, since it appears on the eastern sides of continents, either in coastal or upland locations, as in the United States, South America, Australia, and Africa. Also, *Cfa* and *Cfb* merge with, or even parallel, each other in Europe and in the Caucasus Mountain areas. Furthermore, the *Cs* climate, popularly called *Mediterranean climate*, extends on the western side of the United States northward to Canada, although that location is hardly compatible with the association suggested by "Mediterranean." It becomes apparent, then, that the limits of the *C* climates are too broad, both statistically and areally, to be correctly called "subtropical" and that other names should be used for the various *C* climates. The following names, while indeed long and cumbersome and not too satisfactory, still retain the mesothermal characteristics and, for want of better ones, will serve to distinguish the *C* climates:

Cwa: Mesothermal, hot summer, dry winter
Cwb: Mesothermal, warm summer, dry winter
Csa: Mesothermal, hot, dry summer
Csb: Mesothermal, warm, dry summer
Cfa: Mesothermal, hot summer
Cfb: Mesothermal, warm summer
Cfc: Mesothermal, cool summer

The Mesothermal, Hot Summer, Dry Winter (*Cwa*) Climate

Location and Control Factors

The *Cwa* is the warmest of the *C* climates; were it not for a decidedly cool season, it would be classed as tropical. While much of it is located within the tropic circles, it extends almost to the 40th parallel in China and Korea. In several cases, Köppen does not distinguish *Cwa* from its colder counterpart *Cwb*; he simply labels the regions "*Cw*." This is partly because distinguishing data are inadequate in some sections and undoubtedly because local differences in altitude cause some stations to be classed as *Cwa* while others nearby at higher elevations may be *Cwb*. Where a definite distinction is not made on the Köppen-Geiger map, unless data are available to determine whether the area is *Cwa* or *Cwb*, it will be considered as simply *Cw*.

Much of interior China is *Cwa*. Included are the Shantung Peninsula, the Tsinling Mountains that divide northern from southern China, the hills and lowlands of extreme southern China, and most of the highland masses of Burma and the Indochinese peninsula. A considerable part of the middle Ganges Valley and other isolated sections of India belong in the category. In Africa, the chief belt extends through the uplands of Angola into those of Northern Rhodesia, Nyasaland, and Tanganyika Territory, mostly between the 10th and 17th parallels south. Several discrete areas of *Cw*, both *a* and *b*, appear in east Africa, from the Union on the south to the highlands of Ethiopia in the north, and also as the backbone of Madagascar. In South America, a belt of *Cw* flanks the *Aw* on both the poleward and interior sides and extends northward on the eastern margin of the Andes. It includes some areas above 10,000 feet. This latter is a discontinuous and undifferentiated

Fig. 20–1. Heavy summer rain with a moderately dry winter favors intensive rice cultivation in the Cwa lands of southeastern China. (Courtesy Ewing Galloway, New York.)

belt but extends on northward into Central America and Mexico where it includes both mountains and plateaus.

At moderate elevations in low-latitude situations, the *Cw* climate exists for the same reasons that the *Aw* climate does. During the summer, the doldrum belt invades the area, bringing the rains; in the winter, the doldrum belt has moved to the opposite side of the equator and is replaced by the belt of trades or by the subtropical high-pressure belt, bringing clear weather. Naturally, the lower sun in the winter season means lower temperatures. And like the *Aw* climate, the highest temperatures usually occur before the time of the highest sun (May in the northern hemisphere) because the cloudy weather of the rainy season reduces the intensity of insolation. The chief difference between the two types of climate is largely a reduction in temperature and in the amount of precipitation by reason of altitude. Compare Mexico City (*Cwb*) at 7,500 feet with Veracruz (*Aw*) at sea level and only 200 miles away. Each month in Veracruz averages 17 degrees warmer than Mexico City, and the precipitation is three times as great in Veracruz, although the distribution in each city is the same—wet summers and rainless winters. Much the same sort of condition holds true for similar situations in South America and in southern Africa. It may be stated in general that in all these areas, when the altitude is between 3,000 and 5,000 feet (depending upon latitude), the temperature for the warmest month will be above 72° and the climate is classed as *Cwa*; for altitudes higher than 3,000 to 5,000 feet (again depending upon latitude), the warmest month will be under 72°, and the climate is classed as *Cwb*.

The situation is somewhat different in the *Cw* regions of southeast Asia, although altitude (where applicable as in the Yunnan Plateau of southwest China) may still play an important role, and on the poleward limits in China and Korea the seasonal changes in insolation help to account for the larger annual ranges of temperature there as compared with lower latitudes. But in both India and China, and to some extent in low latitudes in eastern Africa and in northern Australia, prevailing winds play the most significant role. In India, it is the southeast wind from the Bay of Bengal, controlled in part by the great Himalayan chain of mountains but more importantly controlled by the intense summer low-pressure area over the Sind in Pakistan. In the lower Ganges Valley, heavy rains result, diminishing with progress up the Ganges. In China, the winds also prevail from the south or southeast in summer, bringing in *mT* air from the South China and East China seas with resultant heavy precipitation near the coast and decreasing amounts inland, unless mountain terrain increases the amounts locally.

In winter, in both India and China, the winds are essentially reversed in direction, coming down the Ganges mostly as northwesterly winds and over China as northerly and northwesterly winds, with consequent cool, dry air in the Ganges Valley and cold, dry air over China. Similar seasonal reversal of prevailing winds occurs in low latitudes over eastern Africa and over northern Australia. At no other place in the world do prevailing winds have such a remarkable effect upon temperature and precipitation as in India and China.

Temperature

The chief temperature characteristics of most of the *Cwa* climate are the hot summers—not merely 72° or slightly above but extremely hot summers, especially if altitude is not a factor. At Benares (Figure 20–2), which may be considered as typical of the Ganges Valley at least, the mean temperature for the warmest month (May) is above 90°; Hong Kong on the coast of China is appreciably cooler, though still hot with a temperature of 82° for July; in the Yangtze Valley of China, it is 2 or 3 degrees warmer than at Hong Kong. The parts of lowland South America have temperatures about like Hong Kong. But Salta in Argentina and the *Cwa* regions of Africa have warmest-month temperatures of 72°, or only slightly higher; altitude, of course, accounts for this moderate summer heat. In fact, as already indicated, were it not for altitude, most of the *Cwa* areas of Africa would be classed as *Aw*.

Everywhere summer is oppressive because of high relative humidity. The slightest exertion during the daytime in the rainy season is an effort. Where rainfall is heavy, 60 to 80 inches or more, there is seldom any relief from the oppression; but in areas where the rainfall is more moderate, as in interior elevated situations, life is more bearable because not every day witnesses heavy showers. Where winds prevail from the sea for several months, as in India and China, conditions are as hard to bear as in any wet equatorial location: perpetually hot and humid with only slight relief at night. Such a climate, however, is not an unmixed calamity for it means reasonably certain crops.

In all *Cwa* areas, the onset of the cool season brings renewed energy and relief to

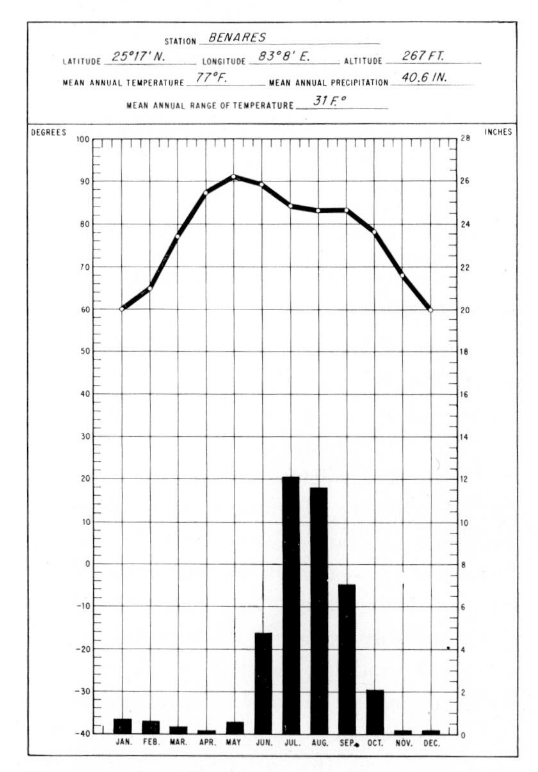

Fig. 20–2. Cwa climate.

the people; the rains slacken, skies clear, and a balmy climate ensues. Winter offers definite compensations to most localities. At Benares, the season from November to February finds temperatures under 70° and as low as 60° in January, and though the days may be warm or hot, the nights are cool—occasionally too cool when a weak cold front enters the Ganges Plain from the west. Most of the time, however, the winds are very light, due to settling of air from aloft rather than to any cold continental air mass; the high mountains to the north are effective barriers to any ingress of cold air from the winter anticyclone of northeast Asia. In China, the story is not quite the same. There are no considerable barriers to the easy sweep of cold, dry *cP* air out of Siberia. The strong winds surge southward and "all China shivers down to the tip of its toes—even to Hong Kong, the world's coldest subtropical city." In North China, the winds frequently bring great clouds of loessal material from the Mongolian steppes, or even from the closer uplands of Shensi Province, sometimes being carried southeastward far out over the Yellow Sea. Since neither the summer monsoon nor the winter monsoon of China is perpetual during either season, each blowing only 50 to 60 per cent of the time, the summer and winter mean temperatures do not reveal all the weather conditions that might be experienced. On the other hand, the lack of strong frontal storms, particularly in winter, means that the weather, from the standpoint of temperature, is steadier than in corresponding latitudes of the United States.

Mean annual ranges of temperature are greatest in northern or central China (40 degrees), more moderate in India (30 degrees) and in South America (20 degrees), and least in Africa (15 degrees).

In many *Cwa* areas where crop cultivation is limited, as in parts of South America and Africa, the winter season may have skies laden with dust or bearing extreme haze from smoke of grassland or savanna fires. Nevertheless, the haze of winter seems preferable to the steam of summer.

Precipitation

The annual totals of precipitation vary considerably in the *Cwa* climate, e.g., from 85 inches at Hong Kong to 28 inches at Salta. As the *w* indicates, winters are dry, and the wettest month receives at least 10 times as much rain as the driest month. The seasonal distribution is most pronounced in India (Figure 20–2) where Aprils are almost rainless but where heavy rain comes in July and August. January shows a slight amount of rain as weak frontal storms (probably from the Mediterranean) move across Afghanistan and onto the Ganges Plain. Hong Kong is almost as extreme as Benares, July (the wettest month) receiving about 14 times as much rain as February (the driest month); and northeast China shows about the same distribution. West central China, however, has a more even distribution owing to the weak winter storms which proceed down the Yangtze Valley; in fact, it is said that cloudy days are more frequent in that area in winter than are clear days.

During the rainy season in both India and China, particularly in the early part from mid-June to mid-July, the sky is continually overcast and some rain falls almost every day, so that clothing and shoes become mildewed. After July, periods of fine weather are interspersed with the rainy periods. Monsoon rainfall, wherever it occurs, is not too dependable. This is not too important in southern China, where rainfall is abundant, because the people are prepared for excessive rains, and it requires a considerable reduction from the normal amount before any hardship is experienced because of the lack of rain. In central and northeast China, the situation is different: the rain comes either too early or too late, or else too much rain comes in too short a time, or it fails altogether, so that some portion of the region is subject to drought or flood; a similar situation prevails in the middle and upper Ganges Valley. Normal rainfall at best is none too heavy, and even a slight diminution during the growing season may cause suffering. The *Cwa* area of Korea seems to have a slightly more dependable rainfall than North China, owing perhaps to its peninsular position.

The region in South America experiences high variability in amounts, in seasonal distribution, and in dependability of precipitation. It is strictly limited in amount in northern Argentina, yet, farther north on the east slopes of the Andes, it is relatively heavy. In the plateau of southeast Brazil, the rainfall is so evenly distributed (the wettest month only six to nine times the driest month) as to exclude much of it, statistically, from the *Cwa* classification. In Africa, most areas experience only moderate precipitation, from 30 to 40 inches; the winters are really dry, some months being rainless. The plateau character of the country means limited available moisture even in summer, but the lower temperature by reason of altitude means that the relatively small annual total of rain is quite effective.

Typhoons, which are an important item in the climatic picture of China, will be mentioned in connection with the *Cfa* climate, since those tropical cyclones affect the coasts to a greater extent than the interiors.

Responses to the Cwa Climate

Natural Vegetation. The types of natural vegetation of the *Cwa* climate are fairly varied. Location with respect to the particular part of the climate and elevation are the principal determinants. Over much of the climate, forest vegetation once prevailed indigenously, but many areas had mixed forest and grasses, while in other areas grasses dominated the scene. In China, the forests were probably mixed in the Shantung Peninsula and middle Yangtze sections, semideciduous in the Yunnan Plateau, and broadleaf evergreen in the southern part. Broadleaf deciduous trees (forest savanna) presumably occupied the major part of the *Cwa* in both India and Africa. While semideciduous broadleaf evergreens and broadleaf deciduous trees occupied some of the eastern section of the *Cw* of South America, much of the central, western, and northwestern extension was grassland savanna with scattered trees and herbaceous plants. Broadleaf evergreens occupied most of Central America and Mexico. The upland of Madagascar was natively a grassland.

Soils. Most of the soils of upland areas of the *Cwa* climate are classed as mountain *pedalfers*.[1]

[1] The term *pedalfer* is applied to a soil which develops in a humid climate, generally under a forest cover. Most of the soluble mineral salts are lost by leaching, and the clay particles are removed mechanically from the surface layer and largely deposited from 1 to 2 feet below the surface. Pedalfers are inherently poor soils, being deficient in humus, lime, and other nutrients, yet they respond to proper treatment, particularly fertilizing and crop rotation, so that they may be highly productive. *Pedo* refers to soil, *al* to aluminum, and *fer* to iron (Latin *ferrum*).

The river valleys and deltas of China and India and the Pilcomayo-Paraná system of South America, on the other hand, are alluvial in nature. The Gran Chaco soils of South America are reddish prairie soils. On the equatorward margins of the *Cwa* in both Africa and South America, some lateritic soils make their appearance, particularly in hilly terrain; the continuous and unscientific utilization of these soils has caused their structure to break down and the process of laterization to be accelerated. Since alluvial soils constitute the more level areas, they are somewhat more intensively used than the upland soils; and since leaching, oxidation, and bacterial action are much more pronounced in warm climates than in cool climates, it follows that soils of the *Cwa* climates, where the summers are hot and wet, are inherently of poor quality. The dense populations of India and China, however, have dictated that the soils be utilized to their limit regardless of their character. Continuous fertilization seems to be the chief answer to their high productivity.

Occupations in the Cwa Climate

Cwa areas include some of the most densely populated regions of the world. Substantial portions of the climate in both China and India are congested, while other areas are moderate to dense. A few of the upland areas are lightly populated, although concentrations of population appear around Rio de Janeiro and São Paulo.

Subsistence agriculture is the dominant occupation. North of the Tsinling barrier in China, wheat, soybeans, barley, and sorghums are the chief crops; south of the barrier and in the central Ganges Valley of India, rice is preeminent. Millets and sorghums are also important in India, usually on the drier margins. In the highlands of southeastern Asia, in the large *Cwa* belt of Africa, and in the Andean portions of South America, hunting, fishing, collecting, forestry, and primitive agriculture are practiced. In South America, cattle are marketed as beef; in India, they are used as draft animals.

Mining assumes some importance along the Andes, in Mexico, in the Katanga of Belgian Congo, in India, and in China. Minor manufacturing is limited to the Ganges Valley and Mexico; very little is carried on elsewhere except as very small-scale enterprises. Trade and transportation are less important than the dense population of India and China would suggest. Few ports exist, however, because the climate extends to the coast in only a few places. Hong Kong is the chief point for exports and imports, but Macao, Canton, and Haiphong have some facilities and handle significant amounts of trade.

The Mesothermal, Warm Summer, Dry Winter (*Cwb*) Climate

Location

The mesothermal, warm summer, dry winter *Cwb* climate is invariably located at either higher elevations than the *Cwa* or on its poleward side. In India, Burma, and China, *Cwb* areas are wedged as narrow strips between the *Cwa* on one side and the Tibetan highlands on the other side. In Africa a few areas are designated as *Cwb*; parts of the Ethiopian highlands (including Addis Ababa) and most of highland Madagascar are *Cwb*. If data were available, undoubtedly the higher *Cw* areas on the Andean flank would be classed as *Cwb*. The city of Mexico, as well as Puebla just east of it, belong to this climate.

Temperature

High altitudes are chiefly responsible for temperatures being lower than in the *Cwa*. Mean annual temperatures are around 60°, while those of the *Cwa* are 10 to 15 degrees or more higher. Highest monthly temperatures (always in May) probably do not exceed 65 or 66° in the northern hemisphere; compare with the January mean of 72° at Pretoria in the Union of South Africa. Either November,

Fig. 20–3. In the Cwb lands of Ethiopia, rough terrain favors grazing. The cattle are largely of a hybrid zebu stock. (Courtesy United Nations.)

December, or January is the northern hemisphere cold month, but it is always July in the southern hemisphere. Some of the coldest-month means are: 53° at Mexico City (Figure 20–4), 59° at Addis Ababa, 52° at Pretoria, and 40° at Darjeeling, India (7,400 feet).

Since the lowest monthly temperatures are not much less than those of the *Cwa* (sometimes they are even higher) and since the greatest monthly averages are not so high, it follows that mean annual ranges of temperature are characteristically less than those of the *Cwa* climate: Addis Ababa 7 degrees, Mexico City 12, Pretoria 20, and Darjeeling 22. Such small ranges of temperature are characteristic of highland climates. With the warmest months only in the middle sixties, few days are ever excessively hot, and with the coldest months usually in the fifties, relatively few nights are uncomfortably cold. *Cwb* areas, therefore, should be classed as temperate, not tropical.

Precipitation

Although precipitation of the *Cwb* climate is distributed seasonally like that of the *Cwa*, it is usually not so great in amount—from 25 to 35 inches approximately. Mexico City has only 23 inches annually (Figure 20–4), Pretoria 26 inches, but Addis Ababa 50 inches and Darjeeling, 123 inches. Each of the two last-named stations is subject to strong monsoonal influence, and altitude is an important additional factor in the heavy rain at Darjeeling. The monsoon effect at the latter station is indicated by the fact that the rainfall exceeds 25 inches each month in June, July, and August, while November, December, and January each receive less than 1 inch. Everywhere in this climate, summer is the season of greatest rain, and winter the season of least; ordinarily the wettest month receives 10 times, or more, the precipitation of the driest month, which is characteristic of the climate. Precipitation controls are the same as those of the *Cwa* climate.

Vegetation, Soils, and Occupations in the Cwb Climate

Both the vegetation and the soils of the *Cwb* areas are similar to those of the *Cwa*. Somewhat more of the *Cwb*, however, are native grasslands. Included are the grass veld of southern Africa, the grasslands of central Madagascar, and other upland sections. The soils of the veld are black or dark gray and are better than most of the *Cwb* soils. Subsistence agriculture, with maize and associated grains—such as kafir and various other sorghums—as the chief products, marks the agricultural economy of the climate in southern Africa, Central America, and Mexico. In parts of South America and Africa, stock raising is important; and in the upper Brahmaputra Valley of Tibet, nomadic herding is practiced. In some of the more remote areas of southeast Asia, Africa, and South America, hunting, fishing, and collecting are the main ways of life.

Mining assumes major importance in the African and Mexican *Cwb*: in Katanga with its great variety of mineral production, in the Johannesburg district with its premier position in world gold production, and in the mountains and plateaus of Mexico with their silver production. Other minerals are produced in lesser amounts within the climate. Johannesburg and Mexico City are important both for a modest variety of manufactures and for a fairly substantial amount of trade.

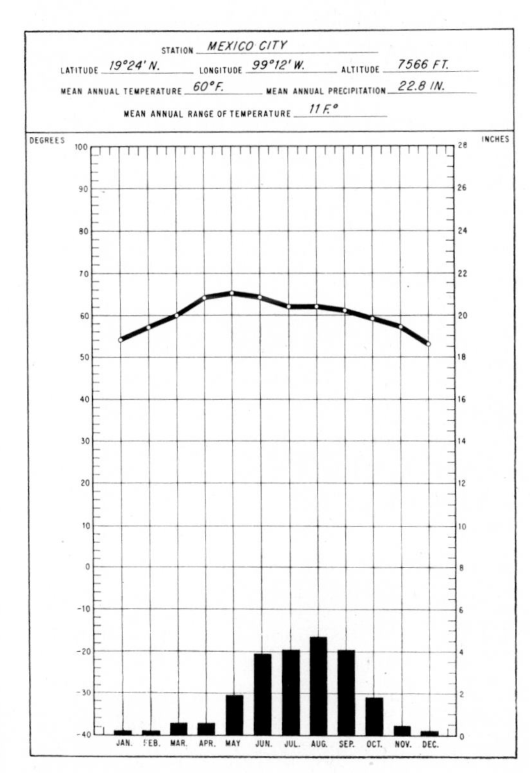

Fig. 20–4. Cwb climate.

The Mesothermal, Hot Dry Summer (*Csa*) Climate

Location

The *Csa* climate is most extensive on the borders of the Mediterranean Sea, hence the name "Mediterranean climate" which has been popularly applied. But, as pointed out earlier, this name is not wholly applicable because the *Cs* climates extend far beyond the shores of the Mediterranean Sea, particularly in western Asia. In North America, they extend into considerably higher latitudes than does the Mediterranean Sea. Furthermore, we generally associate subtropical conditions with "Mediterranean climate"—that is, a climate the coldest month of which will average well above freezing. Yet, the Köppen formula, by its definition of *C*, includes areas having a coldest month as low as 27°. Certainly such a climate is not subtropical.

The *Csa* climate also prevails in most of the Central Valley of California, in the plains of the Columbia and Snake rivers of interior Washington, Oregon, and Idaho, as far north as Bellingham, Washington, and into Vancouver Island; it is found as an isolated area on the eastern edge of the Great Basin at the foot of the Wasatch Mountains in Utah, including Salt Lake City. It makes two appearances in Australia, one in the southwestern part and the other in the vicinity of Adelaide.

It is to be noted that the two *Cs* climates, both of which embrace scarcely 3 per cent of the earth's surface, extend far into the interior of Asia, almost to 80 degrees east longitude. Wherever the two climates are not differentiated, the higher elevations are usually *Csb* and the lower ones are *Csa*. This eastern extension includes the slopes, uplands, and in some places even the high mountains of Turkey, Iran, Afghanistan, West Pakistan, and the Soviet Union. While the latitudinal extent of the Eurasian *Csa* climate is only 1,100 miles, the east-west spread is about 4,800 miles. We shall first consider the climates which have hot summers (*Csa*), later taking up those which have milder summers (*Csb*).

There is considerable diversity of landforms within the *Csa* climatic areas. The great Central Valley of California is singularly flat, except on its edges, where it slopes upward to the encircling mountains, and in a few sections where transverse hills intervene. The Salt Lake oasis is largely flat, but it includes low ranges and the western slope of the Wasatch Mountains. The *Csa* peripheries of the Mediterranean Sea are largely composed of narrow coastlands which are usually backed by a series of parallel or transverse ranges and their associated valleys and, still farther inland, by mountains or plateaus. Some of the latter are contained within the *Csa* climate; some are not. As mentioned before, the *Cs* extensions eastward from the Levant are largely upland and mountainous in character, although valley oases such as Tashkent and Samarkand are also included. The *Csa* areas of Australia are largely lowlands backed by plateaus (southwestern section) and low mountains (Adelaide region).

Climatically, *Csa* areas are located between the dry climates on one hand and either the *Cfb* (in places *Csb*) or the colder microthermal climates on the other. In California the *Csa* area surrounds an area of steppe climate and is separated from other dry lands by the Sierra Nevada. In Asia the *Cs* climates are either flanked on both sides by steppes or are bordered on one side by them and on the other side by microthermal climates at higher altitudes.

Temperature

The *Csa* climate is one of hot summers and mild winters; cold winters are not the rule. Summers, because of a large percentage of clear days with intense insolation, are almost as hot as the bordering steppes; and although 72° for the warmest month is the low limit, most of the region has warm-month means far above that; Algiers has three months above that limit, Athens four, and Sacramento two (Figure 20–6). July or August is invariably the hottest month in the northern hemisphere. Daily maxima frequently reach 95 or 100°, while absolute maxima between 110 and 115° are not unusual. One-third to one-half of the days of July and August are likely to have maxima above 90°.

While prevailingly clear skies lead to the receipt of much heat during the daytime, they also favor rapid terrestrial radiation during the night. This results in a greater diurnal range than would occur on east coasts in the same latitude; diurnal ranges are usually more than 30 degrees in summer.

The coolest month in the northern hemisphere is January. Mean monthly temperatures in winter are usually between 40 and 55°, well above the minimum of 27° allowed by the Köppen formula. Tashkent and Salt Lake City, however, have cool-month means around 30 degrees, which is to be expected for places located on the dry margins of the climate. Daily maxima in winter rise to 55 or 65°,

Fig. 20–5. The olive is probably the most typical plant of the Csa climate and particularly so in southeastern Italy, as here in Matera Province. (Courtesy Standard Oil Company of New Jersey.)

followed by minima not much below 45°, making the daily ranges considerably smaller than in summer. Mean annual ranges are usually between 25 and 35 degrees, although in interiors they may be as high as 50 degrees.

In spite of these mean temperatures above freezing, subfreezing temperatures do occur, although they are of short duration. Occasionally, a *Low* will travel farther south than usual and is followed by a cold front bringing in *mP* air in which the temperatures can well be below freezing; in Europe *cP* blasts from the north and northeast bring intense cold for the latitude. Where fruit and nut crops are produced, the use of smudge pots, or other devices, may be required to prevent or lessen damage by freezing. While some of these cold snaps may be disastrous, most frosts are light, and damage, if any, is small. Inasmuch as the cold air drains onto the valley floors, resulting in an inversion of temperature, the more sensitive crops are usually raised at higher altitudes so as to escape the frosts that might occur. In spite of the fact that the growing season cannot be said to be 12 months long, frost interruptions to a year-round season are generally neither severe nor prolonged; hardier crops survive them, and winter is the most likely growing season, summers being too dry.

Precipitation

Among the chief characteristics of the *Cs* climates are a dry summer and a modest amount of precipitation throughout the winter. The precipitation and to some extent the temperature regimes are the result of the location on the poleward sides of the subtropical high-pressure belts and on the equatorward sides of the prevailing westerlies. It is the seasonal shifting of these belts that is responsible for the summer and winter dif-

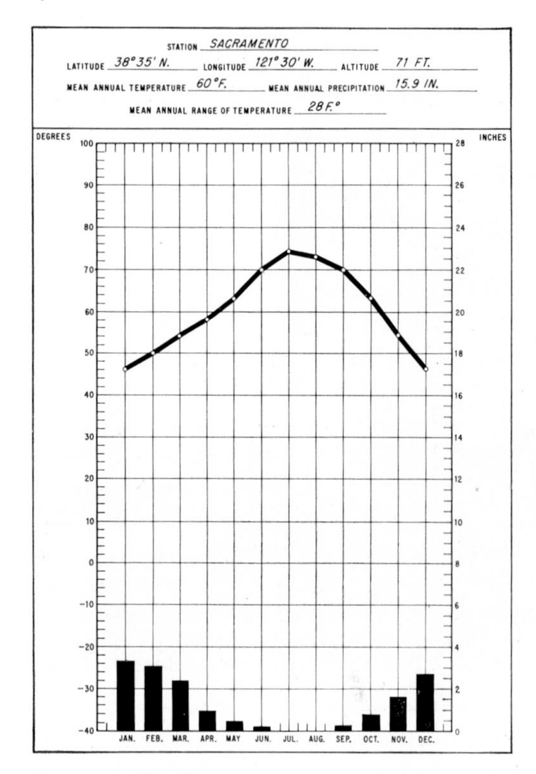

Fig. 20–6. Csa climate.

ferences; it is also the reason that this climate is classed as a transitional type.

In the summer, the subtropical high-pressure belts (horse latitudes) shift poleward into lower middle latitudes as the sun migrates poleward. These high-pressure belts, as masses of subsiding and warming air, are not favorable for the formation of clouds. The skies are clear, the weather is calm, and desert, or near desert, atmospheric conditions prevail; some of the vegetation reflects the fact by drying up, taking on a seared, brown, and dormant appearance. In spite of the summer heat which would favor convection during the daytime, it is only on rare occasions that sufficient moisture is available to produce showers or thunderstorms, except in mountainous terrain. Parts of the Central Valley of California (Figure 20–6) are practically rainless during the midsummer season; the same is true of the interior edges of the climate in Australia and northern Africa.

The summer season is one of bright sunshine and low relative humidity. The driest portions receive 90 per cent or more of the possible sunshine. Coastal sections, however, are subject to fog, a higher relative humidity, and a somewhat lower temperature level. This is particularly true where cool currents parallel the coasts, as along western Portugal and northwestern Africa.

As the winter season approaches, the wind and pressure belts shift equatorward with the sun, bringing *Cs* areas under the domination of the prevailing westerlies and their attendant middle-latitude frontal storms. Some of the storm centers, however, pass to poleward, so that the equatorward locations have less cyclonic influence, and hence less precipitation, than is true of places located farther poleward. That is the reason for the general increase of precipitation with increasing latitude, especially under the same conditions of exposure and elevation.

Higher relative humidity and a correspondingly lower amount of sunshine mark the winter season as compared with the summer. On the average, about 50 per cent of the possible sunshine is received during the winter, although the figure is higher on the dry edges and in interior situations. Nevertheless, the amount of sunshine is relatively high and is sufficient to attract large numbers of winter vacationers from the more humid areas. In addition to the sunshine, the mild winters place both of the *Cs* types among the world's foremost resorts, especially since competing areas are as a whole less accessible or frequently less developed. Although tourists come to the region in summer, it is the mild and rather sunny winter that most attracts them.

The frontal, or cyclonic, precipitation in the *Cs* climates is usually of shorter duration and of smaller magnitude than in the more humid climates to the north and east (*Df* and *Cf* types). More of it comes in the form of short showers and with less prolonged cloudiness than in the latter places. Occasionally, however, storms travel farther equatorward, bringing a two- or three-day period of overcast, cold, and drizzly weather. Orographic precipitation accounts for the greater amounts of rain or snow at higher elevations. In the lowlands, snow is a matter of some concern to the people, but it rarely lasts a day; when attended by large invasions of cold air, the snow, coupled with the cold, may be disastrous to crops. On the uplands, snow may fall frequently; on mountains, it is often heavy and may remain throughout the winter. Melting snows in the mountains are important sources of water for *Cs* regions, both for city water supplies and for irrigation.

Mean annual precipitation varies considerably, but the average is between 15 and 30 inches. On the dry edges, it may be less; on the poleward edges and in orographic locations, it may be much greater. The months of minimum precipitation in the northern hemisphere are usually July, August, and September, although June is sometimes as dry. At Sacramento, July and August are rainless, though most stations record some precipitation (0.1 to 0.3 inch or more) in those months. Wettest months in the northern hemisphere may be any month from November through March. On the dry side of the climate, there is sometimes a delay in the receipt of maximum precipitation: Tashkent receives its maximum, 2.6 inches, in March and April, and Salt Lake City receives its maximum, 2.0 inches, in March.

Fortunately, from an agricultural and vegetational standpoint, the precipitation of the *Cs* climates is concentrated in the winter season when evaporation is at a minimum; were it in summer, the modest amount received would be insufficient for much crop agriculture, since semiarid and even arid conditions would prevail at that season. Winter precipitation, on the other hand, enables substantial crops to be raised, while summer conditions, except where irrigation is practiced, are truly dry. Because of high summer temperatures, low summer rainfall, and only moderate winter rainfall, many climatologists consider the *Cs* climates to be *subhumid* rather than humid.

Responses to the Csa Climate

Natural Vegetation. The indigenous vegetation of the *Csa* climate varies considerably from place to place. Elevation, exposure, slope, soils, and precipitation are the chief factors. As a whole, the wetter, higher, and cooler elevations were generally clothed in open forests; the slopes, by thinner woodlands or scrub (such as chaparral and maquis); and the valleys and lowlands, by open park woodland, chaparral, grasses, or xerophytic associations. Much of the original vegetation has been altered, however, and some exotic vegetation has been introduced. The tallest form of vegetation in most of the climate is more like an open woodland of somewhat stunted trees than like a true forest. Many of the trees are flat-topped, have thick, corky, and uneven barks, and are gnarled in shape; most types have drought-resisting characteristics, such as small, leathery leaves, sometimes degenerated into thorns. Between the trees, shrubs and bushes of numerous types, heights, and colors occupy much of the open ground. Bunch grasses appear in some places, but continuous sods are rare. Evidence of a dry season and adaptability to it is shown in the development of thick and glossy leaves. Other means of retaining water, or of resisting being eaten by animals, have been developed by trees, bushes, and shrubs in the form of spines and spicules and acrid sap. The cork oak, live oak, and olive are good examples of this sclerophyllous woodland.

The *chaparral* of California (more common in the *Csb* than in the *Csa*) and the *maquis* of the Mediterranean borderlands consist of shrubs, scrubby trees, and bushes. Sometimes they are so closely associated as to form a dense growth; elsewhere they are more scattered, with bunch grass, bare ground, and sometimes annuals intervening. During the moist season and after occasional precipitation, the annuals burst forth into bloom and the whole maquis-covered hillside is enlivened, temporarily replacing the usual drab, dusty, and parched appearance.

Although grasslands of limited extent have existed in *Csa* areas, particularly on the steppe and mountain borders and in valleys, only the better ones have been utilized agriculturally or pastorally. The hot dry season and the shallow soils are inimical to good grass growth and hence to good crops. On the dry borders, xerophytic varieties of plants appear; they also take over in some sandy and porous limestone or rocky areas, even though well within the climatic region. It is recognized that most of the *Csa* lands resemble deserts and steppes in the dry season, in both the color of the vegetation and its cessation of growth.

Soils. The topography makes a complicated picture of soils in this climate. Since much of the area is in slope, little opportunity is afforded for residual soils to become mature. The immature soils which result under such conditions are likely to be shallow, stony, porous, and infertile. The limited precipitation is not a very effective agent in vegetational growth. Most slope soils are classed as either gray-brown mountain pedalfers, or as red and yellow subtropical soils.

The best soils are likewise immature, but they are alluvial in nature. The San Joaquin Valley of California and the many small valleys of the Mediterranean area, as well as those of the southern hemisphere, are usually formed by alluvium from the surrounding hills, plateaus, or mountains. Deltas, piedmont alluvial fans, and flood plains are, therefore, the best agricultural areas and are often intensively cultivated.

Occupations in the Csa Climate

While the Mediterranean Sea borderlands have long been known as a cradle of western culture, most of the other *Csa* areas have been occupied more recently. By far the most important occupation is farming; the hot summers with moist winters favor a distinctive combination of farming and pastoral activities. Winter wheat and barley are especially well adapted to the climate; harvesting is accomplished in the early part of the dry season. The hot, dry summer is excellent for the curing of fruit. Grapes, figs, olives, peaches, plums, almonds, walnuts, and citrus fruits are raised. Many localities are devoted to the intensive cultivation of one crop. Irrigation is commonly practiced in lowlands, and terracing conserves land in parts of the Mediterranean area.

Sheep and goats graze and browse in the poor grass and scrublands of the Mediterranean hills. Donkeys, or burros, are the usual beasts of burden. Of all the animals, only the sheep is numerically important in the newer *Csa* areas, and there horses replace donkeys and mules.

Commerce has long been a major activity on the Mediterranean Sea; it is assuming increasing importance in other *Csa* regions. Fishing, also, has long engaged substantial numbers of people, and the canning of fish is an important industry in Iberia, Italy, and Greece. Not only the Riviera of Italy and France but almost all the other *Csa* regions are attracting increasing numbers of vacationists, and the tourist trade is eagerly sought. Neither manufacturing, mining, nor forestry are major activities, yet they are all practiced to some extent. Manufacturing is locally important in the major cities of Spain, southern France, Italy, Australia, and the San Joaquin and lower Sacramento valleys. Mining includes some coal and gold in Australia; silver, lead, copper, and mercury in Spain; sulphur, mercury, and iron in Italy; iron and phosphates in the Atlas Mountain countries; and some oil in the San Joaquin Valley. The only forest product of significance is cork from Iberia.

The Mesothermal, Warm Dry Summer (*Csb*) Climate

Location

Almost the entire west coast of the United States is classed as *Csb*; excluded are the coasts of California south of Los Angeles to the Mexican border, and northwestern Washington. The northern boundary is approximately 47 degrees north latitude, although it does include the Canadian city of Victoria on Vancouver Island. In Europe, only the northwestern part of Iberia plus a narrow strip on the northern side of the *Csa* in Spain and southern France are classed as *Csb*. Small coastal areas in the southern hemisphere are: Chile between latitudes 33 and 39 degrees, the Capetown area of South Africa, areas west and south of Adelaide in both South Australia and Victoria, and the extreme southwestern portion of Australia. Also included in the *Csb* climate are portions of the undifferentiated *Cs* region that extend eastward from Syria far into Asia, as previously indicated in considering the *Csa* climate. Only in the latter case and in small areas of Spain and in the lower portions of the Snake and Columbia River valleys of Washington and Oregon does the *Csb* climate extend more than 50 or 60 miles into the interior of continents.

Temperature

The *Csb* climate is one of moderately warm summers and mild winters; but summers are not so hot, while winters are somewhat cooler than in the *Csa* climate.

Warmest-month temperatures are characteristically between 60 and 70° (Figure 20–8). Stations on the equatorward border, such as Marseilles (72°) and Los Angeles (71°), are a little warmer, while some that have notable marine influence, such as Eureka (57°), are appreciably cooler. Only on the warmer margins of the region do stations have over four months with mean temperatures above 60°. The number of days that reach 90 or 100° is considerably less than in the *Csa* climate, although such temperatures occasionally occur, especially near either the drier or warmer margins of the climate. Diurnal ranges of 25 to 35 degrees result from the generally clear atmospheric conditions that obtain in the summertime. Absolute maxima are not so high as in the *Csa* climate.

January prevails as the coolest month in the northern hemisphere, July in the southern. As was true in the *Csa* climate, likewise in the *Csb* climate, the coldest-month means are usually between 40 and 55°. Since most of the *Csb* locations are near the coast, few of them have mean winter temperatures as low as 32°. In all probability, however, some of the *Csb* stations in interior Asia have winter monthly temperatures that approach the minimum prescribed for the *C* climates (27°). Daily winter maxima of 50 or 60° followed by minima of 37 to 45° are common to the climate. Winter diurnal ranges are therefore between 15 and 20 or even 25 degrees, which is somewhat less than in the summer.

Since most areas are near coasts, mean annual ranges are usually small, somewhere between 10 and 30 degrees. The mean annual range at Eureka is only 10 degrees, but at Santiago, Chile, it is 31 degrees. Compared with the *Csa* climate, with its mean annual ranges of 25 to 35 degrees, the *Csb* climate is therefore more temperate. It is the higher summer temperatures of the *Csa* climate, rather than the somewhat lower winter temperatures of the *Csb* climate, that are responsible for the greater range of the former.

Precipitation

Both the amount and distribution of precipitation are similar to those of the *Csa* climate. On the North American Pacific Coast the amount increases from 15 inches annually in Los Angeles to 33 in Seattle. The amounts vary, however, with exposure and elevation. There is a greater amount in parts of northern

Fig. 20–7. The coastal stretches of the Csb climate in the northern half of California are dominantly forested with redwoods. This scene is about 5 miles from Crescent City. Photograph by W. I. Hutchinson. (Courtesy U.S. Forest Service.)

California (Eureka 37 inches) and Oregon (Portland 44 inches) than in Victoria, while Santa Barbara has slightly more than Monterey (18 inches compared with 17 inches). Apparently the total precipitation increases northward to northern California and Oregon, then decreases somewhat again.

July and August are invariably the driest months in the northern hemisphere, January and February in the southern. Some stations record only traces of rainfall in the dry months. This is true in Santa Barbara, Monterey (Figure 20–8), Valparaíso, and Santiago. Los Angeles and San Francisco usually have no precipitation in July and August. Other stations have from 0.1 to 0.9 of an inch in the driest month. Statistically, they could have as much as 1.5 inches and still be included in the *Csb* climate, provided the driest month was less than one-third the amount of the wettest month.

The wettest months are characteristically in the winter, with December or January usually receiving the greatest amount. The maximum varies from slightly over 3 inches in the drier locations, such as Santiago, Los Angeles, Monterey, and Corunna to over 5 inches in Eureka, Victoria, and Seattle.

Csb precipitation is caused by the same conditions that are responsible for *Csa* precipitation. Since the *Csb* areas are chiefly located on the poleward side of the *Csa*, they are controlled by the subtropical high for a shorter length of time and by the westerly winds for a longer period than in the case of the *Csa*. The summer temperatures of the *Csb* are therefore lower, the effectiveness of the precipitation is greater, and the vegetational cover is better. More cloudiness probably occurs in the northern half of the *Csb* of the United States and Canada than in any other *Csb* area. While the relative humidity, cloudiness, and precipitation are notably higher in the winter than in the other seasons, sunshine and clear days are nevertheless frequent. Again, the northern portion of the United States is considerably cloudier and damper than the southern half of California. Because of the greater amount of sunshine and higher temperatures, the southern part attracts more winter vacationists than the northern part.

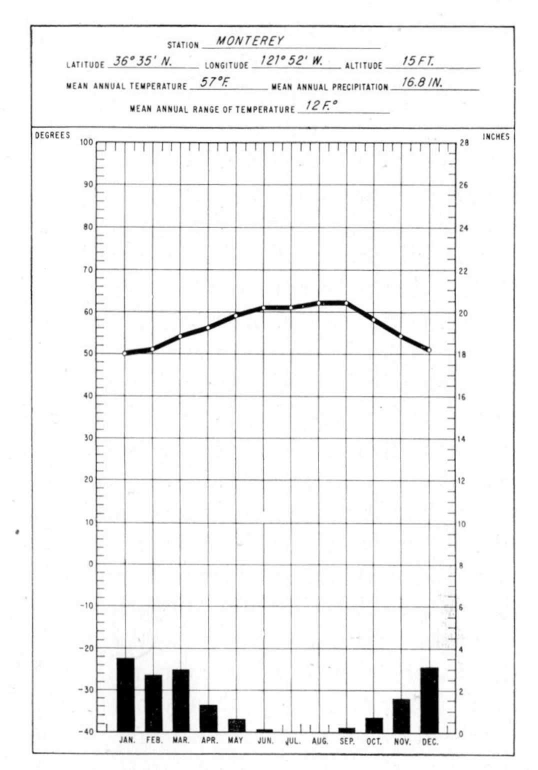

Fig. 20–8. Csb climate.

Fogs are usually associated with coastal *Csb* climate. The causes and characteristics of those fogs were discussed in Chapter 6. Wherever cold currents move close to the coast, and where there is still colder upwelling water along the shore, fogs are frequent; they may occur daily over long periods, notably along the entire Pacific Coast of the United States, especially in the Golden Gate area. It is not to be inferred, however, that fogs occur every day; dense fogs probably do not average much above 40 per year, although light fogs occur more often. Even so, northwest Washington is the foggiest and cloudiest part of the United States; hardly one-fourth of the days may be classed as clear. Fogs are most frequent in the dry season, and although they rarely yield precipitation, except in instances when the fog may condense on trees and drop to the ground as water, it is apparently because of fogs that the famous giant redwoods grow along the coast of northern California. The fogs tend to form near sundown but are dissipated by the sun during the forenoon of the next day. The corresponding coasts of Africa, both north and south, and also the coast of Chile are subject to the same type of fog. In latitudes lower than 40 degrees, bright sunshine prevails as soon as the fogs are dissipated. Farther poleward, even though the fogs may lift and become high fogs, they are likely to thicken into stratus clouds of some depth, perhaps obscuring the sun for several days at a time and yielding some light rain. The fact that the poleward sections have appreciable rain in summer (an inch or more in July along the Washington and Oregon coasts) is more likely attributable to the weak cyclonic disturbances that develop over the north Pacific rather than to fogs, though fogs may be associated with the warm fronts of the disturbances.

Snow on the hills and mountain slopes is not uncommon in winter. In the Columbia and Snake River valleys and in the sinuous Asiatic extension, it is also expected during the cool season. When, on occasion, it is coupled with prolonged continental winds and severe temperatures, it may cause damage to crops. Strong sea winds (*pP* air) at times do considerable structural damage to coastal installations along the Pacific Coast as well as damage to shipping.

Responses to the Csb Climate

Natural Vegetation. As a whole, the *Csb* vegetational cover is superior to that of the *Csa.* Since hills, valleys, basins, and even low mountains, rather than extensive plains, make up most of the terrain, its types of vegetation are quite varied. In the drier sections either chaparral, grasses, or semiarid flora exist; in the wetter and cooler parts, forests of considerable importance are indigenous. At the southwestern tip of Australia where precipitation is more than 40 inches, some of that continent's finest eucalyptus forests, although not dense, are an important source of cabinet wood for that lumber-hungry nation. The other *Csb* areas of the southern hemisphere have less forest cover. In the Pacific Northwest of the United States are found some of the best stands of timber known. Included are the redwood groves of northern California and the pine and Douglas fir forests of Oregon and Washington. Many of the hilly portions of southern and central California have open forest interspersed with a type of oat grass. In the wet season this grass is green, but in summer it takes on a golden cast which contrasts markedly with the green of the evergreen and deciduous trees. Much of this grass has presumably been sown on hillsides to prevent erosion. The numerous exotic varieties of eucalyptus and pepper trees that line the citrus groves and highways, in addition to being planted in groves and yards, make up a substantial part of the trees of California south of the Bay area.

Soils. Most of the *Csb* soils are classed as mountain forest soils, sometimes brown forest soils, but often either lateritic or podsolic.[1] Prairie and alluvial soils, however, appear in some places. Southwestern Australia has red and yellow podsolic soils, while the Adelaide and Capetown regions are composed largely

[1] *Podsol,* like *chernozem,* is a Russian word; it is applied to a soil developed most typically under a subarctic forest cover. High acidity, resulting in part from the slow disintegration of the forest debris, leaches, or dissolves, most of the mineral substances, including iron and aluminum compounds. What is left near the surface is largely silica or sand; hence the ashy gray color and the poor nutritive quality of the topsoil. The aluminum and iron compounds are carried to a lower horizon and there precipitated, giving the characteristic coffee-brown appearance to the horizon under the topsoil. Podsolization is in many respects almost the opposite of laterization. In a laterite, the iron and aluminum compounds remain largely on the surface as insoluble hydroxides, while all other materials, including silica, are carried away. The term *podsol* is often applied to any soil which has developed under a humid forest cover in middle or high latitudes.

of prairie soils of reddish color that have developed under plains and hills topography. The soils of South America are mixed podsolic (usually at higher elevations and on slopes) and mountain chernozem and brown soils (chiefly at lower elevations). Podsolic soils prevail in northwestern Spain. The greater latitudinal extent of the climate in North America, however, leads to a greater variety of soils. Southern California's soils are chiefly mountain soils of brown forest, terra rosa, and rendzina types[1] with inclusions of podsolic and of alpine meadow soils. North of San Francisco, soils of lateritic nature mixed with podsolic and alpine meadow types predominate. Gray-brown pedalfers appear in the Puget Sound area, and limited alluvial soils exist where rivers have deposited fans and in parts of the Willamette Valley of Oregon.

Occupations in the Csb Climate

Inclusion of cooler coastal water areas and extension into higher latitudes impose a greater range of agricultural activities in the *Csb* climate than in its warmer counterpart, the *Csa*. Winter wheat and oats are fairly important grains in most areas, and maize has quite wide distribution in northwestern Iberia. Some barley is produced, but it plays a minor role except in the drier portions, or in areas where other crops will not do better. Sugar beets are significant in the valleys south of San Francisco, and some flax is produced in the same general area. Both citrus and deciduous fruits (apples, plums, peaches, pears, apricots, and grapes) appear in most all parts of the climate, except in northwestern Iberia where citrus fruits are lacking. Invariably the major citrus areas are on the warmer margins of the climate, while the deciduous production is distinctly farther poleward. In contrast to the *Csa* climate, neither the fig nor the olive is very important in the *Csb* climate.

Sheep and cattle vie with each other for importance among the animals. The cooler summer promotes better grazing, and thus both beef and dairy cattle have a prominent place. In California, beef herds are growing in number, while dairying, including cheese making, is expanding in the Willamette-Puget Sound section. The best cattle of Chile and Iberia are in this climate.

Many major cities lie within the *Csb* region. Manufacturing is gaining in most of them; Los Angeles and vicinity hold an especially preeminent position because of the volume and variety of its manufactures. The coastal position of the climate means many important ports, most of them with good harbors and all of them with abundant port facilities, and a large volume of trade, exemplified by such cities as Los Angeles, San Francisco, Portland, Seattle, Valparaíso, Capetown, and Adelaide.

As previously indicated, lumbering is practiced on a vast scale in the Pacific Northwest. It also attains some importance in central Chile and southwestern Australia. As a whole, mineral production is not particularly important, although oil and gas are big-scale operations in the Los Angeles area and some coal and gold are produced in other areas. The taking and canning of fish constitute an important activity in the moderately cool waters of North and South America and southern Africa.

The Mesothermal, Hot Summer (*Cfa*) Climate

Location

The *Cfa*, or mesothermal, hot summer climate is located typically on the east sides of continents in lower middle latitudes—actually from close to the tropic circle to approximately the 40th parallel. The most extensive region is in the United States where it extends westward to the Great Plains steppe climate (about

[1] Rendzinas develop best on soft limestone formations in moderately humid climates and under a tall prairie-grass cover. They are rich in humus and are, therefore, dark in color. The soil of the Black Waxy Prairie of Texas is a typical rendzina.

the 102nd meridian), northward as far as central Illinois, almost to Lake Erie and into southern New England. About one-third of the United States is in this classification. In South America it extends from Bahia Blanca (Argentina) to the vicinity of São Paulo (Brazil). Much of the Argentine *Pampa*, all of Uruguay, a considerable part of the lower Paraná drainage basin, and southeastern Brazil are included.

The most extensive area in the eastern hemisphere is in China, Japan, and Korea. The Tsinling Mountains which separate north China from south China constitute the northern limit of the mainland *Cfa* except for southwestern Korea. The southern Japanese islands of Kyushu and Shikoku and all but the highest and most northern portions of Honshu are also significant areas. The highest mountain zones of the larger Indonesian islands and of New Guinea are tempered enough to fall in the classification. Some of eastern Australia, including the eastern highlands between the tropic of Capricorn and Melbourne, and a small coastal fringe in southern Africa, from southern Mozambique past Durban, are *Cfa*. The remaining areas are found in Europe in the Po Valley, the lower Danube, and along part of the coast of the Black Sea.

It should be noted that *Cwa* and *Cfa* climates are similar as far as the Köppen criteria are concerned; the only difference according to the formula lies in the more even distribution of precipitation in the latter. There is actually, however, one other important difference: the lowland *Cwa* areas, notably in the Ganges Valley, have higher mean temperatures in summer. Furthermore, the coastal *Cfa* situations mean semimarine conditions, reflected in somewhat lower summer temperatures and, of course, in adequate cool-season precipitation. The chief exception to the usual littoral location is found in the United States and, to a small extent, in central Europe where the climate extends for some distance into the interior. In spite of this continental position, the winters are not dry because the rapid succession of frontal storms in the cold season yields sufficient precipitation to make the climate humid.

An important controlling factor in the *Cfa* climate of the western hemisphere is the fact that the summer subtropical belts of high pressure, most intense over the oceans, rule air movement over the lands, just as they do in the *Cs* climates on west coasts. There is a marked difference, however, between east coasts and west coasts in that the high pressure overspreads a portion of the land on east coasts, while on west coasts there is no extension of it over the lands. The flow of air in the latter is distinctly equatorward with little chance for rain unless the air should be lifted orographically; stability is a feature of such air masses because the air is descending. In the southeastern United States, the prevailing air movement is from the southeast in summer; the strong heating of the ground over which the *mT* air is moving means strong convection and, therefore, a high degree of instability, resulting in intense local thunderstorm activity. This same condition is essentially true for South America and Australia. In Asia, there is no summer subtropical high-pressure belt north of the equator; instead, it is located south of the equator. This condition favors enormous volumes of *mT* air moving northward across warm equatorial waters, picking up moisture on the way, and carrying it to the shores of China and Japan. The result here is similar to that in the United States but more intensified because of the large volume of humid air as well as the added orographic effect; consequently, rain is very heavy along these Asiatic coasts in summer.

Temperature

In spite of its dominantly coastal position, *Cfa* climate has definitely hot summers. July is most often the hottest month, although August ties or exceeds it in some locations, e.g., Charleston (Figure 20–10), New Orleans, Savannah, Dallas, Galveston, and Baltimore. Warmest-month temperatures can be as low

Fig. 20–9. On the equatorward margins of the Cfa climate in the United States, "King Cotton" was once supreme. (Courtesy National Cotton Council.)

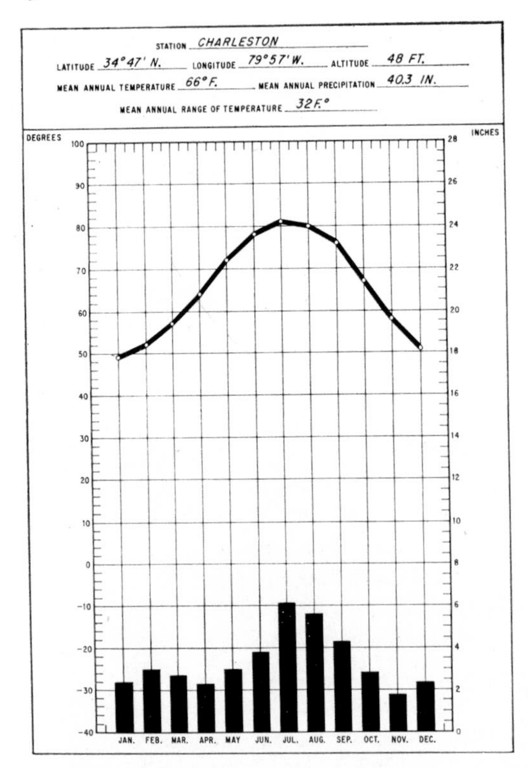

Fig. 20–10. Cfa climate.

as 72°, but they are usually much higher: 76° as far north as Columbus and up into the middle 80s along the Gulf Coast and in parts of China and Formosa. The insular position of Japan means temperatures below 80°, and most stations in the southern hemisphere are still milder (74° at Buenos Aires). Interior locations, logically, have higher summer temperatures than coastal locations in the same latitude. Prevailingly high absolute humidity in summer—the result of the dominance of *mT* air masses—retards nighttime radiation. As a consequence, diurnal ranges of temperature are less than would be found in drier climates with similar temperatures; ranges from 15 to 20 degrees are to be expected except on the dry margins. Maximum daily temperatures are frequently between 90 and 100° every year, and absolute maxima have been observed from as low as 101° at coastal Galveston to more than 115° in Kansas and Missouri.

As a whole, winter temperatures are considered mild, although the latitudinal extent is sufficiently great to make a clear distinction between the poleward and equatorward limits. Mean January temperatures will range from the allowable mean of 27° on the poleward edge to the middle 50s equatorward: 40° at Nashville, 50° at Charleston, and 56° at New Orleans. Although the China region lies closer to the equator than that of the United States, the normal cold blasts of *cP* air from Siberia are likely to cause lower mean winter temperatures there than in the United States. Shanghai with 38°, similarly located to Savannah, averages 14 degrees lower in January (but approximately the same in July). Temperatures in the southern hemisphere in winter are much milder than corresponding latitudes in the northern hemisphere owing to the absence of large land masses to poleward. Thus absolute minimum temperatures are far lower in the northern hemisphere: the lowest recorded temperatures at Brisbane and Buenos Aires are 32 and 23°, while those at Shanghai, Vicksburg, and Springfield (Missouri) are 10, −1, and −29°. As indicated above, it is the invasion of *cP*, or even *cA*, air masses on the rear of winter cyclones that is responsible for the occasional extremely low temperatures in the United States. Such low temperatures usually spell disaster, at least locally, when they invade deeply into the low-latitude sections.

The suddenness with which temperatures plunge after the arrival of the cold front is remarkable. One of the greatest overnight changes in temperature occurred at Oklahoma City, November 11–12, 1911. At 1:00 P.M., the temperature was 83°; at 3:30 P.M., it had plunged to 32°; and by midnight, it was down to 17°. At 3:00 A.M., November 12, it had dropped to 14°. The temperatures of 83 and 17° will probably stand for a long time as the highest and lowest of record there for November 11.

While eastern China has lower mean winter temperatures than our Gulf and Atlantic states, the cold is steadier, the minima are not so low, and damage to crops is not so common. What was stated concerning temperature conditions in the *Cwa* of China is essentially true of the *Cfa* of that country. Cold spells are annual affairs in the southern United States, though the severity varies considerably from year to year.

Mean annual ranges in the United States vary from 25 degrees in the south to 45 degrees on the western and northern limits (48 degrees at St. Louis). It is the cold of winter that creates the greater extremes in the latter locations. For the same reason, the annual ranges in Europe and Asia are around 40 degrees, but in the southern hemisphere they do not exceed 25 degrees.

Precipitation

Although *Cfa* areas in many instances extend in the interior to the steppes, rather high precipitation well-distributed throughout the year is characteristic of the climate. Amounts vary from 30 to 55 or 60 inches annually, part of it falling in the form of snow, especially in poleward, interior, and high altitude locations.[1] As a whole, precipitation increases toward the equator and eastward where *mT* air masses are most frequent: St. Louis 37 inches, Vicksburg 49, and New Orleans 60. Even near the western and drier margins, this increase of precipitation southward is evident: Salina (Kansas) 27 inches, Oklahoma City 33, and Houston 45.

As in the *Cwa* climate, rainfall in *Cfa* areas in both the United States and China is likely to be erratic. Long-enduring droughts, especially on the drier margins, or excessive rains in short periods—though not necessarily characteristic—are not unusual. One of the greatest single falls of rain was near Taylor, Texas, during the night of September 9–10, 1921, when a total of 30 inches was recorded in 15 hours. At Beaumont, Texas, in May, 1923, it was reported that 14 inches of rain fell in 2½ hours. Fortunately, such heavy downpours are rare.

There are considerable differences in the periods of maximum and minimum precipitation within the region. The maximum may occur in any month from March to October: March in Montgomery, April in Little Rock, May in Dallas, June in Kagoshima, July in Charleston (Figure 20–10), August in Pensacola, September in Titusville (Florida), and October in Bermuda. There is not so much variation in the month of least precipitation. February is frequently the driest month (partly because February is the shortest month), but the minimum may also occur in January, November, September, or even August.

While winter is ordinarily the season of least precipitation, there is a great difference in the comparative amounts received in the driest and in the wettest months. In Bermuda, it is very evenly distributed, also along the Gulf Coast of the United States. Toward the poleward and drier margins, a much larger percentage falls in the summer—at Salina, eight times as much in the wettest month as in the driest month. In general, coastal locations have a more even distribution; interiors have a distinct summer maximum.

As already indicated, the warm-season precipitation is largely the result of convectional showers, attended by lightning and thunder. Such showers are most frequent in the United States along the east Gulf Coast (Figure 6–30). Occasional frontal storms add some summer precipitation, particularly in early summer. At those times, the relative humidity becomes high and may remain high for several days, bringing much cloudiness; and when attended by high temperatures, the air is muggy and oppressive. At times, the deep south and much of the area in China may have rather prolonged periods of such enervating weather, especially with persistent southeasterly winds of the *mT* type. It is

[1] The amount of precipitation is higher in Japan where Niigata receives 71 inches, Nagasaki 77 inches, and Kagoshima 85 inches; mountains and insularity largely account for this heavier precipitation.

during such periods, notably in China and Japan, that excessive rains come with resultant floods. Despite the fact that summer is a time of abundant rainfall in our southern states, it is a season of considerable sunshine—much more so than in winter. Under drier conditions, sensible temperatures are appreciably lower, even though the actual dry-bulb temperatures may be high.

Along the Atlantic and Gulf coasts, the annual precipitation is augmented markedly by hurricanes; the Asiatic counterpart is the typhoon. The amount of precipitation caused by these tropical cyclones varies a good deal from year to year, and is often a factor in causing a late summer and early fall maximum. An average of about four typhoons strikes the China coast annually. If they move inland, they soon lose their impact, chiefly because of the mountainous character of the land, but they may swerve toward the northeast and pass over the islands of Japan. Tidal waves, as explained in Chapter 10, are likely to occur, such as the great tidal wave in connection with the Swatow typhoon of August, 1922. The great concentrations of population in the affected areas of Asia mean a great loss of human lives. Typhoons begin earlier (in late spring), last longer (into November or December), and are more numerous than the hurricanes which plague the United States and the West Indies. The heavy rains that occur in Japan are often associated with typhoons, and where there is a mountain backdrop, the rains may be excessive—as much as 30 inches in a 24-hour period. Many hurricanes have struck the United States and have wrought tremendous destruction both by wind and water, one of the most destructive in recent years being that which smote Louisiana in the early summer of 1957. Probably the most memorable from the standpoint of its human toll was the Galveston storm of September, 1900, and the most notable from the standpoint of low pressure was the Florida Keys hurricane of September, 1935.

Another feature of *Cfa* climate in the United States, and not associated to any appreciable degree with that type in other parts of the world, is the tornado, already discussed in Chapter 10. Tornadoes are most likely to occur in the southern part of the region in early spring and in the northern part in late spring or early summer. Through May of 1957, 624 tornadoes had struck some part of the United States, mostly in the southern states. In May of that year, a particularly large number visited Texas, bringing heavy rains and destructive floods to places which had had deficient rainfall for a decade or longer. Though Texas may have needed rain, it seems unfortunate that it could not have been distributed more effectively, but that *is* weather.

Winter precipitation is largely cyclonic, though occasionally local thunderstorms may occur on the equatorward borders. While snow is not to be expected in lower latitudes, it may fall at any time throughout the winter in the higher latitude positions, remaining on the ground for several days before being erased by warm southerly air. Where altitude is a factor, the snow may be heavy and persistent. High humidity, considerable cloudiness, gray days, and chilly to cold conditions commonly attend the warm front of a *Low*, followed by clearing and colder weather as the cold front comes in.

Responses to the *Cfa* Climate

While year-round heat governs the vegetation and soils of tropical climates, and while dryness largely determines their adjustments to dry climates, it is chiefly the appearance of colder weather in addition to the hot humid summers that is responsible for the differences in their reactions to the *Cfa* climates.

Natural Vegetation. Full-fledged tree growth is the dominant response to the combination of temperature and precipitation in the mesothermal, hot summer climate. It is only on the drier margins and in certain soil associations (subsurface limey deposits) where grasses are indigenous. In the United States, a great variety of trees flourishes. Broadleaf hardwoods, mainly deciduous, are most common throughout the northern and central sections;

they are common in the less acid soils of the deep south too. Throughout much of the climate, conifers originally obtained footholds; they tend to predominate in lighter soils of the coastal plains, and they are quite common on the well-drained summits of hill country. It is obvious, therefore, that climate is only one factor in determining the type of forest cover. Mixed broadleaf and needle trees cover many sections, and the southern part of the region has many varieties of broadleaf evergreens to add to the array. Broadleaf evergreens predominate in the Australian and the mild-winter Asiatic sections.

The *Pampas* of Argentina and Uruguay and substantial parts of the dry margin in the United States were originally native grassland. Some of them were tall prairie grasses, some of them were short grasses. While the climax type of such areas is probably grassland, the fact that exotic trees grow much more slowly than in natively forested areas suggests that the grassland areas do not have the best combination of precipitation, soil, and temperature for a good forest cover. Grasses are also found in the so-called "prairies" of southern Louisiana, in the Chunnennuggee Ridge area of Alabama, and in some limestone areas of east central Texas.

Soils. The red and yellow forest soils which prevail over large parts of the southeastern United States are similar to the soils of the wet tropics. With moderately high to high summer temperatures in addition to moist conditions throughout the year, both chemical and mechanical weathering continue much of the time. While leaching is not so great as in the tropics, it is nevertheless great enough to rob the soils, which are chiefly residual, of much of their mineral content. Since much of the region is in slope, erosion is an additional and serious problem. Conservation and reclamation of the soil through control of erosion and through fertilization point toward more intensive use of it and promise more income in a climate where much of the year can be used for growth.

The northern border of the *Cfa* in the United States, with its longer winter, has gray-brown podsolic soils. While they are moderately leached, they are better than the red and yellow soils. Part of their higher value is due to the greater use of fertilizers and also to the lack of soil disintegration and decomposition during the cold nongrowing season; another asset is their quick recovery under a proper crop-rotation regime.

Where grasslands prevailed, they weathered into dark, usually black, soils of much better quality than the forest soils. Thus the Pampas, the drier grassland edges of the North American *Cfa*, and the *Cfa* grasslands on the west slope of the Great Dividing Range of Australia produce well. They are in part chernozem and prairie soils which are little leached and have a high content of humus and desirable minerals. The grasslands of central Alabama and the tall grass prairie of east central Texas are dark calcareous soils of the rendzina type and are highly productive.

Some alluvial lands are of excellent quality; others are sandy and too acid for growth of many crops, but they are often good for truck farming and coniferous reforestation.

Occupations in the *Cfa* Climate

The *Cfa* areas are without question among the most important in the world agriculturally. While subtropical products have long been associated with the *Cfa*, its poleward areas produce temperate crops as well. Outstanding products include enormous amounts of locally consumed rice, plus tea and vegetable fibers in China, Korea, and Japan; cattle, corn, alfalfa, and quebracho in South America; cotton, corn, wheat, fruit, vegetables, naval stores, and tobacco in the United States; wheat and corn in the Po and lower Danube valleys; and sheep, wool, and wheat in Australia. In addition to these, many other products, either of local or widespread importance, could be added to the list. Where the pressure of population is not too great and where the purchasing power is high enough, the milder parts of the climate offer room for expanding cattle industries since little housing is required and animals can graze most or all of the year.

In the southern part of the United States, the *Cfa* is an area of expansion and of considerable promise in agriculture.

High population density and significant population centers including many of the world's largest cities, coupled with a wide variety of agricultural products, important mineral fuel supplies, and an abundance of other minerals afford a surprisingly well-developed manufacturing industry in the United States, Australia, and South America. Lower purchasing power has limited manufacturing in both China and Africa, but Japan is highly industrialized, despite her dearth of minerals.

The world's most important areas producing both coal and petroleum are within, or partly within, the *Cfa* of the United States. Iron, aluminum, sulphur, magnesium, zinc, lead, and building stones are also notable. Australia's chief coal-producing area is in the *Cfa*, around Sydney. The *Cfa* of China produces limited quantities of tungsten, tin, mercury, and antimony; its unworked coal and iron resources are sizable.

With productive and populous hinterlands supplying numerous items of international trade and with many of the world's best harbors serving as media for the exchange of goods, commerce of the *Cfa* climate is phenomenal. Thus many of the world's largest seaports and river ports are within the *Cfa* areas. Port and shipping facilities, as well as the number of people engaged in internal and international trade, are also impressive.

The Mesothermal, Warm Summer (*Cfb*) Climate

Location

Since a major distinction between the *Cfa* and *Cfb* climates is the fact that the latter has no month that averages above 72°, it is not surprising that it is generally found somewhat farther poleward than the *Cfa*. The latitudes of its locations as well as its depths into the continents, however, are by no means uniform. The largest area of mesothermal, warm summer climate is in Europe. In that continent it extends from 40 degrees north latitude in northern Spain and northern Greece to 63 degrees in coastal Norway. Its longitudinal extent is also great, from 10 degrees west to 28 degrees east longitude. Included are the British Isles, except for limited highlands, the Low Countries, and Denmark; most of France, Germany, Poland, Czechoslovakia, Yugoslavia, and Bulgaria; and parts of Spain, Italy, Norway, and Sweden. Parts of the Caucasus Mountains are also *Cfb*. Approximately 30 per cent of Europe is so classified.

In North America, two relatively small areas of *Cfb* occur. One is a narrow coastal strip extending from the northwestern tip of the state of Washington northward past Sitka, Alaska, to 58 degrees. The other is the Appalachian highlands area (mostly surrounded by the *Cfa*), which extends from the southern end of the highlands to Lake Erie and Pennsylvania and also includes parts of coastal New England and most of Long Island.

In the southern hemisphere, the largest *Cfb* area is in eastern and southeastern Australia and includes Tasmania and New Zealand. A small portion of southeast Africa also is included in the climate. In South America, the west coast from about 40 degrees southward to 50 degrees, plus an extension to the east coast, and a small east coastal section south of Rio de Janeiro fit the *Cfb* classification.

While the name "west-coast marine," or "marine west coast," has frequently been applied to the *Cfb* climate, and it could still apply to the west coastal parts of it, the name is scarcely applicable to the climate as a whole as the boundaries are now drawn. The great depth to which the climate penetrates in Europe and its appearance on east coasts, including the interior location in the eastern United States, would seem to preclude such appellations as "west coast" and "marine" in those areas. Therefore, the more general name

Fig. 20–11. No other climate in Europe is better suited to dairying and general farming than is the Cfb type. This scene is near Neuchâtel, Switzerland. (Courtesy Standard Oil Company of New Jersey.)

mesothermal, warm summer climate is applied here.

Temperature

Because temperatures vary relatively little both diurnally and annually and at the same time are not extreme, the *Cfb* is more temperate than any other climate. Diurnal ranges, while somewhat greater in interior locations, are between 15 and 20 degrees throughout the year. They are greater in summer, under clearer skies, than in the winter. Mean annual ranges are about 20 degrees in coastal areas in the northern hemisphere but are as little as 12 or 15 degrees in the southern (12 degrees at Port Elizabeth, Union of South Africa; 15 degrees at Auckland, New Zealand). Interior locations in Europe and those in the eastern United States near the microthermal climates record ranges of 25 to 40 degrees or more, thus reflecting some continental influence (Berlin 36 and Boston 45).

Mean annual temperatures are in the 40s or 50s in most *Cfb* locations. Port Elizabeth is somewhat higher, with a mean annual temperature of 64°; in spite of its relatively high mean annual temperature, it has a small annual range (12 degrees) and its warmest month is 70°.

While summers are warm, they are not so hot as those of the *Csa* and *Cfa* or so warm as the *Csb*, which has more sunshine and lower relative humidity than the *Cfb*. Since the summers are relatively cool, not hot, the higher relative humidity of the *Cfb* areas actually make them feel somewhat cooler than their temperatures would indicate. Summer months average between 55 and 65° near the coast (Figure 20–12) and between 65 and 72° in the interiors. The coasts have negative thermal anomalies of several degrees in the summer; the winter positive anomalies are even greater, however, than the summer departures. Summer diurnal ranges are about 20 degrees.

Whenever summer winds from the interior or from the tropics prevail for several days, they may boost temperatures above 100°. Spells of this sort are rarely prolonged, and severe hot waves are few.

Winter temperatures are often 20, or even 30 or more, degrees higher than places having continental or dry climates in similar latitudes. For example, Paris averages 38° in the three winter months (December, January, and February), but Orenburg, U.S.S.R., 7°. January is generally the coldest month in the northern hemisphere, July in the southern. In the *Cfb* areas, one-half of the days of the winter may have temperatures below 32°, so that frost is not unusual, except that in both Africa and Australia they are much less frequent because of higher winter temperatures. Sustained freezes are not the rule. Growing seasons are from 180 to 200 or 210 days; in the southern hemisphere they are longer.

Just as hot weather may last for a number of days in the summer, so also continental outpourings of *cP*, or even *cA*, air in the winter may cause considerable distress for days or weeks. While it does not occur every winter, occasionally a persistent *High* over interior Europe or Siberia may drop temperatures of western Europe drastically. When easterly winds prevail under such conditions, not only the Continent but even the British Isles may have many days of severe conditions. Central Europe, notably interior Germany, reported the coldest conditions in the prolonged cold wave of the winter of 1956 that had ever been recorded in Germany since official meteorological records were begun in 1776. Temperatures of zero and near zero persisted for several weeks in places which had seldom experienced them. Over 800 lives were lost. Much of the Rhine River and many of its tributaries were reported frozen. The same cold waves penetrated far into the Mediterranean lands, severely damaging olive and citrus trees even as far south as Valencia, Spain. Fortunately such cold outbursts are the exception, but when they occur, they are likely to be disastrous since provisions are not made for such severity.

Precipitation

In locations in higher middle latitudes, cyclonic influences prevail much of the year and result in a fairly even and ample distribution of precipitation. Thus *Cfb* areas rarely lack water. On equatorward margins and in locations close to the coast, there is usually a maximum of precipitation in the winter season; farther inland, as at Paris, Berlin, Vienna, and Breslau in Europe, summer is the season of heaviest rain. While there is no dry period, Paris, Dublin, Utrecht (Figure 20–12), and Edinburgh have a minimum in spring. Some places, such as Melbourne, have a maximum in the fall and spring and a minimum in both summer and winter. In general, it may be said that stations located on west coasts have a distinct winter maximum of precipitation; stations located in interiors have a distinct summer maximum; stations located on east coasts or at some distance from west coasts, but open to some influence of oceans to the west, tend to have an even distribution, or a double maximum. There is, therefore, no uniformity as to times of greater and lesser precipitation; neither is there a deficiency nor an oversupply of precipitation, except that the latter may occur in some highland locations.

Mean annual amounts of precipitation are modest, amounting to 20 or 30 inches on lowlands some distance from coasts. Along west coasts, the amounts are greater, and where there is a backdrop of mountains, the amounts may exceed 150 inches. London has 25 inches and Berlin 23; Brest on the coast of Brittany has 29 inches, and Valencia on the southwest tip of Ireland has 56; Mt. Snowdon, Wales, has over 200 inches. Boston is neither a west-coast station nor is it subject to orographic ascent, but it does have some marine influence and is exposed to a constant procession of warm fronts and cold fronts (though almost completely bypassed during the spring and summer of 1957), so that it receives over 40 inches of precipitation

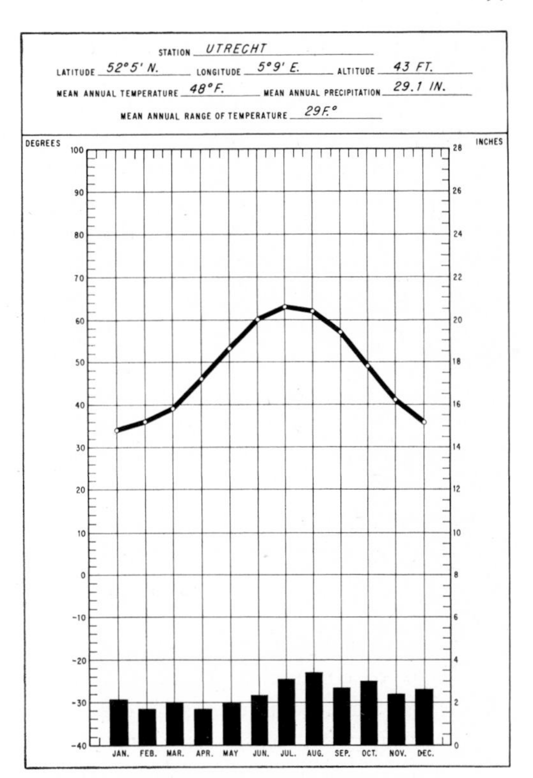

Fig. 20–12. Cfb climate.

annually that is very evenly distributed throughout the year. Auckland, with a distinctly marine location, also exceeds 40 inches, but twice as much is received in the wettest month of winter (July) as in the driest month of summer (January). Puerto Montt (Chile) with 86 inches, Sitka with 81, Bergen (Norway) with 84, and Hokitika (New Zealand) with 116 inches reflect both marine and orographic conditions.

Snow is expected in almost all of the region. In lowland parts of Europe it may lie on the ground a few days, but it seldom lasts long nor is it usually heavy. In the interior and especially in upland areas, such as southern Germany, it may remain for several weeks. The fiorded mountainous coasts of Norway, Canada, Alaska, Chile, and New Zealand have heavy snow of several hundred inches. Some permanent glaciers, or snowfields, exist; in the past, they have been partially responsible for the formation of the fiords.

Although cyclonic storms are not so frequent in the warmer season as in the cooler, they still cause most of the precipitation at all seasons. Cloudy days with drizzly rain, or sometimes snow in winter, are the general rule. Cloudy days occur between 60 and 70 per cent of the time in summer; the sunshine of the summer, therefore, is definitely at a premium. Thunderstorms are unusual in coastal areas, but they are more common as one proceeds into the interior of Europe or into mountainous areas. The marine influence favors stability of the air so that thermal convection is limited, but forced ascent of the humid air by mountains causes rapid cooling and heavy precipitation.

With high relative humidity the rule, advection fogs engulf large parts of coastal *Cfb*; it is so common as to be expected at almost any time. Some coasts have 40 to 60 days a year with dense fogs; visibility is often reduced seriously. Fewer fogs occur in interior Europe, but they still comprise an important element of weather; they are most common in connection with warm fronts, particularly in winter. In addition to fog handicaps, strong winds sometimes lash coastal areas in the winter season as well-developed cyclonic storms invade from the open ocean. Stationary and occluded fronts are frequent factors in prolonging dismal, drizzly weather.

Responses to the Cfb Climate

Natural Vegetation. The combined temperature and precipitation regimes of this climate promote a highly developed plant growth, which is attested by the abundance of large trees. With few exceptions, forests once covered the region. Where the better soils prevailed, the indigenous forests were either deciduous or mixed, while in the acidic podsolic and gray-brown pedalfers of the North German Plain, Les Landes of France, coastal Norway, Canada, Chile, and parts of the Appalachians, conifers thrived. Much of the European forest cover has been altered, but substantial parts of the original Canadian, Chilean, and South Island (New Zealand) stands remain uncut. Broadleaf evergreens predominate in Australia, New Zealand (especially North Island), Tasmania, and at lower elevations in southern Chile.

Exposure to strong winds and the presence of thin soils that resulted from ice-scouring during the glacial period are factors which have prevented the growth of trees in certain limited areas. In their stead, moors, heaths, and bogs composed of shrubs, bushes, brakes, heather, bush cranberries and blueberries, scrub juniper, grasses, and sedges provide scenes that are referred to frequently in English and Scotch literature. The Highlands of Scotland (part of which are classed as *Cfc*), Wales, and northwestern Ireland, as well as portions of the English Lake District, coastal Norway, and South Island abound in one form or another of these stunted growths, most of which have limited utility. Another vegetational exception is the grassland; it appears in north central Spain, in the Hungarian Plain (although most of the Plain is *Cfa*), and in all of the plains of eastern South Island. Some windswept coasts of Norway, South Island, Canada, and Chile are devoid of both soil and vegetation.

Soils. A variety of soils is found in the *Cfb* region. The most common is the gray-brown podsolic soil which ranks as the best of the extensive, mature forest soils. It has developed on most of the lowlands of Europe, in the United States, and in parts of the southern hemisphere *Cfb* areas. Thin, stony, and usually acid soils appear in most of the highland sections. Sandy, acid soils prevail in much of the North German Plain, Denmark, Les Landes of France, and in some mountainous areas. Swamp and bog soils are also common; some of the latter are good. While alluvial soils are not so common as in the *Cfa* regions, wherever they occur they are utilized, if possible. The delta of the Rhine is the most extensive. In South Island, the uplifted alluvial plains are grazed extensively; rocky areas with no soil appear here and there in all of the mountainous fiorded coast.

Occupations in the *Cfb* Climate

This climate has long been regarded as one of the best in which to live and work. A large part of European contributions to manufacturing, agriculture, science, medicine, literature, commerce, and cartography have had their origins in this climate. It is the changing seasons, and particularly the cold season, in contrast to the enervating heat of warmer climes that are considered partly responsible for the mental and physical activity of the people of the region.

The fiorded coast lands of the higher latitudes are very sparsely populated; the hills and plateaus are lightly to moderately populated, and the lowlands are usually heavily populated—actually congested in places. Thus the European region has a very large working manpower. The parts of the United States, Australia, and New Zealand that are placed in this classification are major population areas.

More people are engaged in manufacturing in northwest Europe than in any other area of equal size in the world. It is the livelihood of a large proportion of the population. Almost every known type of manufacturing is practiced there. In other areas where population is moderately dense, manufacturing engages at least 25 per cent of the gainfully employed.

Despite the overwhelming importance of manufacturing, agriculture employs even more people. Many agricultural products that thrive in moderate temperatures and year-round moisture are raised. Wheat, oats, rye, barley, sugar beets, some corn, and numerous fruits and vegetables occupy most of the cultivated land. Animal husbandry of various types, and the raising of fibers such as flax and hemp, are widespread. Outside the United States, sheep herding utilizes much of the poorer lands, but everywhere cattle raising, especially for dairy purposes, is a highly developed industry.

Fishing is favored by the cool coastal waters of the *Cfb* climate. Especially important are the waters around the British Isles, Norway, Canada, Alaska, and, to a lesser extent, those of the southern hemisphere. Although but little whaling is done in the region itself, both Norway and Australia are pushing the industry in southern hemisphere waters.

The large population and productive skill of the Europeans, in addition to their needs for more supplementary and complementary goods, have led to a huge volume of exports and imports. Under normal economic conditions, more of the world's largest ports (based on volume) are concentrated in northwest Europe than in any other area of similar size. The port facilities and media of transportation involve the labor of large numbers of people. Major ports and facilities are present in other parts of the region, too, except in South America and Canada-Alaska where only minor ones appear.

The extraction of minerals is big business in Europe, the eastern United States, Australia, and New Zealand but of less importance elsewhere. Governmental, domestic, and personal services require the time of a good percentage of the inhabitants in most parts of the region; of course, this is true in any region where there is a high state of cultural development. And who will say that climate has not been a significant factor in the cultural development of most of the *Cfb* areas?

The Mesothermal, Cool Summer (*Cfc*) Climate

Location

The mesothermal, cool summer climate stretches poleward from the *Cfb* locations in three places. It extends from 63 to 70 degrees north along coastal Norway; from 53 degrees in the Aleutians northward to 61 degrees in southern Alaska, then back to 58 degrees in

Fig. 20–13. Sturdy and snug buildings make life relatively comfortable in this Cfc setting at Reykjavik, Iceland, where spring merges into autumn and autumn into spring. (Courtesy *Ewing Galloway, New York.*)

insular Alaska; and it makes a very limited appearance in South America and in the Falkland Islands. Small portions of southern Iceland, the Faeroe and Shetland Islands, and several highland areas in Great Britain complete the *Cfc* land areas. All of this type of climate is limited to coastal or insular positions.

Temperature

As would be expected, mean annual temperatures are somewhat lower than at *Cfb* stations. All *Cfc* areas have mean annual ranges around 40 degrees; *Cfb* stations generally exceed 45 degrees. In this cool summer type of climate, there are no more than three months above 50°, but no month averages below 27°. The warmest month hardly exceeds 55°, while the coldest may be well above freezing (37° at Thorshavn, Faeroe Islands). The extremes for Dutch Harbor, Alaska, are 53 and 32° (Figure 20–14).

The unusually small mean annual ranges of temperature for such high latitudes reflect the prevalence of marine influences throughout the year. Warm currents are an aid in the northern hemisphere, e.g., Thorshavn at 62 degrees north has a mean annual range of only 14 degrees. Much cloudiness and high relative humidity usually prevent diurnal ranges from going very high, 12 to 15 degrees being the rule. Only when occasional continental winds or blasts of polar air strike the region do temperatures drop to as low as zero; if such conditions occur, they do not last long. In general, the climate may be classed as uncomfortably cool and raw, with many gray, drizzly days and considerable wind which makes the cool or cold damp air feel very penetrating.

Precipitation

Although excessively heavy rainfall is the exception in this climate, most days are either cloudy, foggy, or rainy, or all three combined; there is little sunshine. Where altitude is a factor, snow may occur in any month; yet snow is not a characteristic of sea-level stations except in winter—even then, it may not last long. Punta Arenas, Chile, located on the coast but somewhat in the lee of hills, receives only 16 inches of equivalent rain in a year; Reykjavik in Iceland receives 34 inches; while the two small island stations of Thorshavn and Dutch Harbor (Figure 20–14) receive 57 and 63 inches, respectively. In Norway, Iceland, and parts of Alaska, at

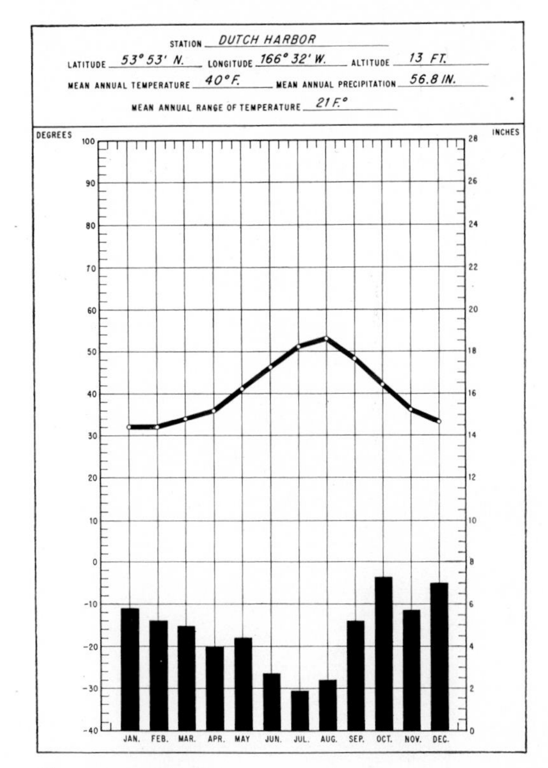

Fig. 20–14. Cfc climate.

relatively short distances from the coasts, altitude may be sufficient to reduce the temperature to a level where most of the precipitation is in the form of snow that does not melt during the brief summer season; thus glaciers proceed from the uplands to the coasts, contributing some icebergs to the shore currents. Obviously, the uplands are not *Cfc* climate but rather polar climate.

In the northern hemisphere, there is a definite winter maximum of precipitation, but there is no month that is ever dry. Punta Arenas, however, has a fairly even distribution with October (spring) being the driest month.

During winter, the seasonal Icelandic and Aleutian *Lows* cause unusually stormy weather. Many of the cyclones of higher latitudes gravitate to these areas, bringing not only rather heavy precipitation but high winds and dreary days—dreary because, even at best, daylight lasts only a few hours at the winter solstice. In the summer, on the other hand, high-pressure areas are more common over the water. With the resulting descending air, though still marine, less precipitation occurs so that occasionally some beautifully clear days may punctuate the otherwise drab recurrence of foggy, sunless weather; hence, summer is necessarily somewhat more pleasant than winter.

Responses to the Cfc Climate

Natural Vegetation. Taiga (northern forest) and *tundra* are the types of vegetation which this climate favors. Forests are more common in slightly elevated and protected situations, while tundra may appear anywhere. Frequently the soil is too thin or too poorly drained to support trees. Moreover, strong winds definitely inhibit tree growth. Therefore, where trees cannot make a stand, a scrubby type of vegetation takes over. If the soil permits, grass will thrive in spite of wind. Areas which cannot support trees, scrub, or grass will be tundra or bare rock. Grass does particularly well in parts of the Faeroes and southern Chile.

Soils. Soils are invariably podsolic, except where small amounts of alluvium have been deposited. Usually the soils are thin, stony, and have little innate value. Bog soils exist, but they have small utilization in so cool and damp a climate.

Occupations in the Cfc Climate

The natural occupation in these fog-enshrouded coasts and islets is fishing. Important fishing fleets set out from Trondheim and Narvik (Norway), from Reykjavik, from the Faeroes, and from ports in Alaska (Juneau, Kodiak, and Dutch Harbor). Almost everyone living along the coast is in some way connected with fishing—taking, canning, salting, or drying fish. Lumbering, gold mining, merchandising, and trading occupy the time of other people. Sheep herding is important in the Faeroes (meaning "sheep") and in southern Chile—in fact, some of the longest fleeces come from these areas. Agriculture is practiced almost under duress. In the Faeroes, barley is the only cereal that can mature. Garden produce, such as cabbage and turnips, does well, but many less hardy vegetables must be produced under glass. Even though population is sparse, small cities do exist and thrive because of the high per capita trade across the seas.

21

The Humid Microthermal Climates

The humid microthermal (*D*) climates have both a distinct summer and a distinct winter season. The length and intensities of the two opposite seasons vary considerably with both latitude and continental location. Summers are longer and hotter on the equatorward margins while winters are longer and more severe in the interiors farther poleward. The spring and fall seasons are also evident, but they, too, vary in length. The *D* climates are located between the constantly cold (*E*) climates poleward and the mesothermal (*C*) and dry (*B*) climates toward the equator. With but one slight exception (the mountainous area of South Island, New Zealand, which is undifferentiated *Df*), the microthermal climates appear only in the northern hemisphere, in high or higher middle latitudes, excluding the Arctic Ocean fringe; consequently, further references to these climates will apply to the northern hemisphere only. They extend completely across Alaska and Canada and also Eurasia, although they have relatively little actual ocean frontage in northwestern Europe. Their latitudinal extent is greater on the eastern sides of the continents than on the western sides.

The humid microthermal climates are frequently referred to as "continental climates" because they cover large interior areas and therefore have little marine influence. The continental influence is shown by relatively large seasonal extremes of temperatures and as a rule by moderate precipitation. By way of contrast, portions of the mesothermal climates are influenced by marine conditions, such as relatively small annual ranges of temperature, especially for their latitudes.

The microthermal (meaning "little heat") climates have an extreme latitudinal spread of some 46 degrees, from 28 to 74 degrees north latitude. Areas of the *D* climates south of 40 degrees in the United States and south of 50 degrees in Eurasia, however, are chiefly mountainous; most of them are detached from the main continental *D* masses. While these several mountainous areas are in the middle latitudes, rather than in the higher latitudes, their elevations reduce their temperatures so that they fall within the limits of the *D* climates.

Statistically the *D* climates must have one or more months with a mean temperature between 50 and 64°, and the coldest month below 27°. Frequently several months are below the latter figure. It is the cold of the winter rather than summer temperatures that distinguishes the *D* climates, although summer, whether short or long, may be definitely warm, or even hot. In general, cold winters, ranging from moderate to severe, are characteristic of the microthermal climates.

Summer is normally the season of maximum precipitation. At that time the absolute humidity is highest, and convectional overturning leads to showers, sometimes heavy ones; this is particularly true in the warmer portions and in areas where monsoon influences are strong, as in northeast Asia. Rainfall reliability, nevertheless, is none too high, and dry years occur fairly often, especially toward the borders of the *B* climates. Total precipitation varies from about 10 inches annually on the dry margins to 40 inches on the *Cfa* border, except in mountainous areas where local orographic precipitation may be higher.

Snow is an important element. It covers the ground frequently and for long periods of time, especially in the colder and less dry portions. The effect of a snow cover is to reflect heat rather than to absorb it; whatever heat is added by day is soon radiated upward at night, and the temperature drops. Ground that is covered with snow for weeks, or even months, is therefore subject to continuously low temperatures. On the other hand, a snow blanket tends to retain whatever heat is in the ground, and thus an early and prolonged snow cover prevents the ground from freezing as deeply as when the ground is bare under the same temperature conditions.

The annual snowfall varies greatly within the *D* climates. Under orographic conditions it may attain 400 inches, as in places in the Sierra Nevada range in the United States. As a whole, it is less on equatorward margins, and it is less in parts of the interior, despite the low temperatures, than is true of the microthermal climates as a whole. The *Dw* climates of Eastern Asia have much less snowfall than the *Df* areas. The lack of snow undoubtedly aids in the deep freezing of the permafrost sections of Siberia.

While absolute humidity is lower in winter than in summer, relative humidity is moderately high in most of the *D* areas. It will be recalled that at low temperatures the ability of air to include moisture is very greatly reduced, so that at a temperature of zero a cubic foot of saturated air can include slightly less than ½ grain of water vapor which is about 1⁄40 of what it could include at 100°. Therefore, whenever the dew point is reached at low temperatures, the amount of snow that falls cannot be very great. Thus snowfalls in the northlands are generally low, a fact that is not commonly recognized.

Middle-latitude cyclonic storms constitute an important control during the winter season. They are more significant in areas of milder winters and less so in areas of severe winters, particularly in eastern Siberia. This is explained partly by the fact that tropical and marine air masses rarely invade these areas in the winter season and partly by the fact that these areas are the scenes of semipermanent high pressures which build up over the cold lands in the winter. They repel rather than attract cyclones; in fact, they are the sources of the great masses of cold air which sweep southward at irregular intervals in great pronglike bursts, the *cold fronts* already discussed in Chapter 9. Such areas have a smaller amount of cloudiness and precipitation than the *D* climates as a whole.

Eight separate types of climate comprise the microthermal group. They can be divided into two categories: (1) those which are moist throughout the year (*Dfa*, *Dfb*, *Dfc*, and *Dfd*) and (2) those in which the mean precipitation of the driest month is less than one-tenth that of the wettest month (*Dwa*, *Dwb*, *Dwc*, and *Dwd*). Inasmuch as the greatest distinction between each pair of climates is in the relatively dry winters of the *Dw* division, each *Df* climate will be followed by its respective *Dw* climate; thus *Dfa* is followed by *Dwa*. As a rule the *Df* climates are both more extensive and more important economically than the *Dw* climates.

The Microthermal, Hot Summer (*Dfa*) Climate

Location

The *Dfa* climate appears almost exclusively in the United States and Eurasia. The largest area is a triangular wedge in the United States, the apex of which is centered at the western end of Lake Erie. The legs of the triangle reach westward to the 100th meridian in South Dakota and in Kansas, respectively.[1] Another small area of *Dfa* includes western Massachusetts, northwestern Connecticut, and a small portion of the Hudson River Valley south and east of the Catskills. The lower elevations of the undifferentiated *Df* area in the Sierra Nevada and Cascade mountains of the western United States are also *Dfa*. In Europe, most of the industrially and agriculturally important Donets Basin of the U.S.S.R. and a very narrow band south of the Carpathians in Romania, including the capital city of Bucharest, have the same classification. As the map indicates, a substantial part of the highlands of central Asia are shown as undifferentiated *Df*. Included in this high mountain mass are the Hindu Kush, the Pamir, the Karakoram, and the Tien Shan. Parts of this area, habitually lower portions, are *Dfa*. The same is true of parts of the Er Rif and Atlas mountains of northwest Africa. Almost one-half of Honshu, the main island of Japan, is *Dfa*. Most of the area is mountainous, however, and is surrounded by *Cfa*, except in the northern part of the island. The *Dfa* areas are not so extensive as either the *Dfb* or *Dfc*, but they are important reservoirs of both agricultural and industrial products.

The *Dfa* climate is characteristically located on the south margin of the *Dfb* climate and north of the *Cfa* climate. Usually it extends to the steppes (*BSk*) of the interior. In the mountainous areas of the western United States and central Asia, the climate appears at certain lower elevations.

A small area of *Dsa*, without definite boundaries, appears on the Köppen-Geiger map in northwestern Iran, northern Iraq, and perhaps in the extreme eastern part of Turkey. As the letters indicate, this area has hot dry summers, but cold winters; its precipitation, like that of the *Csa*, comes mainly in the winter.

Temperature

The summers of the *Dfa* climate are long and generally hot. When the humidity is high, as it frequently is, the sensible temperatures take on a tropical aspect. In reality, the summer-temperature regime is at least subtropical, if not tropical. Tropical-maritime air masses (*gT* or *aT*) invade the corn belt of the United States, much of which is in the *Dfa* climate, with considerable regularity. Even the pressure in the summer time is tropical in nature. The barograph often records little change in pressure from day to day, except for the slight double maxima around 10 A.M. and 10 P.M., and the slight double minima around 4 A.M. and 4 P.M. Occasional invasions of *cP* air masses (summer cool waves) dispel the enervating heat. Hot waves, or *cT* air masses, sometimes *TS* (tropical superior), both out of the southwest, may persist for many days in July or August; drought often accompanies these hot spells.

Because of the length and heat of the *Dfa* summers, the climate is frequently referred to as the "long summer" or "hot summer" phase of the humid continental climate. It is also

[1] While the Köppen-Geiger map is drawn to exclude Detroit and the Twin Cities from the boundaries of the *Dfa* climate, statistically those cities are just within it and are so shown on the map in this book.

Fig. 21–1. Corn and Dfa climate are almost synonymous. This farm is in Iowa. (Courtesy U.S. Department of Agriculture.)

called the corn-belt climate because of the prevalence of the maize crop throughout its extent in the United States. Maize and soybeans are particularly well suited to a long, hot, humid summer. On such a summer night, especially when a low cloud cover is present to keep the heat of the day from radiating from the ground surface, corn may literally grow an inch or two in height. The *Dfa* growing season is from 150 to 180 days, occasionally more.

The hottest month, which is invariably July, averages above 72° (to classify as *Dfa*); in the United States, the figure is usually several degrees higher. Fort Wayne averages 74° in July, Chicago 75°, Des Moines 77° (Figure 21–2), and Omaha 78°. It appears that maximum monthly temperatures are somewhat higher on the dry margins than in the more humid portion of the climate.

Mean maximum July temperatures are often above 90° while absolute maxima (usually in July or August) are above 100°, which is almost everywhere characteristic. In unusually hot summers, numerous places in the American *Dfa* area have recorded as many as 10 successive days with maximum temperatures above 100°. August is characteristically from 3 to 5 degrees warmer than June.

The coldest month in the *Dfa* climate, without exception, is January. It is to be noted that the dry edge of the American *Dfa*, as would be expected, is colder than the more moist sections, reflecting the greater continentality of the former. Less moisture and lighter vegetation, it will be recalled, promote a more rapid radiation of heat from the earth and also favor both a greater diurnal and a greater mean annual range of temperature. The coldest month in Fort Wayne and Detroit averages 26°, Chicago 25°, Des Moines and Omaha 23°, while Sioux City drops to 19° and Huron (South Dakota) to 13°. The two latter, however, are on the northern as well as on the dry side of the climate.

Mean annual temperatures are close to 50°. That for Chicago and Toledo is exactly 50°; Des Moines and Bucharest average 51°, Omaha 52°, Sioux City and Albany 48°, and Huron 46°.

Mean annual ranges reflect both the continental influence and the effect of a fairly high latitudinal position. Except for the *BWk* and *BSk* climates, the *Dfa* climate has a greater range than any other climate mentioned thus far. Generally, mean annual ranges are between 50 and 60 degrees.

Precipitation

Mean annual precipitation in the *Dfa* climate is not great, although the type is classed as humid. The latitude is high enough so that evaporation is not excessive, despite the heat of the summer. The yearly amount of precipitation is usually between 25 and 35 inches. In the United States it decreases in general toward the west and the north. The total in Fort Wayne is 35 inches, in Chicago 32, in Des Moines 29 (Figure 21–2), in Omaha 26, and in Huron only 18 inches.

Precipitation is well distributed throughout the year, but there is a distinct late spring or early summer maximum and a winter minimum. Only on the drier margins is the amount less than 1 inch per month. There is a tendency for a greater proportion of the precipitation to fall in the month of maximum precipitation near the drier (western) margins in the United States than in the more humid parts of the climate. Therefore, while Peoria and Urbana receive 11 and 12 per cent, respectively, of their total precipitation in May and Chicago receives 13 per cent in June, Des Moines, Omaha, Sioux City, and Huron receive over 17 per cent of theirs in June. June is usually the rainiest month, followed by July or May or even August or September. In general, as one proceeds northward and westward, the time of maximum precipitation comes later; this general rule applies to all of the central

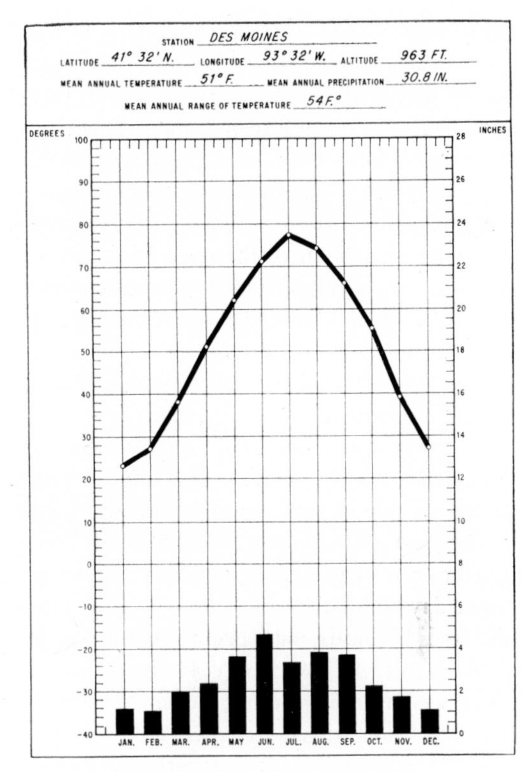

Fig. 21–2. Dfa climate.

United States and Canada, regardless of climatic classification.[1] It should be emphasized, however, that there is a fairly high amount of sunshine in the warm season, higher than at any other season in spite of the rainfall maximum at that time. The intensity of the summer sun, as temperatures prove, is comparable to that experienced in some lower latitudes.

Much of the warm-season precipitation is convectional and of short duration, but occasional frontal storms bring gray days and slow-falling rain that sinks in slowly, doing the soil and crops much good. Severe thunderstorms, squalls, and tornadoes wreak considerable havoc every year, as they do in *Cfa*

[1] This is in contrast to the Pacific Coast of the United States and Canada where the season of maximum precipitation comes earlier as one proceeds northward, e.g., February in San Diego, January in San Francisco, December in Portland, November in Vancouver, and October in Sitka.

and to some extent in the *Dfb* sections of the United States, some years being worse than others. These three climates are among the chief battlegrounds of tropical and polar air masses that form frontal storms occasionally of great violence and excessive precipitation. The greater the difference in the temperature and moisture content of the opposing masses, the more violent the squalls may become along the fronts. These differences are often greatest in the spring and early summer while polar sources are still cold but subtropical latitudes are receiving direct rays of the sun.

As previously indicated, the season of maximum cyclonic activity is winter, with spring and fall in between and with the least development in summer. With their nonperiodic yet frequent appearance, gloomy winter days ensue. More or less snow or rain falls and blizzards may occur. These spells are frequently followed by cold-air invasions, with accompanying crisp, clear weather. Under such conditions, extremely cold weather is likely to occur, with northerly or northwesterly winds.

Compared with the *Dfb* climate, the amount of snow is appreciably less. This is partly because the *Dfa* is located farther south, with the result that the centers of the cyclones frequently pass to the north of such locations. As pointed out in Chapter 9, this means that the warm interfrontal area (zone of convergence, or warm sector) will frequently cover the *Dfa* areas, bringing higher temperatures and rain instead of snow. On the other hand, when the center of the low passes to the south, the *Dfa* areas are first in the northeast and then in the northwest quadrants of the storm. The northeast quadrant usually brings snow, sometimes much of it.

In northern Illinois and Indiana, between one-fourth and one-third of the winter precipitation is usually in the form of snow. The total amount is from 10 to 40 inches annually. The total number of days with snow varies within the climate in the United States from about 30 days on the south to over 90 days in parts of the northern boundary, including southeastern Minnesota and on eastward into western Massachusetts.

Responses to the *Dfa* Climate

A cold winter in which the mean temperatures are below freezing for an average of three months, plus a long, hot summer in which the mean temperatures are above 70 degrees for two or three months produces, when precipitation is favorable, a combination of vegetation and soils rarely surpassed in other parts of the world.

Natural Vegetation. The natural vegetation of the *Dfa* climate is easily designated. About two-thirds of the area in the United States, Romania, and the U.S.S.R. was originally grassland. In the major triangular section of *Dfa* in the United States, only the larger river valleys and portions of northern Illinois, Indiana, and Ohio, southern Minnesota and Wisconsin and eastern Iowa had sizable tracts of broadleaf deciduous trees. These were of the oak-hickory association in the western part and of the beech-maple varieties in the eastern section. Areas between the river valleys in eastern Iowa and parts of Illinois had a mixture of trees and grasses (mostly bluestem).

Much of northeastern Illinois and northwestern Indiana, in addition to most of the western half of the climatic region, however, were level to gently rolling plains of grasses, stretching as far as the eye could see and interrupted by trees only along the watercourses. The grasses of Illinois and Iowa were tall prairie grasses; those of the western drier edge were shorter, some of them being properly classed as steppe varieties.

The Romanian and the eastern Ukrainian sections are part of the extensive short-grass Russian steppes. The *Dfa* sections of eastern New York, New England, and Japan developed a mixed forest of broadleaf deciduous and coniferous trees. The dominant vegetation of the mountainous *Dfa* sections in Asia, and probably where it appears in the western United States, is also grass, with other herbaceous plants growing singly or in groups. It seems reasonably clear that the absence of forests in much of the *Dfa* region is due, in part at least, to three factors: (1) the insufficiency of precipitation, particularly toward the *BS* margins; (2) the erratic distribution of precipitation, both seasonally and annually—again more pronounced on the dry margins; and (3) the type of soil and terrain, calcareous soils and level, or gently rolling, lands being ideal for grasses, regardless of adequate or even abundant precipitation. Only in the eastern United States and Japan, and in favored locations elsewhere, are there extensive forests or woodlands as a climax vegetation. Since vegetation is, with few exceptions, a decisive index of climate, it appears fairly conclusive that parts of the *Dfa* climate may well be considered transitional, that is, on the border between humid and semiarid climates.

Soils. The most extensive *Dfa* soils are classed as *prairie* and *chernozem* types. These two are the finest soils known, with the exception of some of the alluvial soils. They dominate fully two-thirds of the *Dfa* triangle of the midwestern United States and all of the *Dfa* area in Romania and the U.S.S.R.

Prairie soils develop under tall or moderately tall grasses; chernozems form under both tall and some short grass steppes; both have a thick accumulation of humus. Prairie soils occur where there is moderate rainfall and some leaching, but not enough to cause excessive mineral loss. Chernozem soils are structurally and texturally the best known, but they sometimes lack sufficient precipitation to produce good crops. Prairie soils are therefore frequently more intensively utilized. Both soils are dark in color and, if not abused, are easily tilled. In some sections of both chernozem and prairie soils, there is the addition of *loess*, a yellowish and very fine, silty soil that has been blown by wind from drier areas; locally, some loessal deposits are thick. Peat and bog soils form in some lowlands, especially in morainic lake beds.

East of the Illinois-Indiana border and in the vicinity of the Mississippi River, as well as in the New York-New England enclave, gray-brown podsolic forest soils predominate. As indicated previously, these are the best of the forest soils and are much used for agriculture in level terrain. The mountainous *Dfa* areas have a variety of soils, most of them podsolic.

Occupations in the *Dfa* Climate

No other climate is more richly endowed than the *Dfa* for the pursuit of agriculture. Throughout its extent in the United States, cultural features reflect the productiveness of its climate and soils. Corn, wheat, oats, soybeans, cattle, and swine are raised in great abundance; sheep, barley, rye, various hays, fruits, and vegetables are raised in lesser quantities yet are significantly important. Much of the corn, oats, soybeans, and hays are fed to animals, but most of the wheat, barley, rye, fruits, and vegetables, as well as the swine and the beef cattle, are used either locally or on the more populous eastern seaboard. In both Romania and the U.S.S.R., the *Dfa* sections are among the most advanced parts of those countries. The mountainous portions of the *Dfa* climate are generally of less significance economically.

Much manufacturing is carried on in the North American triangle, the eastern United States enclave, and in the U.S.S.R. Many of the manufactures entail the processing of foods and animals raised on the farms. Others include some of the largest iron and steel works in the world, the manufacture of numerous farm implements, vehicles, oils and gasolines, plastics, synthetics, airplanes, clay products, tools, and household appliances; many of the industries, besides the processing of foods, utilize products of the farm, range, and forest. The number of manufacturing establishments is increasing steadily in the United States. Almost every town has at least one factory, and cities are growing rapidly as manufacturing centers. While the rural farm population is decreasing, the cities with their manufacturing population are increasing in size.

Coal is the mineral of widespread distribution within the *Dfa* climate. Much of it underlies Illinois and Iowa, while the Donets coal basin of the eastern Ukraine is the basis of the power for the Soviet Union's most important heavy manufacturing industry. Small amounts of petroleum and natural gas are derived from fields in Ohio, Indiana, Illinois, and Kansas. The southernmost part of the Ploesti oil field of Romania is also within the *Dfa* climate. Because the main areas within the climate are nonmountainous, metals are lacking, although small amounts may be obtained in the *Dfa* mountains of Asia and the western United States.

Flat and bountiful farmland, high purchasing power, much and varied manufacturing, the world's largest railway network emanating from Chicago, relatively cheap Great Lakes transportation, and a fairly dense population have developed an intricate system of commercial exchange in the North American *Dfa* triangle. The parts of Romania and the Soviet Union that are within the climate are also among the most important commercial portions of their respective nations. While agriculture appears areally to be the most important activity within the climate, both commerce and industry vie with it for importance. Neither of these could have attained such importance without good, or even superior, transportation systems. All of the nonmountainous *Dfa* regions are well supplied with rail, air, and highway facilities, and each one has access to major water routes.

While the *Dfa* climate as a whole is not particularly endowed with vacation sites, all the major cities, the Great Lakes and smaller lakes, the uplands of the New York-New England enclave, the mountains of the western United States and central Asia, and the Japanese alps have certain attractions and are to be identified with the tourist trade to some extent. In addition, the American *Dfa* areas are important crossroads between the south and the vacation lands farther north.

The Microthermal, Hot Summer, Dry Winter (*Dwa*) Climate

Location

The *Dwa* climate (sometimes called the "Manchurian climate") appears only in eastern Asia, from 36 to 47 degrees north latitude. It is limited to Manchuria, Korea, and northeastern China. While about one-half of the climate is upland, substantial parts are productive lowlands, including the Manchurian Plain, the northern part of China's Yellow Plain, and the northern part of Korea's chief plain. Major cities within the climate are Peiping and Tientsin, China; Dairen, Mukden, and Harbin, Manchuria; and Pyongyang and Seoul, Korea.

Temperature

Both Mukden and Peiping have three months that average above 70°. July is the warmest month, with 77° at Mukden (Figure 21-4) and 79° at Peiping; August is somewhat warmer than June. While the summer temperatures of the two places are not greatly different, Mukden's winter temperatures are lower than those of Peiping. Mukden has five months that average below freezing with a low of 8° for January, while Peiping has only three. Spring temperatures rise rather rapidly, while the fall months witness a fairly rapid drop. Mean annual temperatures are about the same as in the *Dfa* climate (44° at Mukden, 53° at Peiping). Mean annual ranges are greater than most of the *Dfa* stations (55 degrees at Peiping and 69 at Mukden). Since the summer temperatures are about the same, these ranges reflect the generally colder winters of the *Dwa* climate as compared with the *Dfa* climate.

It is interesting to compare Mukden and Peiping with their latitudinal counterparts in the United States, Albany and Philadelphia: Mukden is 17 degrees colder than Albany in January but 4 degrees warmer in July, while Peiping is 10 degrees colder than Philadelphia

Fig. 21–3. The dry winters and warm, moist summers favor spring wheat rather than winter wheat in this Dwa region of Heilungkiang Province, Manchuria. (Courtesy Eastfoto.)

in January and 2 degrees warmer in July The reason, of course, is found in the strong monsoonal tendency of air circulation in eastern Asia.

Precipitation

It is chiefly the amount and the distribution of precipitation that distinguish the *Dwa* from the *Dfa* climate. The amount in the former is somewhat less than in the latter, totalling approximately 25 inches, e.g., the mean annual precipitation of Mukden and Peiping is only 24 inches, which is similar to that of the dry edge of the *Dfa* climate.

A distinct dry season occurs in the winter, with both December and January receiving

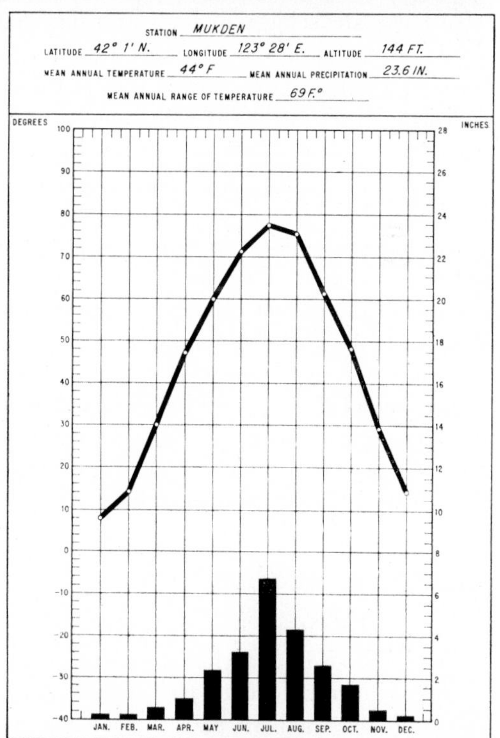

Fig. 21–4. Dwa climate.

only 0.1 inch in Peiping and 0.2 inch in Mukden (Figure 21-4). This amount is less than one-tenth of the amount of the month of maximum precipitation, and for this reason it is designated as *Dwa*. The dry season is the result of the monsoonal winds blowing from the dry high-pressure interior of Asia. Little moisture is available under such conditions. Winter cyclones, which are rare compared with the *Dfa* areas, bring the meager precipitation at that season.

The monsoons reverse in the warm season in response to the development of a low-pressure system over hot interior Asia (Figures 15-2 and 15-3). They bring in from the seas and oceans warm moist air which yields rain convectively, or orographically in hilly areas. Occasionally a typhoon adds an extra amount. Hence, a distinct maximum of precipitation occurs in the summer season. July is the wettest month—6.7 inches in Mukden and 9.4 inches in Peiping. As in central and eastern China, northern China and Manchuria are subject to both droughts and floods. Droughts are more likely than floods because, even at best, *normal* rain is barely sufficient in this region of high summer temperatures. The dry winter season leads to much clearer weather than that experienced in similar latitudes of the United States or Japan. Cloudless skies, in turn, constitute an important reason for lower winter temperatures—the receipt of insolation during the short clear days does not compensate for the great loss of heat by radiation during the long nights. Snowfall in this region is less than in the *Dfa* region, except in some mountainous areas. In northern China, dust storms from the semiarid lands to the west are a common winter occurrence.

Responses to the *Dwa* Climate

Natural Vegetation. It is believed that the rolling lowlands of Manchuria were originally clothed with an excellent cover of short grass, except in the southern part surrounding the Yellow Sea; the eastern Manchurian uplands and the mountains of northern Korea apparently had, and still do have to some extent, mixed deciduous and coniferous forests of considerable value (the most valuable forests of all China and Korea), while the peripheral areas of the Yellow Sea were chiefly forested with broadleaf deciduous trees.

Soils. The soils which have developed under the grass vegetation of the Manchurian Plain, like those under similar conditions in the western part of the American *Dfa* triangle, are of the excellent prairie and chernozem types. Under mixed forest vegetation, the eastern Manchurian uplands and the mountains of northern Korea have developed podsolic soils, with islands of tundra and alpine meadow soils at high elevations. The southern part of Manchuria, surrounding the Yellow Sea, has gray-brown pedalfers, but alluvial soils prevail in the Peiping-Tientsin section.

Occupations in the *Dwa* Climate

As previously indicated, agriculture is one of the important assets of the *Dwa* climate. The Manchurian Plain, long coveted by the Chinese, eventually became their only significant area of surplus food production. The home of the extensively raised soybean, it also produces large quantities of wheat, kaoliang (an important grain sorghum), and various millets. Until 1930, fully one-half of the world's soybeans were raised on the Manchurian Plain. Barley, soybeans, rice, wheat, oats, and susu (kaoliang) are the chief products in Korea. The Peiping-Tientsin region specializes in the production of winter wheat and kaoliang. Animal industries, except for the raising of draft animals, are by no means as important as in the *Dfa* areas.

The variety, production, and reserves of minerals are considerable. For some years the most important iron and coal production of all China was centered at Anshan and Fushun, respectively. Other deposits of both coal and iron are worked to some extent. Oil is derived from oil shales in southern Manchuria. Korea has long exploited gold, and North Korea has important resources of both iron and coal, some of it anthracite. Copper, silver, graphite, and tungsten are also mined in Korea.

Manufacturing assumes some importance in the *Dwa* realm. The Japanese-created iron and steel works at Anshan are well known, and extensive and newer mills are located in the eastern Manchurian uplands. Heavy manufacturing is also carried on in both north China and Korea. The processing of foods, lumber products, clay products, and clothing are leading manufactures in various parts of the climate.

More important than manufacturing, however, is commerce. For centuries, Peiping, Port Arthur, Dairen, and Harbin have been important international trade centers. Newer cities of importance for both trade and manufacturing have been established in recent years. Accessibility to the chief cities is good, either by river, rail, or ocean, although rivers, such as the Yalu and Sungari, are frozen during the winter season.

Lumbering, milling of lumber, and wood products have risen in response to the growth of excellent forests in north Korea and in the eastern Manchurian uplands. Varieties of importance in lumbering include large reserves of spruce, elm, larch, birch, oak, fir, and Korean pine. The trapping of furs is a growing industry in the eastern uplands of Manchuria. Fishing engages the activities of some, both in the chief rivers and in the Yellow Sea and its various tributary waters.

The Microthermal, Warm Summer (*Dfb*) Climate

Location

The microthermal warm summer (*Dfb*) climate extends across North America and Eurasia as two irregular and discontinuous bands that vary in latitudinal width from a few degrees to as much as 27 degrees. The main bands include much of the northern United States and southern Canada and much of central eastern Europe eastward to the headwaters of the Ob River system in central Siberia. These areas are chiefly plains, low mountains, or rolling plateaus. Extensions, sometimes interrupted from the two main masses, occur in several directions. Most of these extensions, especially in lower latitudes, are high mountains or plateaus. Included are the Selkirks, Rockies, Cascades, Sierra Nevada, and Wasatch of the North American continent and the Pyrenees, Alps, Carpathians, Caucasus, east Anatolian uplands, and the high levels of parts of Hindu Kush, Pamir, and Tien Shan of Eurasia. The high mountains of both South Island, New Zealand, and the island of Hokkaido, Japan, are also *Dfb*. It is clear that elevation, rather than latitude, is responsible for the *Dfb* classification of these large mountainous masses that are located farther equatorward than the major *Dfa* and many of the *C* climates. Obviously, the summit levels of most of these mountains are too cold to be classed as anything except arctic, but the areas would be too small to show on the map.

While the *Dfb* climate has the *Dfc* extending along its entire northern boundary, it has several climates on its southern border. Wherever the *Dfa* appears, it is on its southern edge, while in both western North America and Eurasia, the cool steppe borders it for great distances. In other places, the *C* climates either touch it or completely surround it. Naturally, climates on the north of this region are more severe. Those on the south may be warm and moist, or they may be dry.

A small area of *Dsb* climate appears in interior Oregon south of the Columbia River and west of the Snake River. As the letters indicate, this area has the same climatic characteristics as the *Csb* of our western coast, except that its coldest month averages below 27°.

Temperature

Location in higher latitudes means somewhat lower temperatures for the *Dfb* than for any of the other climates already discussed. Both the mean summer and the mean winter temperatures are from 10 to 15 degrees lower than the *Dfa*, though mean annual ranges are about the same. Usually only two or three months average above 60°, and July is the warmest except in a few localities, such as Kars (Turkey) and Nemuro (Hokkaido, Japan) where it is August. Typical of warmest-month temperatures are: Edmonton 61°, Moscow and Duluth (Figure 21–6) each 66°, and Concord (New Hampshire) 70°.

While there may be six or seven months with means above 50° in the *Dfa* climate, there are no more than five in this climate. This shorter period of warmth has earned for this climate the name of "humid continental, short-summer phase." The shorter summer, though with longer days as compared with lower-latitude climates, is ideal for the raising of spring wheat, especially on grass-covered plains in the subhumid sections of the U.S.S.R., the northern United States, and the Prairie Provinces of Canada. The growing season is

Fig. 21–5. Dairying is favored by the cool, moist summers of the Dfb lands in New York State. (Courtesy Standard Oil Company of New Jersey.)

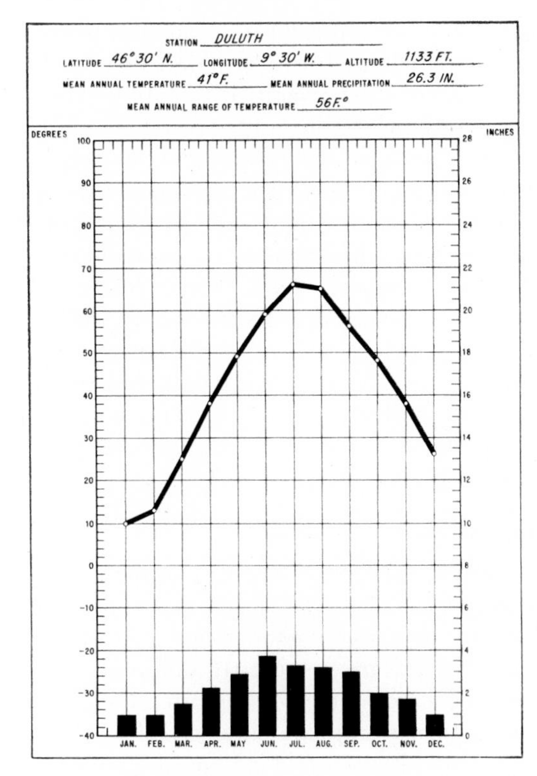

Fig. 21–6. Dfb climate.

only 3½ to 5 months in length, which is a deterrent to the growth of many crops, including corn. On the poleward and dry borders, there has been no month in which frost has not occurred. Frosts late in the spring may kill the young shoots of the wheat plant, and frosts in August may cause damage to the grain while it is in the "milk" stage; normal weather, however, means a sure crop. With minor exceptions, few crops of any sort grow to maturity in any of the climates colder than the two *Db* climates.

Three months, sometimes four or five, have mean temperatures below freezing. The coldest month is almost everywhere January, followed by February and then December. When a fourth month is below freezing, it is March. By April the mean temperatures have climbed to 35 or 45°. October is between 40 and 50°, followed by a sharp drop in November when the long winter sets in. Typical of the coldest month averages are: Winnipeg −4°, Duluth 10°, Moscow 12°, Nemuro and Concord 22°, and Warsaw 26°. Duluth's absolute minimum is −41° and Winnipeg's −46°. One of the lowest temperatures ever observed in the region was at Prince Albert (Saskatchewan), −70°. Mean annual ranges of temperature are very high, seldom being under 50 degrees; Winnipeg's is exactly 70 degrees.

Dfb climates, at times, are subject to bitterly cold winter weather when great masses of *cP* or *cA* air sweep out of the north. The cold blasts of air out of Siberia which swept over Europe in the winter of 1956 have already been mentioned. In January, 1957, the northern and eastern United States and southeastern Canada experienced the coldest weather of record—as low as −55° on the southern margin of the region in upstate New York.

Precipitation

Interior locations, particularly those bordering on the cool steppe, have considerably less precipitation than locations near, or adjacent to, water bodies. Higher elevations also tend to have less than lower elevations, unless recording stations are located on windward slopes. In humid sections of the region, the precipitation ranges from 30 to 40 inches; on the dry margins it will be less than 20 inches. Selected stations across the region in North America, proceeding from west to east, have the following amounts: Edmonton 17, Winnipeg 20, Duluth 26, Sault Ste. Marie 30, and Montreal 40 inches. In Eurasia, the amounts are generally smaller: Oslo (Norway) 23, Warsaw 22, Moscow 21, and Kazan 15 inches, but Nemuro has 38 inches.

North American *Dfb* stations, such as Montreal and Concord, which are located in the more humid sections of the region, have a notably even distribution of precipitation throughout the year. All months at Montreal and Concord average close to 3 inches. Nemuro, on the other hand, has a strong summer and fall maximum and a winter and spring minimum. Almost all the dry margins have strong summer maxima and winter minima of precipitation. Between the most moist and the driest (subhumid) *Dfb* stations are those that have mean annual precipitations of between 20 and 30 inches; seldom do these have less than 1 inch per month, and in most cases they have a summer maximum and a winter minimum. The rainiest month (3 or more inches) at Kiev, Oslo, Warsaw, and Duluth (Figure 21–6) is either June, July, or August; the driest month is January or February.

Compared with the *Dfa* climates, much less of the *Dfb* precipitation is convectional in origin, since the summer months are not so hot and instability is not so great. A larger proportion of the summer rainfall is of frontal origin. More cloudiness and drizzly weather result than in the *Dfa* climates. Winter precipitation is largely cyclonic in origin, but the air masses are usually of polar origin and produce relatively little precipitation. Precipitation (.01 inch or more) falls on an average of 120 days a year at Concord, and it receives 52 per cent of the possible sunshine. On the drier sides of the climate, the relative humidity is much lower than in the more moist portions, hence the amount of sunshine is notably higher, both in the summer and in the winter. The drier parts of the North American *Dfb*, however, are more subject to blizzards, cold waves, arctic continental air masses, and lower temperatures than are the more moist sections.

Lower winter temperatures also mean that a larger percentage of the *Dfb* precipitation is in the form of snow than is true in the *Dfa* climate. Early and persistent snowfalls prevent the ground from warming during the day, and at night the surface quickly radiates what little heat is gained; a prolonged snow cover serves to accentuate the cold and to drop the temperature lower than if the ground were bare. Near Duluth there are usually 75 to 80 days of snowfall each winter. A mantle of snow stays on approximately one-third of the year, from November to late March or April. At Concord eight months may have at least traces of snow. In the last 57 years, one-third of the Octobers and one-fifth of the Mays recorded snow. Although the mean annual snowfall has varied from 29 to 94 inches, the average winter has between 50 and 60 inches. In New England, New York, Michigan, Wisconsin, Minnesota, Ontario, and Quebec, winter sportsmen with increasing interest are making use of this deep and prolonged snowfall. Special weekend trains ply between the nearer *Cfa* and *Dfa* cities and the chief winter sports centers.

Responses to the *Dfb* Climate

Despite the shorter summer and the longer winter which the *Dfb* climate has compared with the *Dfa*, the vegetation and soils of the two climates are not greatly different. The growing season of the *Dfb* is shorter, however, and the major crops are different.

Natural Vegetation. Four main types of vegetation appear in the *Dfb* climatic areas. The more moist sections have mixed forests. Included are most of the eastern United States and Canada (except Ontario and southern Michigan), most of the northwestern part of Europe, much of the U.S.S.R. area around the Caucasus Mountains, and part of Hokkaido. North of these areas and in the western United States and Canada, most of the *Dfb* climate has coniferous forests. On the southern side of the mixed forests, in Ontario, southern Michigan, and the U.S.S.R., as well as in other small areas, deciduous forests prevail. On the drier margins in the United States, Canada, and the U.S.S.R., areas of prairie and steppe grasses appear.

Soils. Podsols, gray-brown pedalfers, and various mountain podsolic soils cover about two-thirds of the *Dfb* climatic zones. On the dry margins of both Canada and the United States, prairie and chernozem soils of excellent quality produce large quantities of spring wheat. Much of the southern periphery of the

Eurasian *Dfb* belt is also chernozem and prairie, or degraded chernozem.

Occupations in the *Dfb* Climate

Two agricultural concentrations are particularly noteworthy in the *Dfb* climate. On the drier margins, both in North America and in Eurasia, spring wheat attains international importance. The more humid parts specialize in dairying and are parts of the world's two most important dairying belts. Included are the eastern half of the *Dfb* area in North America and the north and northwestern parts of the Eurasian *Dfb* belt. Other agricultural products of significance are rye, barley, oats, sugar beets, millets (in the U.S.S.R.), various hays (especially alfalfa and clover), tobacco, flax, hemp, and deciduous fruit (especially in the Great Lakes area). Except for dairy cattle, animals are not so important as in the climates with warmer winters where expensive feed and housing are not required for so long a period. Swine, beef cattle, and sheep, however, are raised to some extent in much of the climate.

Fishing is of major importance in the Great Lakes, on the coasts and offshore waters of the Atlantic provinces of Canada, and on the rivers and coasts of the main Eurasian *Dfb* area.

Both in North America and Eurasia, lumbering, hunting, and forestry are practiced on the northern margins of the climate. In some places, lumbering is of prime importance; in other places, cutover lands offer little of value. The activities of the high mountain mass surrounding the Pamirs of interior Asia are limited chiefly to nomadic herding.

Fairly abundant coal, several fields of petroleum and natural gas, significant deposits of iron ore (including the waning Mesabi of Minnesota and the high-grade Magnitogorsk deposits of the southern Urals of the U.S.S.R.) as well as deposits of nickel, tungsten, and chromite mark the minerals of the *Dfb* climate. In addition, the minerals of the Urals, the Rockies, and the high mountains of the western United States have a great variety which includes most of the known minerals, including those that are radioactive. Without question one of the richest climates, from a mineralogical standpoint, is the *Dfb*.

The population distribution of the *Dfb* varies considerably. The most populous parts are the eastern section of North America and the western part of Eurasia. The Alps of Europe and the island of Hokkaido are also areas of fairly dense population. Wherever population is dense, manufacturing is usually important; the Great Lakes area, the St. Lawrence Valley, and New England have many cities which are industrial. Most of them are important for commerce also. Many cities of the U.S.S.R. owe their importance, in part, to manufacturing. Specialities in manufacturing in the *Dfb* climate include flour milling, pulp and paper processing, the canning of fish, processing of milk and cream products, iron and steel manufacturing, metal products, food processing, and the making of Swiss watches.

Since many mountainous sections as well as uplands near centers of dense population are included within the climate, winter sports activities as well as summer tourist attractions are well developed in some areas. While Switzerland has long been considered a chief playground, the United States, Canada, and even the U.S.S.R. have also developed and expanded a multitude of tourist interests that are based on hilly terrain, lots of snow, accessibility to large cities, and a generally favorable temperature regime.

The Microthermal, Warm Summer, Dry Winter (*Dwb*) Climate

Location

The microthermal, warm summer, dry winter (*Dwb*) climate surrounds the *Dwa* climate except on its southern side. It embraces eastern Manchuria, parts of northern and western Manchuria, and a small part of the U.S.S.R. above the Amur River; the north-

Fig. 21-7. Hunting and trapping are the chief means of livelihood for the members of this "Udarny Okhotnik" cooperative in the Dwb climate of Khabarovsk Territory, Siberia. (Courtesy Sovfoto.)

eastern section of Korea, the extreme southeastern part of the U.S.S.R., and sections north and southwest of Peiping, China. A large part of both the Tsinling Mountains and the Great Snowy Mountains of interior China, between latitudes 28 and 37 degrees north, are also classed as *Dwb*.

Temperature

Temperatures in the *Dwb* climate are similar to those in the *Dfb*. The *Dwb* summer temperature at Vladivostok (Figure 21–8), however, is as warm as the warmest part of the *Dfb*, while the winter temperature is as cold as the coldest part of the *Dfb*. The mean annual range of 64 degrees is therefore appreciably greater than in the average *Dfb* location. The maximum summer temperatures are somewhat delayed compared with the *Dfb* climate; July, August, and September, rather than June, July, and August, are the warmest months. August has the greatest heat (as was true at Nemuro, *Dfb*).

At Vladivostok, five winter months, which are also the driest months, are below freezing (November through March). They average about the same as those of Edmonton. The mean annual temperature is 40°, which is about the same as Duluth. The *Dwb* temperatures are consistently lower, month by month, than those of the *Dwa* climate, but the greatest difference is in the summer. At that time the *Dwb* averages almost 10 degrees cooler.

Precipitation

As the *w* indicates, the *Dwb* climate has a dry winter. Vladivostok, the chief city in the climate, averages a total of only 1.3 inches of precipitation for the five-month period from November to March (Figure 21–8). The driest month is January, which has only 0.1 inch; the wettest month is August, which receives 3.5 inches. This coincides with the month of maximum temperature. July, August, and September receive 55 per cent of the mean annual precipitation, which is only 15 inches.

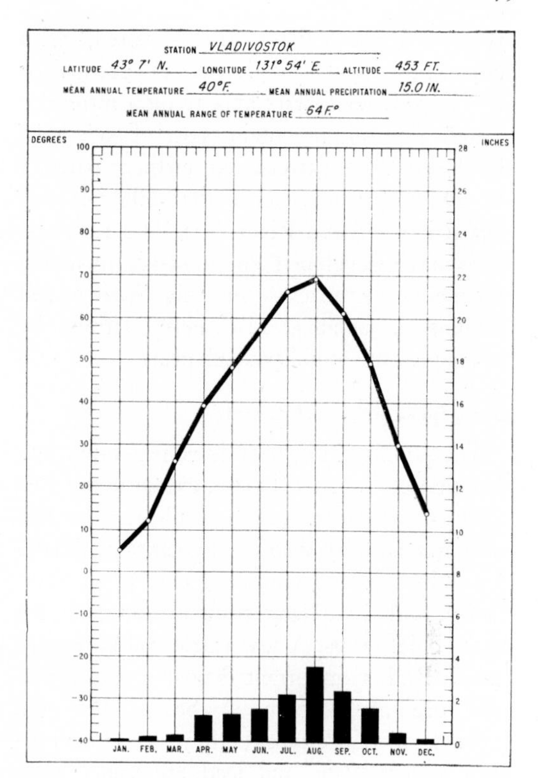

Fig. 21–8. Dwb climate.

This is less than the driest of the *Dfb* stations. The distinct summer maximum of precipitation reflects its convectional and monsoonal origin. The dry winter is the result of the subsiding and outflowing air from the intense Siberian high-pressure area. What little snow falls is usually fine and powdery. Cyclones are relatively infrequent compared with the number in the *Dfb* climate. Winter days in northern Manchuria are much clearer than in *Dfb* areas of the same latitude.

Responses to the Dwb Climate

Natural Vegetation. The eastern and northern part of the *Dwb* climatic area is clothed with a valuable mixed forest, with conifers dominating locally. The western side, plus the two smaller areas near Peiping, were originally, no doubt, partly grasslands and partly broadleaf deciduous forests. The Tsinling and Snowy

Mountains are chiefly grass-covered, with herbaceous plants singly or in groups or patches. Mixed forests appear intermittently.

Soils. Podsolic mountain types of soils have developed in the eastern and northern parts of the *Dwb* climate, except that alluvial soils appear in the Amur River area. The grassland areas have developed chernozems, including mountain chernozems in the Snowy and Tsinling mountains. Elsewhere within the mountains, mountain podsols prevail.

Occupations in the *Dwb* Climate

The population density is considerably less than that of the *Dwa* climate, except in the area southwest of Peiping. As a result, there is less economic activity than in the warmer climate. Agriculture is the dominant enterprise, with barley, millet, kaoliang, peanuts, and soybeans as the chief crops. A few cattle and swine are raised. Rather important deposits of coal and iron ore are known in Shansi Province (southwest of Peiping); they are mined to some extent. Zinc and lead are mined near Vladivostok. Some of the best forests of the U.S.S.R.'s Far East are north of Vladivostok, but they are relatively little used so far. The contiguous forests of eastern Manchuria are being utilized to a greater extent, largely for building purposes. Fishing along the coast and on the Amur and Sungari rivers is chiefly seasonal but of some consequence. Subsistence farming, nomadic herding, hunting, fishing, and collecting of forest products is the livelihood of peoples in the remote sections.

Khabarovsk, on the Trans-Siberian Railroad at the junction of the Ussuri River with the Amur, is a growing city of some recent industrial development. Birobidjan, west of it, is in the heart of a relatively new, intensively and extensively cultivated Jewish settlement. Vladivostok, at the end of the Trans-Siberian Railroad, is an old port and one of the important military bases of the Soviet Union. In spite of five months' averaging below freezing, the port is ice-free the year round, being kept open by icebreakers.

The Subarctic Climates

The *Dfc*, *Dwc*, *Dfd*, and *Dwd* climates are commonly referred to as "subarctic climates" because of their high latitudinal location; all of them represent extreme continental conditions. In them great mean annual ranges of temperature occur, the greatest being in the *Dwd* climate of northeastern Siberia. They are sometimes treated as one climate, but Köppen makes certain distinctions between them.

While most sections of all four of the subarctic climates have a forest cover, it is the *Dfc* that is particularly noted for its extensive stands of northern conifers. The Russian term of *taiga* is commonly applied to such forests; thus this climate is sometimes, and quite properly, called the "taiga climate."

The Microthermal, Cool Summer (*Dfc*) Climate

The microthermal, cool summer climate is the most extensive and also the least severe of the subarctic climates. It is probably second only to the *BWh* and perhaps the *Aw* climate in areal extent. It is the ideal habitat of the great northern coniferous forest, a region of sparse population and slow exploitation, unless unusual circumstances combine locally to make mining, lumbering, or trade profitable.

Location

While it touches the North American Arctic Ocean only in limited places, the *Dfc* climate extends completely across Alaska and Canada from the Bering Sea to the Atlantic Ocean. Except possibly for certain undistinguished high altitudes in the Cascades and Sierra Nevada, the only occurrence of the *Dfc* in the United States is in northwestern Montana, where a tongue extends down from Canada. The southern boundary passes slightly north of Minnesota, Michigan, and Maine. The latitudinal extent in North America is from 47 to 70 degrees. The northern boundary is always the *E* climate; the southern

Fig. 21-9. The long, cold winters of the Dfc climate in Alaska pose serious problems for the railroad. (Courtesy Wide World Photos.)

boundary is the *Dfb*, except along the Pacific, where a narrow fringe of *Cfc* is located seaward. Sizable islands of *E* climate, largely tundra, occur within the *Dfc*, in both Alaska and Canada as well as in Scandinavia.

In Eurasia, the *Dfc* begins very close to the North Atlantic, extending eastward across Scandinavia and Finland far into the U.S.S.R. to 121 degrees east longitude. In Eurasia it is located principally between 60 and 70 degrees north latitude, although near the headwaters of the Ob River it extends as far south as 48 degrees. A large part of the Ob and Yenisei drainage systems and a small part of the Lena are in the *Dfc* climate. Although the *Dfb* borders the *Dfc* along most of its southern extent, the *Dwc* also joins it on the east and southeast. The *Dfc* also appears in the extreme eastern U.S.S.R. There it includes the island of Sakhalin, the area surrounding the mouth of the Amur, the Kamchatka Peninsula, and a considerable area north of the Sea of Okhotsk. While the *Dfc* climate does not extend completely across Eurasia, it is interrupted by only about 40 degrees of *Dfd* and *Dwd*.

Temperature

It is the long, bitterly cold winter as well as the very short summer that distinguishes the subarctic climates. Winter follows soon after summer, and summer follows soon after winter, so that both fall and spring are transition seasons of short duration. Frosts, oftentimes killing ones, occur as late as early June; they resume again in late August or early September. The growing season, therefore, is at best 3 months, but often it is only 2 or 2.5 months. Even in midsummer a northerly wind may drop temperatures below freezing; thus midsummer frosts with their disastrous results to vegetation may occur.

Much of the ground is frozen deeply and may not thaw out from year to year, except on the surface. In some places the ground is reported to be permanently frozen over 200 feet deep, hence the name *permafrost* that is often applied to such areas. When the ground is solidly frozen, transportation is relatively simple; but when two or three feet thaw in the summer, producing a mass of spongy ground and superficial swamps, it is difficult to negotiate much of the taiga.

Long nights and very short days characterize the winter. Under clear conditions it is not surprising that rapid nocturnal radiation causes the temperature to drop quickly. Minimum temperatures of −50 or −60° are known to occur within the *Dfc* climate, and in places like Dawson (Yukon Territory), winter temperatures often remain well below zero for many days at a time. In fact, it was at Dawson in December, 1917, that the mean temperature for the month was −49° and the highest for the month did not go above −35°. Occasionally, the warm sector of a passing *Low* may raise the temperature close to, or above, freezing. On the other hand, large accumulations of *cP* or *cA* air may sweep southward, especially over lowlands, attended by strong winds bearing snow. Such a condition in Canada is called a blizzard; it can cause much suffering to men or beasts who may be exposed to it. A particularly violent type is the buran that blows out of the northeast in Russia and western Siberia; another is the purga from the Siberian tundra.

Six or seven months have mean temperatures below freezing, which means that true winter can begin in October and remain until late April or early May. While Archangel (U.S.S.R.) and Haparanda (Sweden) have no months with means below zero because they are located on large bodies of tempering water, places located in continental interiors such as Fort Vermilion (Figure 21–10) and Fort George (Quebec) may have three or four months with means of zero or below.

The Mackenzie River in Canada and the Ob and Yenesei rivers of Siberia are frozen from five to eight months—a longer period in the lower (northern) parts than in their upper basins. Thawing in early summer, therefore, almost invariably results in floods in the lower basins because the mouths are choked with arctic ice and the ground is still frozen so that it is unable to absorb any water. Any lakes in the drainage basins remain frozen longer than do the rivers, although they freeze later in fall. Consequently, the use of rivers for navigation, significantly so in Canada, is limited to the short period between the thawing of the lakes in early summer and the freezing of the navigable streams in early fall. In northern Siberia, the rivers in some years are navigable for only two months.

July is invariably the month of maximum temperature that ranges between 55 and 60°, and August is above 50°. June may or may not be that warm; thus only two or three months can be said to have even moderate warmth, and summer as known in the United States hardly exists. Absolute maxima, however, may reach 90°—it was once 98° at Fort Vermilion—though the usual daily maxima do not exceed 70 or 75°.

Perhaps it should be emphasized that the northern boundary of the subarctic climates is the isotherm of 50° for the warmest month. Only four or five months at best in the *Dfc* climate are warm enough for vegetal growth, and the so-called "growing season" may not be more than two months, with frost in some years occurring every month. The northern limits of the taiga are mute evidence that the amount of heat received is much less than on the southern boundary.

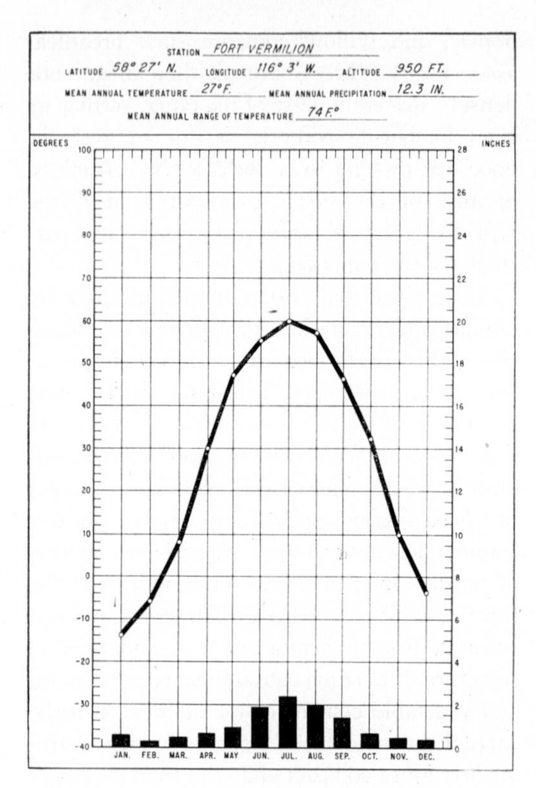

Fig. 21–10. Dfc climate.

Compensating somewhat for the short summer is the length of the summer day. Quick-ripening fruits and vegetables, and even grains, receive the equivalent of a longer growing season. Along the 50th parallel, the sun shines 17 hours a day in the middle of June; it shines 16 hours in the middle of July when the actual acquisition of heat is greater. On the 60th parallel, the duration of sunlight is 18 hours in the middle of July, and beyond the arctic circle there may be continuous sunlight for days, weeks, or even months, depending upon latitude (Tables 12–5 to 12–7). For example, at 70 degrees north latitude, the duration of sunlight is over two months.

Mean annual temperatures are between 20 and 30° in continental interiors. They are slightly higher where there is some marine influence; at both Archangel and Haparanda the figure is 33°. For the same reason, the mean annual ranges at the latter stations are only 48 and 47 degrees, respectively; but they are more likely to exceed 70 or 80 degrees at other places (82 degrees at Dawson).

Precipitation

Low-latitude climates, having no more precipitation than an average *Dfc* station, would be classed as arid or semiarid. Because evaporation and temperatures are not high, its 15- to 20-inch mean annual precipitation is sufficient to produce moderately dense taiga forests. The low precipitation is the result of relatively low temperatures, for the most part continental rather than marine influences, low specific and absolute humidity, and strong anticyclonic control in the winter season.

In most places precipitation is concentrated in the summer, but occasionally it extends into the early fall. At Haparanda there is a delayed season of maximum precipitation, with August through November receiving 2 or more inches monthly. At Archangel and Fort Vermilion (Figure 21–10) both July and August receive 2.4 inches, but at Churchill, Haparanda, and Dawson, September is the rainiest month, with 3.6, 2.4, and 1.7 inches, respectively.

Summer precipitation is mostly frontal in origin (except in mountainous areas) and only slightly convectional. Although summer days may be quite warm, or even hot, there is insufficient heat to cause frequent convective overturning. The occasional thunderstorm is usually cold-frontal in nature. Measurable precipitation may be expected on about one-third of the days during June, July, and August. This relatively large number of rainy days seems incompatible with the fact that in interior Canada, despite the *Dfc's* being the most moist of the subarctic climates, the sun shines approximately from 40 to 50 per cent of the possible time in summer. This seems reasonable, however, since the duration of any rain may be only a few hours, while the duration of possible sunshine during those summer months averages from 16 to almost 24 hours a day.

Winter, with its low temperatures and low absolute humidity, is the season of least precipitation, practically all of which comes in the form of dry, but hard, powdery snow. The snow cover lasts five to seven months. Each winter month receives the equivalent of from 0.5 inch to as much as 1.5 inches of precipitation. January, February, March, or even April may be the month of minimum precipitation.

The persistence of the winter anticyclones over Greenland and Siberia and the fairly regular outbursts of cold air mean that cold frontal conditions extend far south, often beyond the limits of *Dfc* climates. On occasion, however, mild humid air from intermediate latitudes may be forced far north into *Dfc* areas by cold air masses on their left, causing warm frontal conditions with the attendant cloud or fog and frequently heavy snow. While many people still think of the buran of central Asia and the blizzard of North America as belonging to the northern forests, the trees actually act as deterrents to such storms since they break the continuity of the horizontal wind movement and thus lessen its velocity. Although blizzards do occur within the taiga, their best development and greatest frequency are on the steppes and other less vegetated areas of upper-middle latitudes.

Responses to the *Dfc* Climate

Natural Vegetation. The taiga prevails over the vast areas that have the *Dfc* climate. Here and there, however, areas of tundra on one hand and mixed forests on the southern margins make appearances. According to Küchler, much of Siberia east of 65 degrees east longitude has a form of taiga that is deciduous because of the dry winter conditions which cause the needles to be shed wholesale. The great northern coniferous forest proper is composed of relatively few species, with fir, spruce, pine, and larch predominating. Where heavier nonsandy soils (usually clay) exist, deciduous patches are common. Such species as white and yellow birch, alder, choke cherry, poplar, and willow compose such broadleaf associations. Trees are neither huge nor densely spaced in most of the taiga, so that its total lumbering potential is not so great as once was thought to be the case. Nevertheless, because of the vast area covered, it is the world's greatest softwood reservoir, particularly for pulpwood.

Soils. Since soil is determined in part by climate and vegetation, the *Dfc* has developed one of the world's most distinctive, widespread, and poorest of types, called the *podsol.* The net result of 15 to 20 inches of precipitation, with severely cold winters and short, moderately warm summers, is actually an excess of precipitation over evaporation. Active summer leaching follows. A ground covering of needles intensifies the acidic nature of the topsoil, which is usually thin, gray, or ash-colored, low in humus content, and poor in structure. Eluviation causes the meager mineral and vegetable content of the *A* horizon to be carried down to the *B* horizon, which is frequently more compact and solid than the upper surface. Acid as it is, the sandier soil is more easily tilled than the heavier podsols which have a poor texture and tend to become highly compacted if ploughed when too wet.

Many pockets of peaty soils have developed within the taiga. Some of them are acidic in nature, while others are of good quality, having their origin largely in the decay of swamp vegetation. Other islands within the podsols are clay soils of fair to superior quality. A few of them are fairly large and have been developed agriculturally on the southern edge of the *Dfc* in Canada, Finland, and the U.S.S.R. In highland areas, mountain types of podsols, with islands of tundra and alpine meadows, occur. Along the Ob River, young alluvial soils of good potentialities have been formed.

Occupations in the *Dfc* Climate

The taiga is largely unpopulated and little utilized. Hunting, fishing, trapping, collecting, primitive agriculture, and forestry are prac-

ticed here and there throughout the area, usually only on a subsistence basis. On the southern border, in Alaska, Canada, Scandinavia, Finland, and the Soviet Union, farming of a somewhat higher order (hardy grass and root crops) and forestry (both lumbering and pulping) are practiced. Some of the latter interests are large commercial holdings. Lumbering is also carried on at easily accessible places in the far northern part of the U.S.S.R., at Igarka, for example, which is north of the arctic circle.

The old granitic base of the Canadian shield, as well as of Siberia, is rich in minerals. The Urals alone are said to have practically every known mineral; certainly there is a great variety. Important iron deposits are worked in several parts of Canada, in Sweden, and in the Soviet Union. Gold is important in Alaska, Canada, and the U.S.S.R. Minor coal deposits occur in Alaska, Canada, and the Soviet Union. Deposits of copper and zinc appear in several places. Nickel and oil in Canada, platinum, phosphates, and radioactive minerals in the U.S.S.R. are also known. Nomadic herding is practiced in much of the northern periphery of the climate in Eurasia and in northwestern Canada and Alaska. Lapps, Eskimos, Samoyeds, and other groups eke out a living by grazing reindeer, and by hunting, trapping, fishing, and collecting.

Commercial fishing is very important in the offshore and associated waters of much of eastern Canada, Sweden, Finland, Alaska, and the Soviet Union, especially in Sakhalin. Fur also has long been exploited by traders in much of the *Dfc* territory. All the nations within the region have long been engaged both in trapping and in the trading of furs, yet the total number of people involved is relatively small.

The impression may have been given that the entire taiga is a more or less uncivilized domain; that is not the case. The easily accessible parts, particularly Scandinavia, eastern Canada, southern Alaska, and the northwestern part of the U.S.S.R. have fairly high, but not very complex, economies for the most part. But the large and relatively inaccessible interiors, except where mining or some other lucrative enterprise is located, present a wilderness where life is hard, isolated, and rather primitive in the eyes of the average American. Furthermore, the annoying presence of the mosquito and its pestering ally the black fly during the summer is an important deterrent to settlement.

The Microthermal, Cool Summer, Dry Winter (*Dwc*) Climate

Location

This cool summer and dry winter climate is limited to a large area in the eastern U.S.S.R. and northwestern Manchuria, as well as a small section northwest of Peiping. Its latitudinal position is between the 40th and 60th parallels, and it extends from about 90 to 150 degrees east longitude. It is somewhat farther south than most of its moist counterpart, the *Dfc*. Since most of the region lies east of Lake Baikal, Köppen has referred to it as the Trans-Baikalia climate. Unfortunately, reliable published information, other than that furnished by climatic records, concerning both the *Dwc* and *Dwd* climates is definitely lacking; and even the climatic records are relatively few and fragmentary.

Temperature

The temperature conditions in the *Dwc* are similar to those in the *Dfc* but are generally more severe. Rivers flowing into the arctic are frozen from six to eight months, while Lake Baikal is frozen for at least 4½ months, the ice sometimes attaining a thickness of 9 feet. As in other parts of Siberia where the rivers flow northward, the late spring or early summer floods limit navigation to even fewer months.

Seven months with temperatures below freezing—at least three of those well below

Fig. 21–11. The short, cool summers and the long, cold, and dry winters do not favor a luxuriant forest cover in the Dwc lands of the upper Lena River Valley in Siberia. The long and moist summer days, however, permit forage and hardy grain crops. (Courtesy Fotokhronika Tass.)

zero—reflect a long and severe winter. Low as temperatures are, however, they are alleviated somewhat by the dry conditions and by rather quiet air settling in the great Siberian anticyclone, so that temperatures do not seem so low as in the moist *Dfc* climate. Along the east Siberian coast, however, raw weather with many foggy days is to be expected. The coldest month at Okhotsk (Figure 21–12) is January with −11°; March and November are only slightly above zero. By April, the temperature is climbing toward the freezing mark. July and August are moderately warm, about like New York in October or San Diego in January, but both June and September are transitional months, while winter begins in October to last until May. Lucky are the *Dwc* gardeners when the growing season is more than two months, for any month—even from June to September—may have frosts. Absolute minima dip to −70° in the interior near Lake Baikal. Absolute maxima may go to 90° or higher, but from 65 to 80° is the rule. Although mean annual ranges are about the same as the two *Dw* climates already discussed, the latter have summer temperatures 10 to 20 degrees higher.

Precipitation

Location on the eastern, or leeward, side of the large Eurasian continental block, coupled with strong anticyclonic and monsoonal control in the winter, leads to dry conditions that persist during the colder half of the year. Okhotsk (Figure 21–12) has nine months, each with less than 1 inch of precipitation, even though the city is located on the coast where marine influences would cause not only more precipitation but also a warmer winter if it were not for the strong, offshore, seasonal monsoonal influence. The greatest dryness occurs in January, February, and March, when only 0.1 inch falls in the average year. November, December, and April are only 0.1 inch better off. The November-through-April precipitation totals only 0.9 inch. June, August, and September have over an inch of precipitation each, with the latter two months getting the greatest concentration. Together they receive over 50 per cent of the year's precipitation of 7.5 inches. The late summer maximum of precipitation at Okhotsk is partly convectional but mostly cyclonic; by November the strong anticyclonic influence has again set in, and little precipitation results.

Snow falls from early October, or even earlier, into May and, occasionally, later. Most of it is fine, hard, and powdery. Since there is so little of it, the ground is often only very lightly covered, or it may be swept bare by the wind and piled into small drifts. Clear

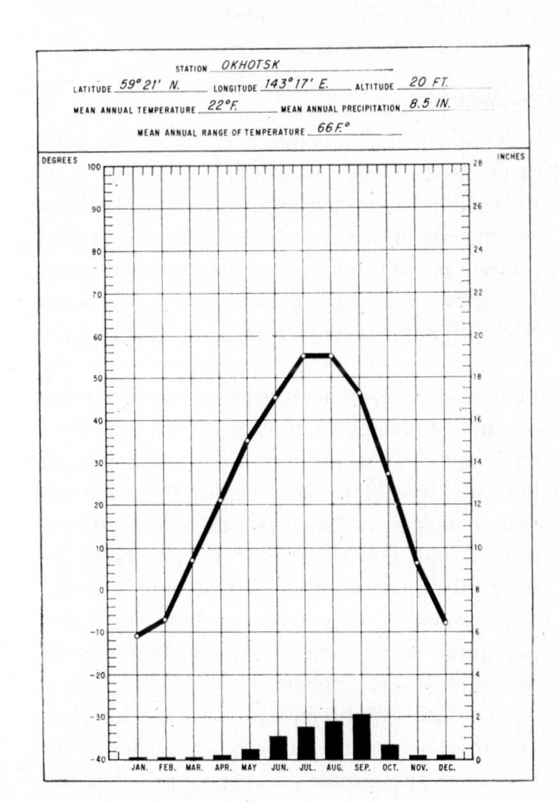

Fig. 21–12. Dwc climate.

conditions prevail throughout the winter over the interior; and even in the summer, cloudiness is only about 50 per cent throughout most of the region.

Responses to the Dwc Climate

Natural Vegetation. Reflecting the dry winter conditions, most of the vegetation north of 50 degrees and somewhat east of Lake Baikal is coniferous, but deciduous rather than evergreen. Surrounding Lake Baikal and to the west, however, where somewhat more winter precipitation falls as the *Dfc* is approached, the true evergreen coniferous forest prevails. South of Lake Baikal much of the area is short grassland that merges eventually into the dry Gobi steppe and desert. The Khingan Mountains of the northwestern part of Manchuria have mixed forests, but both the needleleaf and broadleaf trees are deciduous, so that they are bare in the drier parts of the year. Patches of tundra appear in the northern part of the climate.

Soils. The topography of the *Dwc* climate is composed mostly of mountains and plateaus. In spite of the dry winter, there is sufficient leaching to cause mountain podsols to form under a coniferous forest; they cover most of the climatic region. The area south of Lake Baikal has chernozem, chestnut and brown soils, which are the product of grassland decomposition. Some alluvial soils exist along the Amur, and tundra soils have developed in limited areas.

Occupations in the Dwc Climate

In the truest sense, the *Dwc* climate offers little to man; what he obtains from it and from the lithosphere he must wrest arduously and ingeniously. Were it not for the Trans-Siberian Railroad and the new railroad north of Lake Baikal, it is safe to say that the entire area would be much less densely populated. Irkutsk, Chita, Ulan Ude, and Komsomolsk, all located on the railroads, are cities of some size and importance. Irkutsk is a city of 300,000;

Komsomolsk is a new but rapidly growing city of over 70,000 and is reported to have the largest steel mills and shipyards of the Soviet Far East. Ulan Ude is a meat packing center. Away from the railroads and chief rivers, which are virtually the only transportation media, and thus away from the towns and cities also, population is sparse indeed.

Minerals are varied around Lake Baikal. Tin, zinc, gold, tungsten, arsenic, molybdenum, coal, iron, and radioactive minerals are all exploited. The northern part of the climatic region includes a portion of the Aldan (River) shield, which has important gold exploitations. Both the mining and the reduction of minerals are factors in the locations of major towns and cities. Lumbering, fishing, trapping, hunting, and meager agriculture, both for town supplies and subsistence, mark the livelihood of additional households. Much of the southern boundary has agriculture of a fairly good order, including stock raising, particularly near Lake Baikal. West of Lake Baikal, along the Trans-Siberian Railroad, a narrow belt of steppe is rather intensively sown to barley, wheat, oats, and rye. South of Lake Baikal, toward the Mongolian Desert, nomadic herding is practiced.

The Microthermal, Extreme Subarctic (*Dfd*) Climate

Location

In the northeastern part of Siberia, between latitudes 60 and 72 degrees north, are two relatively small areas having a microthermal, extreme subarctic (*Dfd*) type of climate. One is a lens-shaped area west of the Lena River; the other is north of Kamchatka Peninsula, in the extreme northeastern part of Asia. Although small portions of the climate reach the Arctic Ocean, the oceanic influence is negligible for the climate is not only subarctic but distinctly continental in nature, having a very cold winter with a great mean annual range of temperature.

Temperature

In order to classify as *Dfd*, the coldest month must average below −36°. At Viliuisk (Figure 21–14), five months have means below zero. Both April and September are substantially warmer. Since Viliuisk is located rather far in the interior, at about the same latitude as Okhotsk, its summer temperatures are somewhat higher than at the latter place which is located on the ocean. The three warm months (June, July, and August) average above 50°, with a July mean of 65°, the same as Peoria in September or Duluth in August. Duluth in August is considered delightfully cool to vacationers from the corn belt, or farther south, thus the short *Dfd* summer in Asia should be considered only moderately warm at best. The average temperature for the year is about midway between zero and freezing, 15°, and the mean annual range is very great, 103 degrees; both these figures emphasize the severity of the climate.

Precipitation

Although *Dfd* precipitation is low in terms of inches (9 inches at Viliuisk), the high latitude and low temperatures produce little evaporation; conditions therefore remain moist. No month is without precipitation, but both February and March have only 0.3 inch. Seven months (October through April) have considerably less than 1 inch each, while the months of May through September range from 0.9 to 1.6 inches each (Figure 21–14). Precipitation is concentrated in the warm season, but only June, July, and August receive more than 1 inch each. Since August is the rainiest month and February and March have the least precipitation, there is a definite seasonal retardation of both maximum and minimum precipitation as compared with a number of the less severe climates.

Most of the precipitation is cyclonic in origin. The semipermanent anticyclonic influence in the winter prevents some cyclones from invading the climate, as is true of the

Fig. 21–13. This bleak scene in northeastern Siberia near the arctic circle gives some notion of the difficulty of living in the Dfd climate. Permanently frozen subsoil precludes a forest cover along the coast. (Courtesy *Ewing Galloway, New York.*)

Dwc climate, yet they bring enough to maintain fairly humid conditions. Neither snow cover nor cloudiness is so great as in the *Dfc* climate.

Responses to the *Dfd* Climate

Natural Vegetation. Low total precipitation and proximity to dry-winter areas on at least one side have produced different vegetative types within the two *Dfd* regions. West of the Lena, deciduous conifers are the major association, while the eastern zone is mostly tundra, with deciduous taiga on its western border.

Soils. Again, soils closely reflect climate and vegetation. Podsols have developed west of the Lena, but tundra soils with some podsols

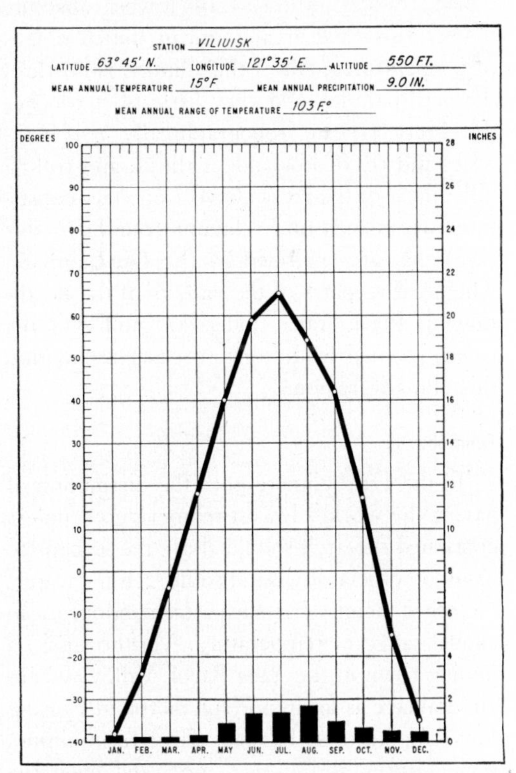

Fig. 21–14. Dfd climate.

occupy the eastern area. Locally some black soils, especially along the Vilyui River, have developed from valley grasslands.

Occupations in the *Dfd* Climate

Human adjustments to the *Dfd* climate are difficult because little is offered naturally. Population is very sparse and towns are few. Nomadic herding, hunting, fishing, a little gold mining, and a bit of forestry occupy the time of the inhabitants, part of whom are Mongolian; these people also occupy the tundras.

The Microthermal, Extreme Subarctic, Dry Winter (*Dwd*) Climate

Location

Sandwiched between the two sections of the *Dfd* climate, and north of the *Dwc*, is the climate that has the reputation of being the world's worst in terms of lowest absolute temperatures and greatest mean annual ranges of temperatures. All of this climate is north of the 60th parallel, and very little of it touches the Arctic Ocean. Topographically, it is composed mostly of lowlands, valleys, and rolling hills; most of the areas above 2,000 feet constitute two islands of *E* climate (tundra in this case) that are encircled by the *Dwd* climate. The greatest parts of the valleys of the north-flowing Lena, Yana, Indigirka, and Kolyma rivers (named from west to east) are within this climatic region.

Temperature

To the *Dwd* climate goes the distinction of having the world's lowest temperatures, unless continued research should show the icecaps to have the official lowest absolute temperatures. No other vicinity has such a tremendous mean annual range of temperature. Verkhoyansk, a mining town in the Yana River Valley, where air drainage is an important factor in causing low temperatures, has had absolute minimum temperatures as low as −90°, and even the mean January temperature is −58° (Figure 21–16). July, on the other hand, has a mean of 60°. That makes the mean annual range 118 degrees. Oimekon, relatively nearby, claims an absolute minimum temperature of −113°. Obviously the inducements to live under such conditions must be great! While Yakutsk on the Lena River (−46°) is not quite so cold as Verkhoyansk in January, characteristically the coldest month in the *Dwd* climate, it is warmer in July (66°); its mean annual range is 112 degrees.

Three months attain means of 50° or more, with July, the warmest month, reaching 60° or more. Therefore, the warm-month temperatures of the *Dwd* climate are fully as high as the *Dc* climates. It is the extreme winter cold, as previously indicated, which distinguishes the *Dwd* from all other climates. Five months (November through March) average below zero, and two more (April and October) are below freezing. May and September are only slightly above freezing at Verkhoyansk and just above 40° at Yakutsk. Mean annual temperatures are scarcely above zero, averaging 3° at Verkhoyansk and 12° at Yakutsk. Because the winter air is so dry, it is said that the low temperatures are not so difficult to bear as one might presume.

Precipitation

The mean annual precipitation varies noticeably from place to place. Yakutsk receives 14 inches annually, while Verkhoyansk receives only 5 (Figure 21–16). The greater part of the precipitation falls during the six-month period from May through October; about one-half of it is received in June, July, and August, the three warmest months. The amount is from about 1.0 to 2.5 inches per month. Minimum precipitation occurs in the winter months, with the least amount in February, 0.2 inch at Yakutsk but 0.1 inch at Verkhoyansk; the three coldest months receive only 10 per cent and 15 per cent of the annual rainfall; hence the reason for characterizing this as a dry winter climate.

Fig. 21–15. Sparse forests of semideciduous coniferous trees are characteristic of the bitterly cold Dwd climate of northeastern Siberia. In spite of the dry winters, snow remains on the ground throughout most of the winter; the season here is April. (Courtesy Fotokhronika Tass.)

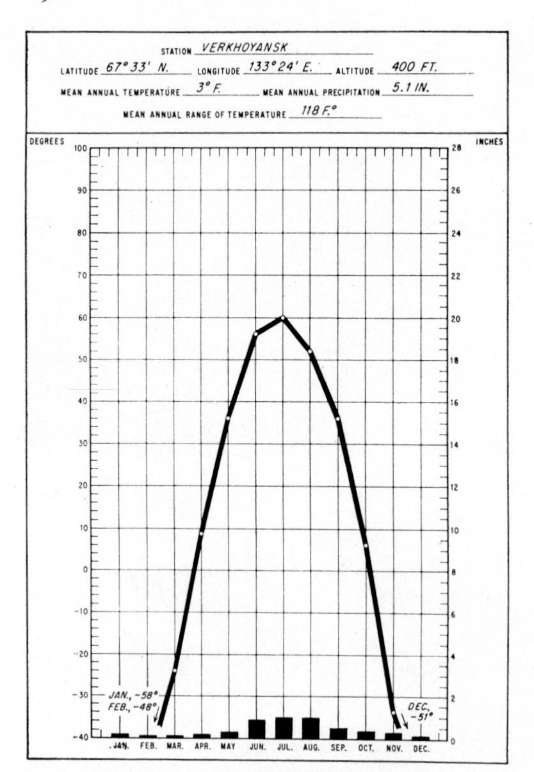

Fig. 21–16. Dwd climate.

No part of Siberia is more subject to control by the cold-season anticyclone. Clear, bright, and severely cold weather prevails; absolute humidity is quite low. The lack of clouds favors rapid radiation and cooling after sundown. In addition to this, the winter nights are very long and the days are very short, so that each night seems to vie with others in an attempt to reach low temperatures. Snow blankets are not usually deep, but they last from five to seven months; they, too, aid in nocturnal radiation and cooling.

Precipitation, both in winter and in summer, is largely frontal. Seldom is there sufficient and continuous summer warmth to produce much convection; even the unusual thunderstorm is of cold-frontal origin.

Responses to the *Dwd* Climate

Natural Vegetation. As is true in the *Dwc* climate, here also the taiga sheds its needles in response to the dry winter; also the forest is more open than in the more moist subarctic climates. Locally there are islands of tundra climate.

Soils. As would be expected, podsols have formed under *Dwd* climatic conditions, except in limited areas in the Lena River Valley where immature alluvial soils are found. In spite of the small mean annual rainfall, considerable leaching and high acidity characterize the extremely poor soils that are natively of little value.

Occupations in the *Dwd* Climate

It is largely under duress or because of the glitter of minerals that people migrate to the *Dwd* climate. Not much forestry, agriculture, commerce, or industry can be sustained under present conditions, except perhaps under subsidy. A small amount of precarious mixed farming is reported in the alluvial soils of the Lena, but elsewhere farming is merely gardening to supply local mineral towns and for home use. Much of the gardening is carried on underground with the use of artificial lighting. Some nomadic herding extends southward from the tundra proper into the *Dwd* climate and the islands of tundra it includes. Some towns, such as Verkhoyansk, are said to exist because of the exploitation of minerals, such as gold, coal, and zinc.

22

The Polar, or Arctic, Climates

All polar, or arctic (*E*), climates have one characteristic in common: no month has a mean temperature above 50°. Here much of the similarity ends. This particular isotherm of 50° for the warmest month is commonly accepted as the boundary between polar and other climates, not merely because it seems to mark the cold or poleward limit of trees, but more importantly because it definitely marks the cold-temperature limit of agriculture; the few inhabitants must depend for food upon sources other than local crops. To be sure, large-scale agriculture is also limited in the colder portions of the *D* climates; but in these climates, subsistence crops, such as cabbage, turnips, potatoes, and certain cereals, are possible in most years.

Although latitude is the most significant control, other physical factors exert important controls, particularly land versus water, altitude, cyclonic storms, winds, and even ocean currents. Thus, polar climates do not comprise a single, well-defined unit in the same way that other types of climate do: *BSh* or *Cfa*, for example. Polar climates cover large areas, not only in high latitudes but also at high altitudes in middle and low latitudes. Over such a vast area as that embraced by the *E* climates, the physical controls cause distinctly different climates to exist. For example, parts of the arctic prairies of Canada have very limited precipitation but severe winter cold; coastal stretches of Baffin Island and Labrador have much heavier precipitation but definitely less severe winters; and the Aleutian Islands experience typically marine conditions in that they have relatively small mean annual ranges of temperature with much cloud, fog, and rain. The *E* climates that occur in other than high latitudes tend to take on the characteristics of the lower lands which surround them, that is, they tend to have the same ranges of temperature and the same distribution of precipitation as the lowlands, though the degree of cold or warmth and the amount of precipitation may vary markedly.

With the expansion of air transportation and the consequent need for weather stations and landing strips in the arctic, with the increased demand and search for minerals, and with the constantly enlarging requirements for scientific data in the fields of geophysics and meteorology, there is good and sufficient reason for a close study of polar climates, especially in high latitudes. Accordingly, more attention will be given to the *E* climates than the sparse density of population would otherwise seem to warrant.

Although the new Köppen-Geiger map makes no distinction between the many and varied types within the *E* climates, they will be discussed under two categories: (1) those in which the mean temperature of the warmest month is above 32° and (2) those in which the mean temperature of the warmest month is below 32°. In spite of the fact that the word *tundra*, by definition, refers to the treeless, often boggy, plains of Siberia, that word has been used by most plant geographers to designate a type of natural vegetation composed of diminutive trees (if any), quick-flowering herbaceous plants, including a variety of grasses, and certain types of mosses and lichens; because such vegetation is typical of the lands having the climate under the first category, the term *tundra* will be applied to this type of climate. In the second category, plant life is almost wholly lacking because the mean monthly temperatures are perpetually below freezing, so the name *frost* or *icecap* will be applied to this climate. In the tundra climate there are permanent human habitations; in the frost climate, permanent homes are not found, although it is not impossible with modern types of clothing, shelter, and supply for people to live there indefinitely as they have done, and are doing, on the icecaps of Greenland and Antarctica.

The Tundra Climate

Location

Most of the northern fringes of North America and Eurasia are tundra. Numerous Arctic Ocean islands are also chiefly tundra. Only the southwestern and southeastern fringes of Greenland and very limited edges of Antarctica, however, are warm enough to be tundra. Most of the *E* portions of Alaska (except the glacial areas), parts of the high mountains of western North America, substantial portions of the Andean system of South America, and the greater part of the huge *E* area of Tibet and associated countries also have tundra climate. Throughout the middle and high latitudes, other upland areas, generally too small to be shown, are either tundra or icecap. In the mountainous areas of the middle and low latitudes, high elevation, of course, is responsible for the temperature regimes that are necessary to produce the tundra. The climate, in its temperature characteristics, is transitional between the icecap of perpetual ice and snow and the subarctic climates with short summers and long severe winters.

Temperature

Tundra temperatures are varied. Stations such as Pt. Barrow, Upernivik (Greenland), and Sagastyr (the U.S.S.R.) have from three to six months averaging below zero; even though these are all located on coasts, they have more severe winters than the majority of tundra stations. Many other places, coastal or otherwise, have no months with means below zero. While continentality, therefore, has its effect in some coastal areas, marine effects are evident in others; furthermore, conditions that would seem to reflect marine influences are found in some interior middle-latitude mountain tundra locations. It is well known that mountain stations usually have relatively small mean annual ranges of temperature, and for this reason some climatologists prefer to group all high mountains together, thus classifying them as having highland climates, rather than distinguishing the various climates that appear at various altitudes. But, recognizing that there are differences with elevations, Köppen classifies

Fig. 22–1. Life in this tundra phase of the E climate in northern Sweden is less rigorous than in some of the extreme continental situations farther south and east. (Courtesy Swedish Information Bureau, Inc.)

all major climates occurring in the larger mountainous areas according to his formula.

The Siberian tundra stations have colder winters than the North American arctic frontier because of the prevalence of offshore winds from the intense Siberian *High*. While high pressure prevails over the northwestern part of North America in the winter, it is not so high nor so large as the Siberian *High*, hence temperatures are not quite so low. There are six months of subzero means at Sagastyr, from November through April, and there are comparable means at Pt. Barrow. While Upernivik on the western side of Greenland has three months (January, February, and March) averaging below zero, Angmagssalik on the eastern side has no month averaging below February's 13°.

Coldest months are always January or February, or, in places that have three months of the same minimum, March is tied with January and February for lowest temperatures. Ben Nevis, Scotland, (24°) and Berufjord, Iceland, (30°) are examples. Of the eleven stations that were compared in this study, only four had months with mean temperatures below zero; the other seven had coldest months ranging from 2 (Pikes Peak) to 37° (Evangelists Island, southern Chile). It is evident, then, that many tundra stations have much

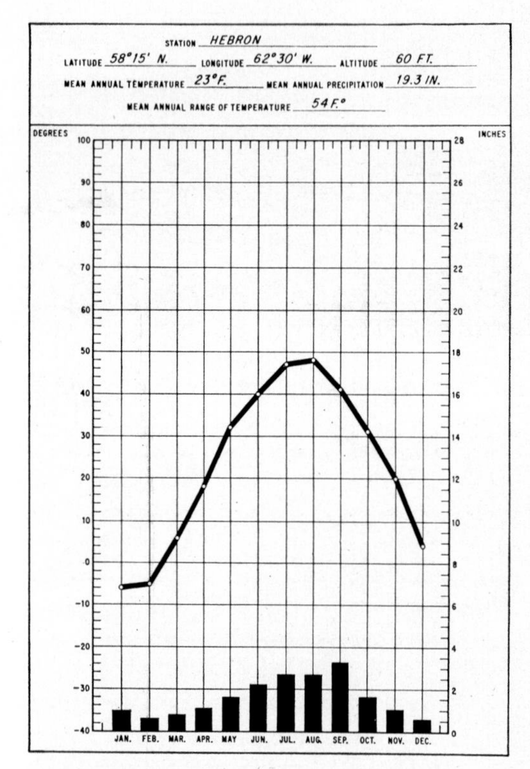

Fig. 22–2. E (tundra) climate.

smaller ranges of temperature than others. Absolute minima of −50 to −70° appear in the coldest parts of the tundra.

Warmest month temperatures in the tundras are similar, regardless of exposure or elevation. They range from 40° at Pt. Barrow and Pikes Peak to 48° at Berufjord, Hebron (Figure 22–2), and Vardō (Norway). July or August is the warmest month in the northern hemisphere, January or February in the southern. The coldest tundra localities have only one month with means of 40° or more, but the majority have three or four, while Evangelists Island has ten months that average at least 40°. Absolute maxima reach over 80° in the tundra of southern Alaska and in parts of Canada on rare occasions, but the usual "summer" day reaches around 50 or 55°. Even then cold spells frequently drop the maxima to 40°, and nights may drop to, or below, freezing.

Days, in the tundras of the high latitudes, are long in the warm season. While the longest summer day at the arctic circle is only 24 hours (summer solstice), the length of continuous day increases up to six months at the north pole. Since a considerable part of the world's tundra is north of the arctic circle, inhabitants there witness weeks or months of continuous daylight, depending upon their latitude (see Table 12–7). The angle of the sun is never high at such latitudes, however, and therefore the amount of heat received is slight. Yet, while the sun is continuously above the horizon, there is little cooling, and the diurnal range is only 5 to 15 degrees. In the winter, the arctic regions are in darkness almost as much as they are in light during the summer;[1] because of the very short days, again depending upon latitude, the winter diurnal temperature range is also small.

Mean annual ranges of temperature are as varied as winter temperatures. Evangelists Island has a range of only 10 degrees, reflecting marine influence. Ben Nevis and Berufjord record only 17 and 18 degrees, respectively. Others have 20, 30, or 40 degree ranges. Those having mean annual ranges of 50 degrees or more are: Upernivik 50, Hebron 54, Pt. Barrow 59, and Sagastyr 77 degrees.

A summary of the eleven stations reveals several interesting facts: (1) Sagastyr, with the lowest mean annual temperature (1°), has the greatest mean annual range (77 degrees) and the least precipitation (3.3 inches); (2) Evangelists Island has the highest mean annual temperature (43°), the lowest mean annual range (10 degrees), and next to the greatest mean annual precipitation (117 inches); (3) Hebron and Angmagssalik, as more temperate stations, are between the extremes. Angmagssalik, for example, has a mean annual temperature of 28°, a mean annual range of 31 degrees, and a mean annual precipitation of 36 inches.

[1] Because of the refractive effect of the earth's atmosphere, the period of continuous sunlight is somewhat longer than the period of continuous darkness at any particular latitude.

Precipitation

Pronounced differences of mean annual precipitation are found in the tundra. In addition, some reporters claim that there are great variations from year to year in some places. A comparison of the eleven stations from many parts of the world's tundras shows a range from 3 inches at Sagastyr, just north of Verkhoyansk, to 171 inches at Ben Nevis, in the Grampian highlands north of Glasgow, Scotland. Since the amounts are so varied, there cannot be said to be a representative amount. The greatest number of tundra stations, however, receives around 20 to 45 inches: Hebron has 19 inches, Vardō 26, Pikes Peak 30, St. Paul Island (Alaska) 32, Angmagssalik 26, and Berufjord 46 inches. Some, but not all, of the inland stations receive somewhat less than these amounts.

Maximum precipitation often occurs in the warmest season, but it frequently occurs in the fall and sometimes in the spring. Not infrequently, on the other hand, the greatest amount may be in the coldest part of the year. Examples of a summer maximum are Upernivik, Evangelists Island, Hebron (Figure 22–2), and Sagastyr; examples of a winter maximum are Berufjord, Vardō, Ben Nevis, and Angmagssalik. The winter maxima of the latter four, all of which are in the North Atlantic, are the result of the unusually large number of cyclones which converge in the area of the semipermanent (winter) Icelandic *Low*. Still other places have either a fairly even distribution or no particular distribution pattern. Places receiving from 20 to 45 inches of precipitation have a maximum monthly precipitation from 3 to 6 inches.

Minimum precipitation also varies as to season. Frequently late winter or spring has the least amount. This is true at Pt. Barrow, Angmagssalik, St. Paul Island, Vardō, Hebron, Ben Nevis, and Sagastyr. Minimum-month precipitation varies from nothing at Sagastyr to 7.8 inches at Ben Nevis, but again, places receiving 20 to 45 inches mean annual precipitation have minimum monthly precipitation ranging from 0.5 to nearly 3 inches.

Except in the mountainous locations, which are chiefly in the middle latitudes, both orographic and convectional precipitation are lacking in the tundra climate. Snow and rain are frontal or cyclonic in origin. Although rain falls in the months that average above freezing, temperatures may drop with little warning even during the warm season, bringing snow. As in the *Dw* climates, so in the major tundra areas the snow is dry and compact; it takes less of it than of fluffy snow to equal an inch of rain. On treeless plains, high winds and severe blizzards are common and drifting of snow follows. Drifting frequently means much bare ground and deeper freezing than where snow remains.

Whereas cyclones and their attendant inclement weather spread over arctic lands nonperiodically, especially in the winter, there is a considerable anticyclonic control with its clear, crisp, and bitterly cold weather. The arctic summers are cool, occasionally warm, damp, often foggy and cloudy, and the least agreeable season; it is at this season that the ground is thawed, muddy, and difficult to traverse. Shallow swamps and muddy ponds cover large areas, making either devious routes or semiaquatic vehicles necessary. Spring and autumn are said to be more stormy than winter.

Responses to the Tundra Climate

Natural Vegetation. The temperature regime, regardless of the variations, lacks sufficient warmth to produce full-sized trees. Cold and dry winds at a time when water is not available from the frozen ground is another deterring factor. As previously stated, most vegetation requires a mean temperature of 40° before it starts growth, or resumes growth. Since the majority of tundra stations have only three or four months with means above 40°, the growing season is quite limited. Furthermore, frosts may occur at any time.

Just as the tundra climate is transitional

between the *D* and frost climates, so likewise the tundra vegetation is transitional between the taiga and the icecap. There are numerous descriptions of tundra vegetation. Some botanists and geographers recognize as many as five types; a three-fold classification is more common and will serve here. The southern margins of the arctic tundras, and those that appear just above the timber line on middle-latitude mountains, have the highest tundra forms. They consist of taiga varieties of trees in stunted size and form, often bushy; hence the name *bush tundra*. Between the small trees and bushes are annuals, so that most of the ground is covered with vegetation in the warmer season. The true tundra, which apparently covers somewhat more territory than the bush tundra, is the *grass tundra*. It appears in the colder and drier parts, chiefly poleward of the bush tundra. The grass tundra is synonymous with alpine meadows in the high mountains. It consists of grasses, small shrubs, and perennials of numerous varieties, sphagnum bogs, and fluffy lichens that are called "reindeer moss." After a fresh warm-season rain, the flowering shrubs and perennials are unbelievably beautiful. Their revival is somewhat like that of the foliage in deserts and steppes. The driest, windiest, and barest tundra is the *desert tundra*. It appears along the rocky coasts, in patches in the interior, and just below the icecaps of mountain summits. Some desert tundras are extensive, having patches of grass, or even bush tundra, in protected places.

Soils. Where tundra soils appear, they are usually thin and impoverished. Even those that are natively good are of no potential value agriculturally because of the lack of heat. Mechanical weathering, leaching, and more or less desiccation because of strong winds characterize most areas. There is little humus except where bogs are present. Symbolic of the tundra soil is a subsurface layer that is more or less fluid and sticky after the upper surface freezes. Because of the liquid within it and the pressure exerted by the frozen ground above it, the layer is subject to expansion. Blowups, called *hummocks*, are thus formed. The warmth of buildings also sometimes prevents freezing beneath them, with the result that the soil under the buildings often heaves upward, doing considerable damage. This phenomenon also occurs in the *Dc* and *Dd* climates of Siberia.

Occupations in the Tundra Climate

The tundra habitat offers little of much potential value. Commercial agriculture and forestry are impossible; industry and commerce have little prospect; mining is limited, but coal, oil, phosphates, nickel, and iron ore are produced. Most of the inhabitants—Eskimos, Lapps, Indians, and others—depend for an existence primarily upon hunting, trapping, fishing, reindeer herding, and the collection of a few tundra articles. Gaining a livelihood is usually difficult. Alpine meadows have some utility as pasturage for flocks and cattle in the warmer season but as a whole offer only limited use, chiefly recreational, during the winter.

The Icecap, or Frost Climate

Location

The icecap, or frost climate, has developed over the huge fields of ice that cover most of Greenland and Antarctica. The ice sheets themselves attain great depths in both cases; the depths of 1,000 to 10,000 feet are the result of millenia of concentration of snow, since the yearly precipitation is not great. The icecaps are remnants of the Ice Age, at which time much of the high latitudes of the northern hemisphere was buried under continental glaciation. The icecap climate also occurs at the highest elevations in many of the mountainous areas of middle latitudes, shown as *E* on the map; in some high-latitude lowland areas, also shown as *E;* and over the more or less perpetually frozen waters in the vicinity of the north pole. In some winters, much of the Arctic Ocean is frozen over.

Fig. 22–3. "Operation Deepfreeze" describes completely this scene in Antarctica, a land of perpetual frost (E) climate. (Courtesy U.S. Navy.)

Temperature

No other climate has such low mean annual temperatures or a complete absence of at least a somewhat warm season. No monthly mean is above 32° in the icecap climate, although from three to five months average above zero.

Eismitte (Greenland) has about as severe a year-round climate as records detail. Nine months (September through May) have means below zero; and February is almost as cold as January at Verkhoyansk; it suggests that the absolute minimum at Eismitte may not be far from that of Verkhoyansk. The three above-zero months at Eismitte (June, July, and August) have means of 4°, 12°, and 1°. A "warm-month" mean of only 12° will hardly reflect many days with maximum temperatures above freezing! Lower temperatures than those of Eismitte, however, may eventually

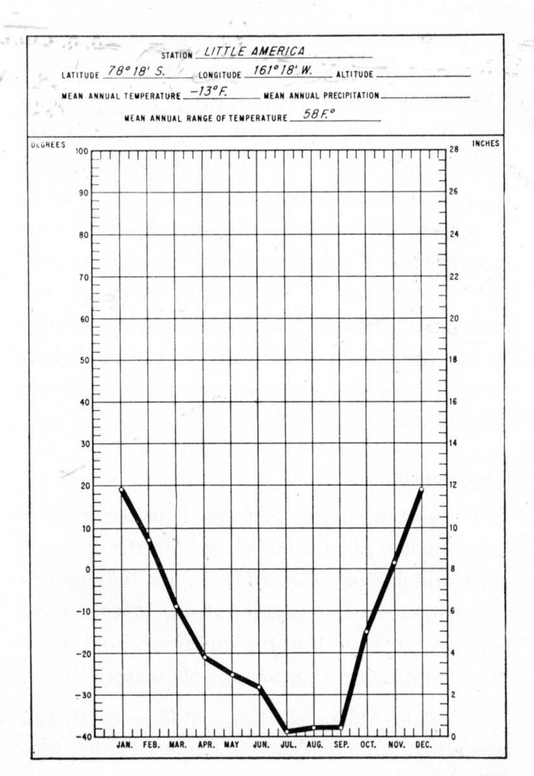

Fig. 22–4. E (frost) climate.

be substantiated from the south pole. Limited data there suggest that the interior of the antarctic continent may enjoy the reputation of having the lowest summer temperatures and also the lowest mean annual temperatures of any place in the world.

The severity of the climate at Eismitte is summarized, in part, by White and Renner[1] thus: "At times, the cold air is so still as to seem absolutely motionless, at other times, howling gales roar across the ice: a wind velocity of 174 miles per hour has been recorded. In really cold periods, one's face when exposed, withers in a few seconds as if scorched by a flame; a deep breath will sear the lungs, and a wet mitten or shoe may necessitate amputation of hand or foot. The snow squeaks under foot, moist cordwood explodes like dynamite, rubber hose breaks like glass, the touch of bare skin to metal pulls off patches of skin, brake-fluid freezes, and even high-test gasoline sometimes refuses to burn."

Little America (or Framheim, Antarctica) is more moderate than Eismitte, while McMurdo Sound (Antarctica) is still more moderate. Little America (Figure 22–4) has eight months with means below zero (May through October); they range from −9 to −39°. July is the coldest month. The four months above zero (November through February) range from 0 to 19°, with both December and January sharing the maximum. McMurdo Sound has seven months with means below zero, and the five warmest months range from 4 to 25°.

Precipitation

Meager amounts, perhaps equivalent to 5 or 10 inches of rain but mostly in the form of hard spicules of snow and ice, are believed to characterize icecap precipitation. Winter temperatures are too low for the air to hold more than infinitesimal amounts of water vapor. Few statistics are available, although limited data are being gathered both in Greenland and Antarctica. Unfortunately most observations have been maintained for only a few months, so that reliable mean annual precipitation data are unavailable. No other extensive climate has such limited information.

Probably, most of the precipitation is cyclonic in origin. At least cyclones are numerous enough in Greenland to raise the question whether a semipermanent anticyclone could exist there in the winter, as was once thought to be the case. Some snow originates simply from descending air as it touches, or approaches, the intensely cold icecap, reaching its dew point and condensing as fine ice or snow particles. Violent blizzards occur during the winter and sometimes in the warm season. During such storms, wind velocities of hurricane force, or greater, are attained, and temperatures are frequently far below zero. Visibility is reduced to zero, and all life seeks shelter for the duration, which may be an hour or several days.

Activities in the Frost Climate

In addition to the monotony of the severe climate, the lack of both vegetation and soil seriously hampers human activities. Permanent human habitations are practically nonexistent, although military installations have been built and Eskimos dwell seasonally on the icecap in some cases. Land animals, other than a few like polar bears and penguins, rarely inhabit the major icecaps, but strictly aquatic life is abundant. Perhaps the most typical occupation of icecap areas is the catching of whales, some seals, polar bears, and fish. Small amounts of coal, copper, and cryolite are known to exist. Although the exploitation of areas of icecap climate has not been developed to any great extent, recent exploration by many nations, largely for scientific purposes, indicates that regions of extreme polar climate may eventually be sources of important economic products, as well as sites for strategic air bases.

[1] C. Landon White and George T. Renner, "Human Geography," p. 265, Appleton-Century-Crofts, Inc., New York, 1948.

Appendixes

APPENDIX A

Selected References

Group 1: Weather

Albright, John G.: "Physical Meteorology," Prentice-Hall, Inc., Englewood Cliffs, N.J., 1939.

Berke, Jacqueline: "Watch Out for the Weather," The Viking Press, New York, 1951.

Berry, Frederic A., et al.: "Handbook of Meteorology," McGraw-Hill Book Company, Inc., New York, 1945.

Blair, Thomas A.: "Weather Elements," 4th ed., Prentice-Hall, Inc., Englewood Cliffs, N.J., 1957.

Botley, Cicely Mary: "The Air and Its Mysteries," D. Appleton-Century Company, Inc., New York, 1940.

Brands, George J.: "Meteorology," McGraw-Hill Book Company, Inc., New York, 1944.

Brooks, Charles F.: "Why the Weather?" 2d ed., Harcourt, Brace and Company, Inc., New York, 1935.

Brunt, David: "Weather Science for Everybody," Thomas Nelson and Sons, New York, 1942.

Byers, Horace R.: "General Meteorology," McGraw-Hill Book Company, Inc., New York, 1944.

Byers, Horace R., et al.: "Thunderstorms," Government Printing Office, Washington, 1949.

Clayton, Henry H.: "Solar Relations to Weather and Life," The Clayton Weather Service, Canton, Mass., 1943.

Clayton, Henry H.: "Sunspot Changes and Weather Changes," The Smithsonian Institution, Washington, 1946.

Donn, William L.: "Meteorology, with Marine Applications," 2d ed., McGraw-Hill Book Company, Inc., New York, 1951.

Eaton, Elbert Lee: "Weather Guide for Air Pilots," The Ronald Press Company, New York, 1943.

Finch, Vernor, et al.: "Elementary Meteorology," McGraw-Hill Book Company, Inc., New York, 1942.

Geiger, Rudolf: "The Climate near the Ground," (translation by Milroy N. Stewart, et al.), 2d ed., Harvard University Press, Cambridge, 1950.

Grant, Hugh Duncan: "Cloud and Weather Atlas," Coward-McCann, Inc., New York, 1944.

Haurwitz, Bernhard: "Dynamic Meteorology," McGraw-Hill Book Company, Inc., New York, 1941.

Haynes, Benarthur C.: "Techniques of Observing the Weather," John Wiley and Sons, Inc., New York, 1947.

Hewson, Edgar W., and Richard W. Longley: "Meteorology, Theoretical and Applied," John Wiley and Sons, Inc., New York, 1944.

Holmboe, Jörgen, et al.: "Dynamic Meteorology," John Wiley and Sons, Inc., New York, 1945.

Humphreys, William J.: "Physics of the Air," 3d ed., McGraw-Hill Book Company, Inc., New York, 1940.

Humphreys, William J.: "Ways of the Weather," The Jacques Cattell Press, Lancaster, Pa., 1942.

Humphreys, William J.: "Weather Proverbs and Paradoxes," The Williams and Wilkins Co., Baltimore, 1937.

Johnson, John Clark: "Physical Meteorology," John Wiley and Sons, Inc., New York, 1954.

Kimble, George H. T.: "Our American Weather," McGraw-Hill Book Company, Inc., New York, 1955.

Krick, Irving P., and Roscoe Fleming: "Sun, Sea and Sky," J. B. Lippincott Company, Philadelphia, 1954.

Laird, Charles, and Ruth Laird: "Weathercasting," Prentice-Hall, Inc., Englewood Cliffs, N.J., 1955.

Miller, Denning: "Wind, Storm and Rain," Coward-McCann, Inc., New York, 1952.

Mills, Clarence A.: "Climate Makes the Man," Harper & Brothers, New York, 1942.

Murchie, Guy: "Song of the Sky," Houghton Mifflin Company, Boston, 1954.

Namias, Jerome, et al.: "An Introduction to the Study of Air Mass and Isentropic Analysis," 5th ed., American Meteorological Society, Boston, Mass., 1940.

Neuberger, Hans Hermann, and F. Briscoe Stephens: "Weather and Man," Prentice-Hall, Inc., Englewood Cliffs, N.J., 1948.

O'Connor, James F.: "Practical Methods of Weather Analysis and Prognoses," Government Printing Office, Washington, 1952.

Perrie, D. W.: "Cloud Physics," John Wiley and Sons, Inc., New York, 1950.

Petersen, William F.: "Man, Weather, Sun," C. C. Thomas, Springfield, Ill., 1947.

Petterssen, Sverre: "Weather Analysis and Forecasting," 2d ed., McGraw-Hill Book Company, Inc., New York, 1956.

Riehl, Herbert: "Tropical Meteorology," McGraw-Hill Book Company, Inc., New York, 1954.

Riehl, Herbert, et al.: "The Jet Stream," (Meteorological Monograph, No. 7), American Meteorological Society, Boston, 1954.

Sausier, Walter J.: "Principles of Meteorological Analysis," University of Chicago Press, Chicago, 1955.

Shields, Bert A.: "Meteorology and Air Navigation," 2d ed., McGraw-Hill Book Company, Inc., New York, 1942.

Sloane, Eric: "Almanac and Weather Forecaster," Duell, Sloan and Pearce, Inc., New York, 1955; also Little, Brown and Company, Boston, 1955.

Sloane, Eric: "Weather Book," Little, Brown and Company, Boston, 1954.

Spilhaus, Athelstan F.: "Weathercraft," The Viking Press, New York, 1951.

Starr, Victor P.: "Basic Principles of Weather Forecasting," Harper & Brothers, New York, 1942.

Stewart, John Quincy: "Coasts, Waves and Weather, for Navigators," Ginn and Company, Boston, 1945.

Talman, Charles Fitzhugh: "The Realm of the Air," The Bobbs-Merrill Company, Inc., Indianapolis, 1931.

Tannehill, Ivan R.: "Hurricanes," Princeton University Press, Princeton, N.J., 1950.

Tannehill, Ivan R.: "Weather around the World," Princeton University Press, Princeton, N.J., 1943.

Taylor, George Frederic: "Elementary Meteorology," Prentice-Hall, Inc., Englewood Cliffs, N.J., 1954.

Trewartha, Glenn Thomas: "An Introduction to Climate," 3d ed., McGraw-Hill Book Company, Inc., New York, 1954.

United States Navy, Bureau of Aeronautics: "Aerology for Pilots," McGraw-Hill Book Company, Inc., New York, 1943.

Wenstrom, William H.: "Weather and the Ocean of Air," Houghton Mifflin Company, Boston, 1942.

Wylie, Charles C.: "Astronomy, Maps and Weather," Harper & Brothers, New York, 1942.

Group 2: Climate

Ahlmann, Hans W.: "Glacier Variations and Climatic Fluctuations," American Geographical Society, New York, 1953.

Baxter, William Joseph: "Today's Revolution in Weather," International Economic Research Bureau, New York, 1953.

Blair, Thomas A.: "Climatology," Prentice-Hall, Inc., Englewood Cliffs, N.J., 1942.

Brooks, Charles E. P.: "Climate through the Ages," McGraw-Hill Book Company, Inc., New York, 1949.

Clements, Frederic E.: "Climatic Cycles and Human Population in the Great Plains," Carnegie Institution, Washington, 1938.

Conrad, Victor, and L. W. Pollak: "Methods in Climatology," 2d ed., Harvard University Press, Cambridge, 1950.

Hadlow, Leonard: "Climate, Vegetation and Man," Philosophical Library, Inc., New York, 1952.

Hare, Frederick Kenneth: "The Restless Atmosphere," Hutchinson's Library, New York, 1953.

Haurwitz, Bernhard, and James M. Austin: "Climatology," McGraw-Hill Book Company, Inc., New York, 1944.

Kendrew, Wilfred George: "Climatology," 3d ed., Oxford University Press, New York, 1949.

Kendrew, Wilfred George: "The Climates of the Continents," 3d ed., Oxford University Press, New York, 1942.

Koeppe, Clarence Eugene: "The Canadian Climate," McKnight and McKnight Publishing Company, Bloomington, Ill., 1931.

Miller, Arthur Austin: "Climatology," E. P. Dutton and Company, Inc., New York, 1953.

Shapley, Harlow: "Climatic Change: Evidence, Causes, and Effects," Harvard University Press, Cambridge, 1953.

Stetson, Harlan T.: "Sunspots in Action," The Ronald Press Company, New York, 1947.

Tannehill, Ivan R.: 'Drought, Its Causes and Effects," Princeton University Press, Princeton, N.J., 1947.

Ward, Robert DeCourcy: "The Climates of the United States," Ginn and Company, Boston, 1925.

Group 3: Miscellaneous

Bull. Am. Meteorol. Soc., (Monthly, except July and August), American Meteorological Society, Boston.
"Centenary Proceedings, 1950," Royal Meteorological Society, London, 1950.
Clayton, Henry H.: "World Weather Records," The Smithsonian Institution, Washington, 1927—.
"Climate and Man," *Yearbook Agr.*, 1941, U.S. Department of Agriculture.
"Climatic Summary of the United States" (Supplement to "Summaries of Climatological Data for the United States, 1930), U.S. Weather Bureau, Washington, 1931—.
"Climatological Data for the United States by Sections," U.S. Weather Bureau, Washington, 1914—.
"Compendium of Meteorology," American Meteorological Society, Boston, 1951.
"The Cooperative Weather Observer," U.S. Weather Bureau, 1951.
J. Meteorol., (Bimonthly), American Meteorological Society, Boston.
"Meteorological Abstracts and Bibliography," American Meteorological Society, Boston, 1950.
Meteorol. Monographs, 8 nos. through 1954, American Meteorological Society, Boston.
"Meteorology and Atomic Energy," (Prepared for the U.S. Atomic Energy Commission), U.S. Weather Bureau, 1955.
Monthly Weather Rev. with Supplements, U.S. Weather Bureau.
"Smithsonian Meterological Tables," 6th ed., The Smithsonian Institution, Washington, 1951.
"Summaries of Climatological Data for the United States," (Bulletin W), U.S. Weather Bureau, 1930.
Thiessen, Alfred H.: "Weather Glossary," U.S. Weather Bureau, 1945.
Visher, Stephen S.: "Climatic Atlas of the United States," Harvard University Press, Cambridge, 1954.

Group 4: An Index of Information Published on the Back of the Daily Weather Map as of September 20, 1956

Subjects are listed by titles, subtitles, and dates of the most recent printing. Only a small supply of the more recent issues is maintained; and, therefore, the Weather Bureau can fill only limited requests. Bound volumes of back copies of the Daily Weather Map are available in many libraries and some Weather Bureau offices. The interested student will want to subscribe to the Daily Weather Map, thereby being able to maintain an up-to-date file of information. All inquiries should be addressed to: Chief, U.S. Weather Bureau, Washington 25, D.C.

Agriculture

Drought and Its Effects: 11-21-52
Freezing Temperatures and the Motorist: 10-20-50
Frost without Freeze: 3-28-56
Fruit-frost Service: 2-15-56
Growing Degree Days: 4-9-51
Raisins and Rain Warnings: 9-11-52
Spring Weather and the Home Garden: 3-24-53
Weather and Alfalfa Seed: 2-10-50
Weather and Corn: 7-15-55
Weather and Cotton: 10-17-50
Weather and Cranberries: 11-13-46
Weather and Forest Fires: 7-22-54
Weather and Fruit-growing: 8-22-47
Weather and Livestock: 3-10-53
Weather and Plant Diseases: 5-14-53
Weather and Sheep: 3-16-55
Weather and Sugar Beets: 3-27-47

Aviation

Aeronautical Climatology (Low Ceilings and Visibilities): 1-4-56
Aeronautical Climatology (Thunderstorms): 2-1-56
Altimeters: 5-18-56
Ceiling: 8-3-55
Flying Weather Forecasts: 11-3-54
Flying Weather Information: 7-6-55
Fronts: 11-23-55
Ice on Aircraft: 12-1-54
Jet Stream: 1-5-55
Mountain Wave: 3-2-55
Severe Weather Forecasts: 3-7-56
Storm Detection Radar: 4-6-55
Thunderstorms—Part One: 5-4-55

Thunderstorms—Part Two: 6-3-55
Tips on Weather for VFR Flight: 10-5-55
Turbulence: 2-2-55
Visibility: 9-7-55
Weather and Aviation
How Knowledge of the Weather Aids Flying: 6-8-55
Beginnings of Aviation Weather Service: 3-31-53
World War I and Post War Years: 7-1-53
Meteorology and Civil Aeronautics: 10-15-53
Recent Developments in Aviation Weather Service: 12-17-53
Weather Reports from Pilots: 12-15-55

Climatology

Degree Days in the United States: 11-4-49
Highest Temperatures (July 8–14, 1954): 7-16-54
Highest Temperatures (July 8–16, 1954): 7-29-54
It Isn't the Heat, It's the Humidity: 7-19-55
Local Noon Temperature and Humidity Data (January): 2-7-55
Local Noon Temperature and Humidity Data (July): 8-4-55
Some Extremes of Snowfall: 3-18-55
Weather in Death Valley: 1-31-52
Weather on Mount Washington: 4-22-55
World's Heaviest Rains: 8-24-55

Cooperation

Cooperation with Canada: 2-2-50
Cooperation with Cuba: 1-9-50
Cooperative Weather Program in Mexico: 12-9-49
Cooperative Weather Program with the Bahamas: 12-19-49
Our Voluntary Weather Observers: 7-22-52
Weather Service Cooperation with Bermuda: 5-9-50

Forecasting

Forecasting at Fresno: 7-11-47
Forecasting Storms on the Great Lakes: 10-7-53
How to Forecast Your Own Minimum Temperatures: 11-25-49
Structure of Fronts: 5-11-55

Hurricanes

Emergency Radio in Hurricanes: 8-28-52
Hurricane Fatalities: 9-3-54
Hurricanes as Shown on Daily Weather Maps: 9-9-54
North Atlantic Hurricane of September 8th to 16th, 1944: 10-31-44

Industrial Meteorology

Industry's Own Weather Men
Airline Meteorologists: 12-2-49
Company Meteorologists: 1-10-52
Consultant Meteorologists: 12-7-49
Public Utilities Battle Winter Weather: 3-10-48
Some Uses of Weather Forecasts and Reports: 12-14-49
Weather and Electric Power Flashovers: 1-24-51
Weather and Natural Gas: 2-20-52
Weather and Public Utilities: 3-4-53
Weather in Business: 12-12-49

Instruments

Infrared Absorption Hygrometer: 5-21-54
Instruments Used in Weather Observing: 12-10-53
Rotating Beam Ceilometer: 5-20-54

Operations

Automatic Teletypewriter Weather Observer: 6-25-56
International Weather Reports: 6-5-53
Ocean Vessel Weather Stations: 6-25-52
Radiosonde and Rawinsonde Basic Reporting Network: 7-24-56
Solar Radiation Network: 9-8-55
Winds above the Clouds: 5-7-54
Winds Aloft Basic Reporting Network: 7-25-56

Recreation

Weather and Vacations: 7-16-47
Weather and Winter Sports: 12-8-54

River Services

Man's Control of a River Begins with the Weather: 11-28-47
Minute Men in Storm and Flood: 12-8-49
Ohio River Flood of March 1945: 4-27-45
River and Flood Forecasting Service
  Hydrologic Cycle: 4-22-48
  Monthly Flood Expectancies at Selected Stations: 12-16-49
  Reports and Warnings: 6-22-46
Snows—Warnings—Floods: 7-31-52

Tornadoes

Local Storm Warning Networks: 6-22-55
These are Tornadoes: 6-24-55
Tornadoes: 6-24-55

Transportation

Ice Reporting on the Great Lakes: 2-28-56
Weather and Bananas: 6-19-46
Weather and Harbor Traffic: 2-5-46
Weather and Intracoastal Boating: 12-13-50
Weather and the Motorist: 11-9-46
Weather and the Railroads: 11-17-46

APPENDIX B

Conversion of Fahrenheit to Centigrade

Degrees of the Fahrenheit scale are shown in *tens* in the first vertical column and in *units* across the top; all other numbers are the equivalent degrees of the centigrade scale.

DEGREES FAHRENHEIT	DEGREES CENTIGRADE 0	1	2	3	4	5	6	7	8	9
110	43.3	43.9	44.4	45.0	45.6	46.1	46.7	47.2	47.8	48.3
100	37.8	38.3	38.9	39.4	40.0	40.6	41.1	41.7	42.2	42.8
90	32.2	32.8	33.3	33.9	34.4	35.0	35.6	36.1	36.7	37.2
80	26.7	27.2	27.8	28.3	28.9	29.4	30.0	30.6	31.1	31.7
70	21.1	21.7	22.2	22.8	23.3	23.9	24.4	25.0	25.6	26.1
60	15.6	16.1	16.7	17.2	17.8	18.3	18.9	19.4	20.0	20.6
50	10.0	10.6	11.1	11.7	12.2	12.8	13.3	13.9	14.4	15.0
40	4.4	5.0	5.6	6.1	6.7	7.2	7.8	8.3	8.9	9.4
30	−1.1	−0.6	0.0	0.6	1.1	1.7	2.2	2.8	3.3	3.9
20	−6.7	−6.1	−5.6	−5.0	−4.4	−3.9	−3.3	−2.8	−2.2	−1.7
10	−12.2	−11.7	−11.1	−10.6	−10.0	−9.4	−8.9	−8.3	−7.8	−7.2
+0	−17.8	−17.2	−16.7	−16.1	−15.6	−15.0	−14.4	−13.9	−13.3	−12.8
−0	−17.8	−18.3	−18.9	−19.4	−20.0	−20.6	−21.1	−21.7	−22.2	−22.8
−10	−23.3	−23.9	−24.4	−25.0	−25.6	−26.1	−26.7	−27.2	−27.8	−28.3
−20	−28.9	−29.4	−30.0	−30.6	−31.1	−31.7	−32.2	−32.8	−33.3	−33.9
−30	−34.4	−35.0	−35.6	−36.1	−36.7	−37.2	−37.8	−38.3	−38.9	−39.4
−40	−40.0	−40.6	−41.1	−41.7	−42.2	−42.8	−43.3	−43.9	−44.4	−45.0
−50	−45.6	−46.1	−46.7	−47.2	−47.8	−48.3	−48.9	−49.4	−50.0	−50.6
−60	−51.1	−51.7	−52.2	−52.8	−53.3	−53.9	−54.4	−55.0	−55.6	−56.1

Conversion of Centigrade to Fahrenheit

Degrees of the centigrade scale are shown in *tens* in the first vertical column and in *units* across the top; all other numbers are the equivalent degrees on the Fahrenheit scale.

DEGREES CENTIGRADE	DEGREES FAHRENHEIT 0	1	2	3	4	5	6	7	8	9
40	104.0	105.8	107.6	109.4	111.2	113.0	114.8	116.6	118.4	120.2
30	86.0	87.8	89.6	91.4	93.2	95.0	96.8	98.6	100.4	102.2
20	68.0	69.8	71.6	73.4	75.2	77.0	78.8	80.6	82.4	84.2
10	50.0	51.8	53.6	55.4	57.2	59.0	60.8	62.6	64.4	66.2
+0	32.0	33.8	35.6	37.4	39.2	41.0	42.8	44.6	46.4	48.2
−0	32.0	30.2	28.4	26.6	24.8	23.0	21.2	19.4	17.6	15.8
−10	14.0	12.2	10.4	8.6	6.8	5.0	3.2	1.4	−0.4	−2.2
−20	−4.0	−5.8	−7.6	−9.4	−11.2	−13.0	−14.8	−16.6	−18.4	−20.2
−30	−22.0	−23.8	−25.6	−27.4	−29.2	−31.0	−32.8	−34.6	−36.4	−38.2
−40	−40.0	−41.8	−43.6	−45.4	−47.2	−49.0	−50.8	−52.6	−54.4	−56.2
−50	−58.0	−59.8	−61.6	−63.4	−65.2	−67.0	−68.8	−70.6	−72.4	−74.2

Conversion of Inches to Millimeters

Inches are shown in *units* in the first vertical column, and in *tenths* across the top; all other numbers are equivalent millimeters. To find equivalents for 10 to 99 inches in multiples of 1 inch, multiply all numbers by 10; to find millimeter equivalents for values over 99 inches, multiply the inches by 25.4.

	MILLIMETERS									
INCHES	0	0.1	0.2	0.3	0.4	0.5	0.6	0.7	0.8	0.9
0	0.0	2.54	5.08	7.62	10.16	12.70	15.24	17.78	20.32	22.86
1	25.4	27.9	30.5	33.0	35.6	38.1	40.6	43.2	45.7	48.3
2	50.8	53.3	55.9	58.4	61.0	63.5	66.0	68.6	71.1	73.7
3	76.2	78.7	81.3	83.8	86.4	88.9	91.4	94.0	96.5	99.1
4	101.6	104.1	106.7	109.2	111.8	114.3	116.8	119.4	121.9	124.5
5	127.0	129.5	132.1	134.6	137.2	139.7	142.2	144.8	147.3	149.9
6	152.4	154.9	157.5	160.0	162.6	165.1	167.6	170.2	172.7	175.3
7	177.8	180.3	182.9	185.4	188.0	190.5	193.0	195.6	198.1	200.7
8	203.2	205.7	208.3	210.8	213.4	215.9	218.4	221.0	223.5	226.1
9	228.6	231.1	233.7	236.2	238.8	241.3	243.8	246.4	248.9	251.5

Conversion of Millimeters to Inches

Millimeters are shown in *hundreds* in the first vertical column and in *tens* across the top; all other numbers are equivalent inches. To find equivalents for 1,000 to 9,900 millimeters in multiples of 100 millimeters, multiply equivalent inches by 10. For higher values, divide millimeters by 25.4, or multiply by 0.039; for a useful approximation, multiply millimeters by 0.04.

MILLI-	INCHES									
METERS	0	10	20	30	40	50	60	70	80	90
0	0.0	0.39	0.79	1.18	1.57	1.97	2.36	2.76	3.15	3.54
100	3.94	4.33	4.72	5.12	5.51	5.91	6.30	6.69	7.09	7.48
200	7.87	8.27	8.66	9.06	9.45	9.84	10.24	10.63	11.02	11.42
300	11.81	12.21	12.60	12.99	13.39	13.78	14.17	14.57	14.96	15.35
400	15.75	16.14	16.54	16.93	17.32	17.72	18.11	18.50	18.90	19.29
500	19.68	20.08	20.47	20.87	21.26	21.65	22.05	22.44	22.84	23.23
600	23.62	24.02	24.41	24.80	25.20	25.59	25.98	26.38	26.77	27.16
700	27.56	27.95	28.35	28.74	29.13	29.53	29.92	30.32	30.71	31.10
800	31.50	31.89	32.28	32.68	33.07	33.46	33.86	34.25	34.65	35.04
900	35.43	35.83	36.22	36.61	37.01	37.40	37.80	37.99	38.39	38.98

Conversion of Inches to Millibars

Whole inches are shown in the first vertical column and *tenths* of inches across the top; all other numbers are equivalent millibars. For other equivalents, multiply inches by 33.86, or divide by 0.0295.

	MILLIBARS									
INCHES	0	0.1	0.2	0.3	0.4	0.5	0.6	0.7	0.8	0.9
25	847	850	853	857	860	864	867	870	874	877
26	880	884	887	891	894	897	901	904	908	911
27	914	918	921	924	928	931	935	938	941	945
28	948	952	955	958	962	965	968	972	975	979
29	982	985	989	992	996	999	1,002	1,006	1,009	1,013
30	1,016	1,019	1,023	1,026	1,030	1,033	1,036	1,040	1,043	1,046
31	1,050	1,053	1,057	1,060	1,063	1,067	1,070	1,074	1,077	1,080

Conversion of Millibars to Inches

Millibars are shown in multiples of 30 in the first vertical column and in multiples of 5 across the top; all other numbers are equivalents in inches. For other equivalents, multiply millibars by 0.0295, or divide by 33.86.

MILLI-	INCHES					
BARS	0	5	10	15	20	25
860	25.40	25.54	25.69	25.84	25.99	26.13
890	26.28	26.43	26.58	26.72	26.87	27.02
920	27.17	27.32	27.46	27.61	27.76	27.91
950	28.05	28.20	28.35	28.50	28.64	28.79
980	28.94	29.09	29.23	29.38	29.53	29.68
1010	29.83	29.97	30.12	30.27	30.42	30.56
1040	30.71	30.86	31.01	31.15	31.30	31.45

APPENDIX C

Temperature of the Dew Point (Pressure 30.0 Inches)

Dry-bulb temp.	Sat. vapor pressure	Depression of the wet bulb																												
		1	2	3	4	5	6	7	8	9	10	11	12	14	16	18	20	22	24	26	28	30	32	34	36	38	40	42	44	46
10	0.0631	5	−2	−10	−27																									
12	0.0699	7	2	−6	−19																									
14	0.0772	10	5	−2	−12	−33																								
16	0.0850	12	7	1	−7	−20																								
18	0.0933	14	10	5	−2	−13	−37																							
20	0.1026	16	12	8	2	−7	−21																							
22	0.113	19	15	11	5	−2	−12	−36																						
24	0.124	21	17	13	9	2	−6	−20																						
26	0.126	23	20	16	12	7	−1	−11	−32																					
28	0.150	25	22	19	15	10	4	−4	−17																					
30	0.164	27	25	21	18	14	8	2	−7	−25																				
32	0.180	30	27	24	21	17	12	7	−1	−12	−42																			
34	0.195	32	29	26	23	20	16	11	5	−3	−17																			
36	0.211	34	31	29	26	23	19	15	10	3	−6	−25																		
38	0.228	36	33	31	28	25	22	18	14	8	1	−10	−36																	
40	0.247	38	35	33	30	28	25	21	18	13	7	−1	−14																	
42	0.266	40	38	35	33	30	27	24	21	17	12	6	−3																	
44	0.287	42	40	37	35	32	30	27	24	20	16	11	4	−24																
46	0.310	44	42	40	37	35	32	29	27	23	20	15	10	−7																
48	0.334	46	44	42	40	37	35	32	29	26	23	19	14	1	−41															
50	0.360	48	46	44	42	40	37	34	32	29	26	22	18	8	−13															
52	0.387	50	48	46	44	42	40	37	34	32	29	26	22	13	−2															
54	0.417	52	50	48	46	44	42	40	37	34	32	29	25	18	6	−20														
56	0.448	54	53	51	49	47	44	42	40	37	34	32	29	22	12	−5														
58	0.482	56	55	53	51	49	47	45	42	40	37	35	32	25	17	4	−30													
60	0.517	58	57	55	53	51	49	47	45	43	40	38	35	29	21	11	−8													
62	0.555	60	59	57	55	53	52	50	47	45	43	40	38	32	25	16	−3	−45												
64	0.595	62	61	59	57	56	54	52	50	48	46	43	41	35	29	21	11	−10												
66	0.638	64	63	61	60	58	56	54	52	50	48	46	44	38	32	26	17	2												
68	0.684	67	65	63	62	60	58	57	55	53	51	49	46	42	36	29	22	11	−11											
70	0.732	69	67	65	64	62	61	59	57	55	53	51	49	44	39	33	26	17	2											
72	0.783	71	69	68	66	64	63	61	59	58	56	54	52	47	42	37	30	22	11	−11										
74	0.838	73	71	70	68	67	65	63	62	60	58	56	54	50	45	40	34	27	18	3										
76	0.896	75	73	72	70	69	67	66	64	62	60	59	57	53	48	43	38	31	23	12	−10									
78	0.957	77	75	74	72	71	69	68	66	64	63	61	59	55	51	46	41	35	28	19	5									
80	1.022	79	77	76	74	73	72	70	68	67	65	63	62	58	54	50	44	39	32	24	13	−7								
82	1.091	81	79	78	77	75	74	72	71	69	67	66	64	60	57	52	48	42	36	39	20	7	−43							
84	1.163	83	81	80	79	77	76	74	73	71	70	68	66	63	59	55	51	46	40	34	26	15	−4							
86	1.241	85	83	82	81	79	78	76	75	73	72	70	69	65	62	58	54	49	44	38	31	22	9	−27						
88	1.322	87	85	84	83	81	80	79	77	76	74	73	71	68	64	61	57	52	47	42	35	27	17	0						
90	1.408	89	87	86	85	83	82	81	79	78	76	75	73	70	67	63	59	55	51	45	39	32	24	11	−17					
92	1.499	91	89	88	87	86	84	83	81	80	79	77	76	73	69	66	62	58	54	49	43	37	29	20	4					
94	1.595	93	92	90	89	88	86	85	84	82	81	79	78	75	72	68	65	61	57	52	47	41	34	26	14	−9				
96	1.696	95	94	92	91	90	88	87	86	84	83	82	80	77	74	71	67	64	60	55	51	45	39	31	22	8				
98	1.803	97	96	94	93	92	90	89	88	87	85	84	82	79	76	73	70	66	63	58	54	49	43	36	28	17	−2			
100	1.916	99	98	96	95	94	93	91	90	89	87	86	85	82	79	76	72	69	65	61	57	52	47	41	33	25	12	−22		
102	2.035	101	100	98	97	96	95	93	92	91	89	88	87	84	81	78	75	72	68	64	60	56	51	45	38	31	21	4		
104	2.160	103	102	100	99	98	97	95	94	93	92	90	89	86	83	80	77	74	71	67	63	59	54	49	43	36	27	16	−9	
106	2.292	105	104	102	101	100	99	98	96	95	94	92	91	88	86	83	80	77	73	70	66	62	58	53	47	41	33	24	9	−56
108	2.431	107	106	104	103	102	101	100	98	97	96	95	93	91	88	85	82	79	76	72	69	65	61	56	51	45	39	30	20	0
110	2.576	109	108	106	105	104	103	102	100	99	98	97	95	93	90	87	84	81	78	75	72	68	64	60	55	50	43	36	27	14

APPENDIX D

Relative Humidity in Per Cents (Pressure 30.0 Inches)

Dry-bulb	Depression of the wet bulb																													
temp.	1	2	3	4	5	6	7	8	9	10	11	12	13	14	15	16	18	20	22	24	26	28	30	32	34	36	38	40	42	44	46
10	78	56	34	13																											
12	80	59	39	19																											
14	81	62	44	26	8																										
16	82	65	48	31	14																										
18	84	68	52	36	20	5																									
20	85	70	55	40	26	12																									
22	86	71	58	44	31	17	4																								
24	87	73	60	47	35	22	10																								
26	87	75	63	51	39	27	16	4																							
28	88	76	65	54	43	32	21	10																							
30	89	78	67	56	46	36	26	16	6																						
32	89	79	69	59	49	39	30	20	11	2																					
34	90	81	71	62	52	43	34	25	16	8																					
36	91	82	73	64	55	46	38	29	21	13	5																				
38	91	83	75	66	58	50	42	33	25	17	10	2																			
40	92	83	75	68	60	52	45	37	29	22	15	7																			
42	92	85	77	69	62	55	47	40	33	26	19	12	5																		
44	93	85	78	71	63	56	49	43	36	30	23	16	10	4																	
46	93	86	79	72	65	58	52	45	39	32	26	20	14	8	2																
48	93	86	79	73	66	60	54	47	41	35	29	23	18	12	7	1															
50	93	87	80	74	67	61	55	49	43	38	32	27	21	16	10	5															
52	94	87	81	75	69	63	57	51	46	40	35	29	24	19	14	9															
54	94	88	82	76	70	64	59	53	48	42	37	32	27	22	17	12	3														
56	94	88	82	76	71	65	60	55	40	44	39	34	30	25	20	16	7														
58	94	88	83	77	72	66	61	56	51	46	41	37	32	27	23	18	10	1													
60	94	89	83	78	73	68	63	58	53	48	43	39	34	30	26	21	13	5													
62	94	89	84	79	74	69	64	59	54	50	45	41	36	32	28	24	16	8	1												
64	95	90	84	79	74	70	65	60	56	51	47	43	38	34	30	26	18	11	4												
66	95	90	85	80	75	71	66	61	57	53	48	44	40	36	32	29	21	14	7												
68	95	90	85	80	76	71	67	62	58	54	50	46	42	38	34	31	23	16	10	3											
70	95	90	86	81	77	72	68	64	59	55	51	48	44	40	36	33	25	19	12	6											
72	95	91	86	82	77	73	69	65	61	57	53	49	45	42	38	34	28	21	15	9	3										
74	95	91	86	82	78	74	69	65	61	58	54	50	47	43	39	36	29	23	17	11	5										
76	96	91	87	82	78	74	70	66	62	59	55	51	48	44	41	38	31	25	19	13	8	3									
78	96	91	87	83	79	75	71	67	63	60	56	53	49	46	43	39	33	27	21	16	10	5									
80	96	91	87	83	79	75	72	68	64	61	57	54	50	47	44	41	35	29	23	18	12	7	3								
82	96	92	88	84	80	76	72	69	65	61	58	55	51	48	45	42	36	30	25	20	14	10	5								
84	96	92	88	84	80	76	73	69	66	62	59	56	52	49	46	43	37	32	26	21	16	12	7	3							
86	96	92	88	84	81	77	73	70	66	63	60	57	53	50	47	44	39	33	28	23	18	14	9	5	1						
88	96	92	88	85	81	77	74	70	67	64	61	57	54	51	48	46	40	35	30	25	20	15	11	7	3						
90	96	92	89	85	81	78	74	71	68	65	61	58	55	52	49	47	41	36	31	26	22	17	13	9	5	1					
92	96	92	89	85	82	78	75	72	68	65	62	59	56	53	50	48	42	37	32	28	23	19	15	11	7	3					
94	96	93	89	85	82	79	75	72	69	66	63	60	57	54	51	49	43	38	33	29	24	20	16	12	9	5	1				
96	96	93	89	86	82	79	76	73	69	66	63	61	58	55	52	50	44	39	35	30	26	22	18	14	10	7	3				
98	96	93	89	86	83	79	76	73	70	67	64	61	58	56	53	50	45	40	36	32	27	23	19	15	12	8	5	2			
100	96	93	89	86	83	80	77	73	70	68	65	62	59	56	54	51	46	41	37	33	28	24	21	17	13	10	7	4	1		
102	96	93	90	86	83	80	77	74	71	68	65	62	60	57	55	52	47	42	38	34	30	26	22	18	15	11	8	5	2		
104	97	93	90	87	83	80	77	74	71	69	66	63	60	58	55	53	48	43	39	35	31	27	23	20	16	13	10	7	4	1	
106	97	93	90	87	84	81	78	75	72	69	66	64	61	58	56	53	49	44	40	36	32	28	24	21	17	14	11	8	5	3	
108	97	93	90	87	84	81	78	75	72	70	67	64	62	59	57	54	49	45	41	37	33	29	25	22	19	16	12	10	7	4	2
110	97	93	90	87	84	81	78	75	73	70	67	65	62	60	57	55	50	46	42	38	34	30	26	23	20	17	14	11	8	6	3

APPENDIX E

Table of Sines

Lat.	Sine	Lat.	Sine	Lat.	Sine	Lat.	Sine
0	.00000	23	.39073	46	.71934	69	.93358
1	.01745	24	.40674	47	.73135	70	.93969
2	.03490	25	.42262	48	.74314	71	.94552
3	.05234	26	.43837	49	.75471	72	.95106
4	.06976	27	.45399	50	.76604	73	.95630
5	.08716	28	.46947	51	.77715	74	.96126
6	.10453	29	.48481	52	.78801	75	.96593
7	.12187	30	.50000	53	.79864	76	.97030
8	.13917	31	.51504	54	.80902	77	.97437
9	.15643	32	.52992	55	.81915	78	.97815
10	.17365	33	.54464	56	.82904	79	.98163
11	.19081	34	.55919	57	.83867	80	.98481
12	.20791	35	.57358	58	.84805	81	.98769
13	.22495	36	.58779	59	.85717	82	.99027
14	.24192	37	.60182	60	.86603	83	.99255
15	.25882	38	.61566	61	.87462	84	.99452
16	.27564	39	.62932	62	.88295	85	.99619
17	.29237	40	.64279	63	.89101	86	.99756
18	.30902	41	.65606	64	.89879	87	.99863
19	.32557	42	.66913	65	.90631	88	.99939
20	.34202	43	.68200	66	.91355	89	.99985
21	.35837	44	.69466	67	.92050	90	1.00000
22	.37461	45	.70711	68	.92718		

SD

APPENDIX F

The Koeppe Classification of Climate[1]

The difficulties encountered in making a classification of climates that is both simple and accurate are manifold—in fact, the task is an impossible one. Probably the simplest scheme would be to use the terms *continental* and *marine*. It would not be difficult to select all the localities, or regions, which have a typical continental or marine climate. What to do with the broad transitional zones becomes the difficult problem. Furthermore, it might be desirable to distinguish between continental and marine climates in high and low latitudes; here again one would find transitional zones that would be hard to classify. Whatever the types selected, they should be based upon purely climatic considerations that could satisfactorily embrace the controls of climate as well as the elements of climate.

Places situated in different continents, but in the same latitude, at the same altitude, and in corresponding locations in the continental mass, are generally found to have the same climatic characteristics. As an example, consider the coastal stretches of our Pacific Northwest, southern Chile, and South Island of New Zealand. These coasts are in latitudes of 40 to 45°, backed by relatively high lands and exposed to the same sort of marine influence because of their location in the belt of the prevailing westerlies. The climates of these coasts are so strikingly similar to each other and so different from the climates of places in other situations, that they may be considered a type to which could be given the name *cool marine*. It should be evident, however, that there is no abrupt change from one type of climate to another, e.g., from Washington to British Columbia, or from Oregon to California; instead, there is always a transitional zone of varying width where the climate partakes of some of the characteristics of the types on all sides.

The types of climate described later are distinct in climatic characteristics, they are based largely upon geographic position, they are essentially genetic in that each type has one or more distinct reasons for its occurrence, and they have names that are definitely climatological. The number of types is limited to 16, which seems to be the lowest number consistent with what the types are intended to show. Obviously, the greater the number of types shown, the greater are the refinements and the more accurate is the system. It is also obvious that multiplicity of types leads to complexity and to difficulties in learning for the student.

The regional boundaries in the Koeppe classification are determined by the climatic data available; in areas where the data are few, or lacking, the boundaries are determined either by relief or by natural vegetation. The boundary lines are purposely "smoothed," even at the expense of accuracy, since to show all the deviations that the climatic data might indicate would serve to complicate what is intended to be simple.

The criteria used to distinguish the types are those which seem to be significant not only from the standpoint of what is commonly accepted by climatologists but also from the standpoint of natural vegetation and crops, e.g., the poleward limit of the humid subtropical climate is the isotherm of 40° mean temperature for the coldest month. In the United States, this isotherm is approximately the same as the northern limit of the 200-day frost-free period which seems to mark, in general, the northern limit of commercial cotton production. The criteria as given with each type of climate are not to be considered absolute in every case. For instance, there are places in Honshu, Japan, in the cool littoral climate, where the mean annual ranges of temperature will exceed 40 degrees (by definition, cool littoral climates have mean annual

[1] Based on a wall map in full color, entitled "Climatic Regions of the World," by Clarence E. Koeppe; published by A. J. Nystrom & Co., Chicago. The map which appears as Plate III, following page 216 is reproduced by permission of the publisher.

ranges of temperature under 40 degrees). Again there are some places in the dry summer subtropical climate where the mean temperature for the coldest month will be under 45°, although by definition the coldest month in this type of climate is 45° or above.

Each of the temperature and precipitation charts across the bottom of the map is the composite of some 10 to 12 stations well distributed within the region. In this way, small differences, or peculiarities, in some part of the region are eliminated, and only the broad essential characterictics are shown, e.g., in the humid subtropical region, the month of maximum precipitation may be almost any month of the year, yet the composite graph shows what is highly significant for the region as a whole—that the summer half year receives considerably more precipitation than the winter half year, a fact of great importance agriculturally.

In tropical climates, altitude is ignored in drawing the regional boundaries. This does not mean that altitude is not an important factor in tropical climates. On the contrary, altitude as a control of climate is as significant in the tropics as in higher latitudes. The chief importance of altitude in the tropics, however, lies in the fact that it reduces the general temperature level without greatly altering any of the other characteristics. There are, of course, many small differences in weather and climate between highlands and lowlands lying in the same latitude in the tropics: differences in absolute and relative humidity, in insolation and radiation, in winds and storms, in ranges of temperature—both diurnal and annual—and in other conditions. The student has already been introduced to those differences in his study of Chapter 13. To go into a detailed account of all the meteorological aspects of mountains as opposed to lowlands is far beyond the scope of a simple study of regional climates.

As an instance of a mountain climate in the tropics, consider Bogotá, Colombia, at an altitude of 8,700 feet. This station has all the average characteristics of temperature and rainfall of the wet equatorial climate, except for the fact that the mean temperature for the coldest month is only 57° and the wet equatorial climate should have a temperature of at least 70° for the coldest month. If we assume that altitude reduces temperature on the average about 3 degrees for each thousand feet of altitude, we find that Bogotá would, if it were at sea level, have mean monthly temperatures some 26 degrees higher than actually are experienced there. Otherwise, Bogotá has characteristics like any other equatorial station: small annual range of temperature, abundant rainfall throughout the year, and a tendency for a double maximum of both temperature and rainfall. The same sort of thing is true for all other tropical climates.

In higher latitudes, however, factors other than altitude tend to make mountain and plateau climates essentially different from lowland climates in the same latitude; therefore, climates on mountains and plateaus outside the tropics are given separate classifications from the adjacent lowlands. It should also be constantly borne in mind that there are many areas, particularly in high-altitude situations of intermediate and subpolar latitudes, that do not have the same characteristics as the type with which they are classified, but such areas are too small to warrant separate delineations on the map, e.g., the Cascade Mountains of Washington are classified as moderate subpolar, although it is readily granted that the zones above the timber line have distinctly polar characteristics.

In the following pages is a summary of some of the important facts concerning each type of climate. Paragraph A, in each case, attempts to give the geographical position of the type as well as some of the significant controls. Paragraph B summarizes the principal characteristics of the type with some explanation of, or reason for, the characteristics. Paragraph C mentions the Köppen letter combinations which seem to correspond most nearly to the type; it should be noted, however, that in certain instances there may be other letter combinations that may extend into the margins of the type, but not significantly enough to need listing. The purpose of giving the Köppen type is to aid the student who might want to use the descriptive material in Chapters 18 to 22 to supplement these summaries.

Wet Equatorial Climate

A. This type lies astride the equator except in east Africa (Kenya); it averages approximately 10 degrees of latitude in width. It is, essentially, the doldrum belt of low pressure with its strong convection and converging trade winds. The poleward boundary in the southern hemisphere is pushed northward almost to the equator on the west coasts of South America and Africa by the cold Peruvian and Benguela currents, respectively.

B. All months are warm and moist; no month is under 70° except where altitude reduces the temperature; mean annual ranges of temperature do not exceed 5 degrees. There is usually a double maximum of temperature corresponding to the two preiods of highest sun, but each is delayed about one month; this is reflected in a double period of maximum rainfall, also slightly delayed. North of the equator, the two high-sun periods occur after the vernal equinox and before the autumnal equinox; thus the periods of maximum temperature and rainfall are likely to be only four or five months apart. The same condition occurs in the southern hemisphere except that the two maxima occur after the autumnal equinox and before the vernal equinox. Rainfall is abundant, 60 or more inches per year, and no month receives less than 2 inches—more likely 6 to 8 inches. Rain occurs almost every day—slightly less often on the poleward margins during the period of low sun. Thunderstorms accompanied by squalls are common; otherwise, the air is relatively calm, unless a tropical cyclone visits the area. The weather has the characteristic doldrum pattern of much cloud and rain, light variable winds or calms, and constantly high relative humidity; relief comes only at night, unless one is fortunate enough to reside at a considerable altitude. Quito, Ecuador, like Bogotá, is a type of high-altitude situation where the mean temperature is in the upper 50s—perpetual May for one accustomed to the weather of Boston.

C. Mostly *Af* and *Am;* some *D* in equatorial high-altitude positions.

Trade Wind Littoral Climate

A. This type is found on east coasts of continents lying in trade-wind belts; it extends from points near the equator to the tropic circles, or slightly beyond as in Florida, Mexico, the Union of South Africa, and Queensland in Australia; it includes all the trade-wind islands: Hawaii, the West Indies, the Philippines, and eastern Madagascar. The regions are always relatively narrow longitudinally on the continents, and there is always sufficient altitude to cause orographic rainfall. The trade winds blow with considerable steadiness and force during autumn and winter; they tend to be more variable and erratic in late spring.

B. Mean annual temperatures are 70° or higher, unless reduced by altitude. The wide north-south extent causes annual ranges of temperature to be as high as 20 degrees toward the poleward sides. Rainfall is more than 35 inches, and along those coasts—which are freely exposed to the influences of the trade winds, and especially in areas backed by considerable elevations of land—the rainfall is not only heavy, it is also fairly evenly distributed throughout the year. This feature is entirely lacking in the interior or on the leeward coasts of the tropics, except along the equator. In fact, along the coast, no month is rainless, and there may be a distinct double maximum near the equator, as along the Kenya and Tanganyika coasts of Africa; usually, however, there is a late summer, or autumn, maximum of rainfall and a spring minimum. In the eastern hemisphere, where the summer monsoons disrupt the trade winds, there is a tendency for a distinct dry season and a very wet season. Tropical cyclones (hurricanes and typhoons) are frequent. Rainfall is likely to be highly variable from year to year. Some trade-wind areas have excessively heavy rain, several stations in Hawaii receiving more than 200 inches per year.

C. *Af*, *Am*, and *Aw* along east coasts; some *Cw*, as in the vicinity of Hong Kong.

Wet and Dry Tropical Climate

A. On the poleward sides of the wet equatorial climate, the two annual periods of maximum solar heating are so close together that the two corresponding periods of high temperature merge into one. There is likewise only one period of heavy rainfall, at the time the doldrum belt invades the region. At other seasons the region is under the influence of the trade winds, during which period there is little or no rain. No general rule can be stated concerning the latitudinal extent of the wet and dry tropical climate. It is farthest poleward in India, along the west coast of Mexico, the west coast of Madagascar, and in northern Argentina; it is closest to the equator along the west coasts of South America and Africa, due in part to cold ocean currents; otherwise, it extends to about 15 or 20 degrees north or south latitude. The west coasts of Mexico and middle America are situated to the leeward of the trade winds in winter; in summer, however, the southeast trades of the southern hemisphere cross the equator, are deflected to the right, and become onshore southwesterly winds as they sweep toward the continental low-pressure area of northwest Mexico—thus the winter droughts and summer rains in that area.

B. The mean annual temperature is 70° or higher, unless reduced by altitude; mean annual ranges are generally under 15 degrees except in regions of strong monsoon influence, as in the middle Ganges Valley where ranges are as high as 33 degrees. The highest temperature usually comes before the time of the highest sun and just before the summer rains; in most places in the northern hemisphere the highest temperature comes in May. Clouds reduce the temperature during the rainy season, but there is often a secondary maximum of temperature after the rains have subsided. Rainfall is more than 35 inches, and it is excessively high in monsoon areas, especially if aided by orography (as along the southwest coasts of India and Burma and in northeast India). There is always a distinct winter dry period which may last for six months, and at some stations, two to four months may be completely rainless. Bombay is an example of extreme conditions: 95 per cent of its annual total of 80 inches comes in the four-month period of July to October.

C. Mainly *Aw* in lowlands near the equator; *Am* along the southwest coasts of India and Burma; *Cw* in highlands of the tropics (Ethiopia, parts of southern Africa, and southern Brazil) and in higher latitudes (northern India and northern Argentina).

Semiarid Tropical Climate

A. This type of climate occurs in those portions of the tropics where the mean annual rainfall is much reduced, usually under 35 inches but more than 12 to 15 inches. Ideally, as in the Sudan of Africa, it is found on the poleward sides of the wet and dry tropical climate, and some portion of it always adjoins the desert. It occurs also in the interiors of continents, especially on plateaus, as in central and southern Africa, northeast Brazil, Mexico, Peru, and Bolivia. It is farthest north in Pakistan and farthest south in the Union of South Africa. The lowest latitudes are in Ethiopia, Kenya, Tanganyika, northeast Brazil, Peru, and Angola of Africa (the two latter are due, in part, to cold ocean currents). In summer, the region is on the poleward edge of the doldrums; in winter it is under the influence of the drying trade winds, or the horse latitudes. Monsoons have some influence, as in Pakistan, peninsular India, central Burma, northern Australia, and eastern Africa. Altitude accounts for the islands of semiarid climate in the deserts of Sahara and south

Arabia. Cold, upwelling waters along the shore of northern Venezuela, coupled with the fact that the trade winds tend to be offshore or parallel to the coast in summer, account for the limited rainfall of that area. Rain-shadow effects of mountains largely explain the semiaridity in Mexico, Peru, Bolivia, northeast Brazil, Tanganyika, southern Africa, peninsular India, and Burma. Offshore summer trade winds limit the amount of rain in Yucatan. The amount of rainfall required to fit the semiarid tropical climate depends upon temperature: the higher the temperature, the more rain required.

B. Mean annual temperatures are 70° or higher, unless reduced by altitude; mean May temperatures exceed 90° in northwest India; mean annual ranges are somewhat higher than in corresponding latitudes of the wet and dry tropical climate (as much as 40 degrees in northern Pakistan). There is a tendency for a double maximum of temperature, with the highest temperature coming before the period of the highest sun as in the wet and dry tropical climate. Maximum rainfall comes, of course, in summer; as many as five months may be rainless; the doldrum regime is limited to two to five months. Weak frontal storms bring some winter precipitation (occasionally snow) to Pakistan and northwest India. As is characteristic of dry lands, the precipitation is highly variable from year to year.

C. *BSh* in the Sudan, Ethiopia, Southwest Africa, Bechuanaland, part of the Union of South Africa, northwest and peninsular India, and northern Australia; both *BSh* and *Aw* in Mexico; *BSk* in Peru and Bolivia; and largely *Cw* in other areas.

Tropical Desert Climate

A. The tropical desert is found on the poleward sides of the semiarid tropical climate, especially in Africa, Australia, Pakistan, and to some extent in the area around the Gulf of California. It also occurs along coastal Somalia where both the summer and the winter monsoon winds parallel the coast; along the west coasts of Peru, northern Chile, and southern Africa as the result of being on leeward coasts in trades as well as being washed by cold ocean currents; and in northern Argentina in the rain shadow of the Andes. Most areas lie in the horse latitude belts of high pressure or in trade-wind belts, either in interiors or along leeward coasts.

B. The mean annual temperature is 70° or higher, unless reduced by altitude; mean annual ranges are usually large for the latitude (yet only 8 degrees at Swakopmund on the coast of Southwest Africa); diurnal ranges are the highest of all tropical climates because of the strong heating by day and the rapid radiation by night. The highest temperature ever observed under standard conditions was 136° at Azizia in Libya. Rainfall is under 15 inches on the equatorward sides, the maximum occurring in summer, and under 10 inches on the poleward sides, where there is a tendency for a winter maximum, as in the northern portions of the Sahara and Arabian deserts and in the southern parts of the Atacama and Kalahari deserts. Some desert areas are almost rainless, as at Iquique in the Atacama.

C. Generally *BWh*, although some *BSh* on poleward margins.

Dry Summer Subtropical Climate

A. This type of climate, often referred to as "Mediterranean" climate, is found along west coasts in lower middle latitudes (on the poleward sides of the tropics). Places having this type of climate lie in the horse latitude belts of high pressure in summer and are, therefore, practically rainless in that season except in mountainous sections. In winter, the regions lie in the belts of prevailing westerlies; the marine influence is then marked, so that winter is the rainiest season; cyclonic storms are also important as they bring warm-front precipitation with winds from the south as well as cold-front precipitation with winds from the west and northwest. This climate is usually confined to coastal fringes because of the prevalence of mountains near the coasts. It follows the shores of the Mediterranean as well as the southern and eastern shores of the Black Sea.

B. The mean temperature of the coldest month is above 45°; coastal areas tend to have mean July or August temperatures under 70°, but interiors may have mean temperatures above 80°. Mean annual ranges of temperature are from about 12 degrees near the coasts to as much as 40 degrees in interiors. Rainfall is generally light, rarely exceeding 35 or 40 inches except in orographic situations, such as the west end of the Caucasus at Batum where the average exceeds 90 inches, or along the eastern Adriatic where it may be 100 inches. The dry limit is marked by about 10 inches. Although winter is normally the season of maximum precipitation, occasionally there is a double maximum in late autumn and early spring, as in parts of the Mediterranean. Some summer months may be rainless, and usually the wettest month has at least three times the rain of the driest month. Cold fronts in winter may bring temperatures below freezing; in summer, the descending air in the subtropical high-pressure belts favors clear weather—but cold ocean currents may cause fogginess along coasts, particularly in the less rainy months. Tropical cyclones occasionally bring substantial showers in August and September to the coast of southern California. Rainfall is likely to be erratic and variable from year to year, sometimes only 30 per cent of normal, at other times as much as 250 per cent of normal.

C. Generally *Csa* or *Csb;* equatorward margins are likely to be *BSh.*

Humid Subtropical Climate

A. Humid subtropical climate is found on east margins of continents in lower middle latitudes; it also occurs in certain insular locations not far from east coasts, such as in southern Japan, eastern Formosa, and North Island of New Zealand. It has a rather wide latitudinal extent in North America, South America, and Asia, but it is limited to coastal areas in the Union of South Africa and eastern Australia where highlands rise close to the coasts. Subtropical belts of high pressure over the oceans in summer favor inflow of *mT* air, as in the United States, South America, and Australia; the summer monsoon (southerly) also brings *mT* air into China. Strong cyclonic activity in winter favors stormy weather in the United States—a rapid succession of warm fronts and cold fronts; storms of this sort are less pronounced in other areas, but cold *cP* air from Siberia dominates the winter weather of China, although occasional weak winter cyclones move down the Yangtze Valley, bringing milder weather with clouds and rain (or snow) to that area.

B. The mean temperature of the coldest month is above 40° (approximately the northern limit of the 200-day growing season in the United States). The warmest month is usually above 80° in the United States and China, in the 70s in Japan, South America, southeast Africa, and Queensland, and under 70° in New South Wales, Victoria, and North Island. Mean annual ranges are least in strictly marine situations (15 degrees at Auckland, New Zealand) and greatest in continental situations (40 degrees or more in China). Annual precipitation ranges from 35 to 60 inches or more, and no month is dry, although in China the wettest month may have from four to eight times as much rain as the driest month. The maximum precipitation comes as a rule in summer, although there are many exceptions: March in Montgomery, April in Little Rock, May in Dallas, June in Kagoshima, Japan, August in Pensacola, and September in Titusville, Florida; and in Auckland, the maximum comes in July (the coldest month), the result of insularity. In late summer and early autumn, hurricanes are frequent along the Gulf and Atlantic coasts, and typhoons along the China coast and in Japan; tropical cyclones rarely visit the region in Australia, and they are almost unknown in South America and Africa. Tornadoes are characteristic of the humid subtropical climate in the United States, being most frequent in spring.

C. *Cfa* in the United States, South America, Japan, and coastal China; *Cfb* in southern Africa, Australia, and New Zealand; and *Cwa* in western interior China.

Cool Marine Climate

A. The cool marine climate occurs on west margins of continents and on the poleward sides of the dry summer subtropical climate; it is everywhere limited to the coastal situation, although of fairly wide extent in the British Isles; it involves Tasmania and the western part of South Island, New Zealand. The greatest north-south extent is found in Europe, from 42 to 63° north. It is exposed to the prevailing westerlies at all seasons but is modified by intense cyclonic activity, especially in winter. Warm ocean currents and their eastward drifts tend to keep the winter temperatures well above the average for the latitude.

B. The mean temperature of the coldest month is between 30 and 45°, and the warmest month will have temperatures from 60 to 65°. Mean annual ranges of temperature are small, from 15 to 35 degrees, reflecting marine conditions. At times there is considerable snow; rainfall is moderate to heavy (35 to 60 inches) with a definite winter maximum, but with no month rainless; some coastal stations may receive as much as 3 inches in the driest month. Some protected stations have under 25 inches annually, e.g., Hobart, Tasmania, but windward slopes of the mountains of western Tasmania may have 110 inches. Henderson Lake on the west side of Vancouver Island receives over 200 inches, the maximum coming in late fall and early winter.

C. Mainly *Cfb*.

Cool Littoral Climate[1]

A. Cool littoral climate is located in middle to higher middle latitudes where marine and continental influences are in balance. The best representation is in western Europe between the cool marine to the west and the humid continental to the east. Other areas are Honshu Island of Japan, highlands of moderate elevation in southeast Australia, and the eastern part of South Island. In Europe, cyclonic activity is strong in winter, convection in summer; in Honshu there is a distinct winter and summer monsoon reversal tending to weaken marine influences in winter but favoring large indrafts of *mT* air in summer; in Australia, elevation coupled with the absence of large continental masses to poleward favors an equable climate; and South Island is shielded from the extreme marine conditions of the strong prevailing westerlies.

B. The mean temperature of the coldest month is usually under 40°, though somewhat higher in Australia; temperature of the warmest month is from 60 to 70°, though well into the 70s in Honshu; mean annual ranges are mostly under 40 degrees, though slightly greater in Honshu. Precipitation in Europe is from 25 to 30 inches and up to 50 or 60 inches in other areas. It is evenly distributed in Europe, shows a distinct summer maximum in Honshu, but a winter maximum in South Island. Moderate snowfall is characteristic of the winter season, except at lower elevations in Australia; snowfall is extremely heavy on the west side of Honshu when the northwest winds out of Siberia sweep across the Sea of Japan. Precipitation is generally reliable, except in Australia; and at times it may be excessive in Japan—as much as 30 inches in a 24-hour period. Typhoons frequent the east shores of Honshu from June to November.

C. *Cfb* in Europe, Australia, and South Island; *Cfa* in Honshu.

Humid Continental Climate

A. The humid continental climate extends from the interiors of continents in middle latitudes to east coasts and covers as much as 25 degrees of latitude. It is not found, however,

[1] The wall map of "Climatic Regions of the World" shows cool littoral climate along the east coast of Korea, in Hokkaido, and the northern part of Honshu. This is an error; each of these areas should be shown as having humid continental climate. The error has been corrected on the map used in this book.

in central Asia because of aridity or extreme cold, nor does it occur in the southern hemisphere because of the absence of large land masses. The region occupies almost one-half of the United States, and it extends as a great wedge from central Europe, narrowing down to a point in western Siberia. The climate is humid because there is sufficient precipitation in most years to permit successful agriculture, though severe droughts are not uncommon; it is continental because of large annual ranges of temperature. It occurs farther north in Europe and farther south in China than it does in North America. Cyclones with their warm and cold fronts dominate the weather of North America and Europe, particularly in winter; local thunderstorms are frequent toward the equator in summer. Monsoons constitute an important weather control in eastern Asia: warm and moist southerly winds (*mT* air) in summer, cold and dry northerly winds (*cP* air) in winter. Tornadoes are likely to strike any portion of the east central United States, occurring most frequently in late spring and early summer.

B. The coldest month averages below 40°, and it may average as low as 0°. At least three months go above 60°, and the warmest month may exceed 80°. Mean annual ranges are more than 40° and may go as high as 70° toward the colder and drier margins. The east coast of Asia is from 10 to 15 degrees colder in winter than corresponding latitudes in North America but 3 to 5 degrees warmer in summer. Temperature conditions in Europe are similar to those in the eastern part of the North American region. Precipitation is moderate and seldom excessive. It is not less than 16 inches on poleward sides, as in North Dakota, nor less than 22 inches on equatorward sides, as in Oklahoma. In Nova Scotia, precipitation exceeds 50 inches, which includes about 75 inches of snow; no lowland station in Europe has more than 30 inches total precipitation; in Asia amounts range from 25 to 40 inches. There is a definite summer maximum in most of the region, decidedly so in Asia; but New England and southeastern Canada experience a rather even distribution.

C. The Köppen-Geiger map shows a very complicated distribution of climates in the region classed as humid continental: *Cfa* in the southern part of the North American region and *Dfa*, *Dfb*, and *Cfb* in the northern part; *Cfb* in central Europe; *Dfb* in Russia and western Siberia; *Cwa* and *Cfa* in northern China; and *BSk*, *Dwa*, and *Dwb* in the remainder of eastern Asia.

Semiarid Continental Climate

A. This climate occupies the interiors of continents in middle latitudes. The aridity is due to mountains cutting off supplies of moisture or to great distances from the sources of moisture supply. Equatorward margins of this climate touch the semiarid tropical climate in Mexico, Argentina, and Australia; they also adjoin in northern Pakistan. The difference between the two types is only a matter of temperature. There are many differences within the semiarid continental climate, but the region is a unit from the standpoint of limited moisture and large annual ranges of temperature. The greatest extent of the type, both east-west and north-south, is found in Eurasia. It is most characteristically associated with plateaus, but it occupies lowland areas in the lower Rio Grande, east central Argentina, southern Russia, and eastern Australia.

B. Temperature limits are the same as in the humid continental climate. Extremes are greatest in Montana, central Siberia, and northeast China, and least in the Rio Grande area, Argentina, and Australia. Precipitation is from 6 to 16 inches on poleward margins and from 12 to 22 inches on equatorward margins.[1] Summer is the season of maximum precipitation, except in areas close to west coasts where there is likely to be a double

[1] There is a slight error on the wall map of "Climatic Regions of the World," in the third line of the legend for semiarid continental climate: it should read "between 12″ and 22″ on equatorward margins." The error has been corrected on the map in this book.

maximum as in the northwestern United States, parts of Argentina, Spain, and Turkey. Such areas are transitional between the marine climate of west coasts and the continental climate of the interiors. There is a pronounced summer maximum in northwest China where monsoons play an important part. Precipitation everywhere within the region is highly variable from year to year; some areas may look like deserts after a prolonged dry period. Severe dust storms are common in winter, as in China and on the Great Plains of the United States.

C. *BSk* on poleward margins; otherwise *BSh*.

Intermediate Desert Climate

A. Intermediate desert climate exists largely as "islands" within the semiarid continental climate; but it also occurs in the rain shadow of the Andes, as in Patagonia (southern Argentina). With the exception of decreased precipitation, the type is much like the semiarid continental climate.

B. Temperature limits are the same as in the humid continental climate, but diurnal and annual ranges of temperature are greater. Precipitation is under 6 inches on poleward sides and under 12 inches equatorward; seasonal distribution is essentially the same as in the semiarid continental climate, though even more variable and erratic.

C. *BWk*, except in North America and southern Iran, where it is *BWh*.

Marine Subpolar Climate

A. This climate occurs as narrow strips along west coasts in higher middle latitudes on the poleward margins of the cool marine climate, where mountains rise abruptly and where the marine effect is pronounced. It also extends as a tongue southward along the west flanks of the mountains of British Columbia. Besides being found in Alaska, Canada, Iceland, and Norway, it occurs as a very limited coastal area in southern Chile, but too small to be shown on the map.

B. Winters are mild for the latitude, only one to four months being under 30°; summers as such hardly exist for no more than two months may be above 60°, but at least one month must be above 50°. Mean annual ranges are small for the latitude, ranging between 20 and 30 degrees. Precipitation varies widely within the region, seldom below 35 inches but in some places totaling 80 or more inches. There is always a distinct autumn maximum, usually October; yet rainfall is adequate in summer. Cyclonic activity is frequent and persistent, particularly in fall and winter. Fogs are common along coasts, especially so in the Aleutian Islands of Alaska.

C. *Cfb* to the south; *Cfc* to the north.

Moderate Subpolar Climate

A. This type occurs at high altitudes in middle latitudes as in the Cascades, Sierra Nevada, Rockies, Andes, Pyrenees, Alps, Carpathians, and Caucasus; in semimarine situations in high latitudes, as in parts of Alaska, Sweden, and Finland; and in one insular location, Newfoundland. It is called *moderate* because the winters are not nearly so cold as in the great continental interiors of subpolar latitudes, owing either to its position between the strictly marine and strictly continental climates of higher middle latitudes or to considerable elevation in lower latitudes. Mountains, it should be recalled, have temperature conditions similar to marine climates.

B. Temperature limits are the same as those of the marine subpolar climate, but mean annual ranges of temperature are greater in moderate subpolar climate, except in lower-latitude mountains (about 10 degrees in southern Peru but 35 degrees in the Swiss Alps). Moderate precipitation is the rule in higher latitudes, from 25 to 40 inches, well distributed but with a summer maximum; moderate to heavy precipitation occurs on windward sides of mountains, with a tendency for a cooler season maximum. Snowfall is heavy in such situations, notably on the west slopes of the Sierra Nevada.

C. Mostly *Dfc*.

Extreme Subpolar Climate

A. Extreme subpolar climate, as its name suggests, occurs in nonmarine situations in moderately high latitudes, extending from the arctic circle, or slightly beyond, to 48° in North America but to 30° in Asia because of high altitudes. It is not shown in the southern hemisphere. The climate is dominated by cold-air masses during most of the winter but is occasionally punctured by invasions of warm air from the south; there is considerable cyclonic activity, though weak, in the summer. The summer monsoon, coming from the southeast, penetrates into eastern Asia, hence the distinctly dry winters and the pronounced summer maximum of rainfall of that region.

B. Long and cold winters are to be expected, with five or more months below 30° (as many as seven months in northern Canada and northeastern Siberia). Summer temperature limits are the same as in the other subpolar climates. Mean annual ranges of temperature are the greatest of any climate, e.g., 75 degrees in the Mackenzie Valley of Canada and over 100 degrees in northeastern Siberia. Precipitation is everywhere moderate to light, being less than 10 inches in parts of Siberia. Summer is the season of greatest precipitation, since the extreme cold of winter means little available moisture, but precipitation in the form of snow remains on the ground throughout the winter. Total snowfall in parts of eastern Canada may amount to 8 feet or more, and sometimes as much as 5 feet or more will be on the ground at one time.

C. In North America, *Dfb* along the southern margins, otherwise *Dfc;* in Europe and western Siberia, *Dfc;* in eastern Siberia, *Dfd*, *Dwc*, and *Dwd;* in Mongolia, *Dwc* and *BSk;* in Tibet, *E*.

Polar Climate

A. Polar climate extends generally beyond the polar circles and to slightly lower latitudes in the northeastern parts of Canada and Siberia; any other areas exist because of altitude. It is largely tundra or icecap. The climate is considered a type only because the summers—if they exist—are too cold and too short for any appreciable forest cover; successful agriculture is impossible.

B. There is no month with a mean temperature above 50°. Other temperature conditions, as well as precipitation and other climatic conditions, vary greatly from one area to another; therefore, no other distinctions are made, partly because of insufficient data and partly because so little use is currently being made of the areas other than for scientific purposes and for military and commercial aviation.

C. *E* (tundra or icecap).

APPENDIX G

Climatic Data for Selected Stations

Stations are arranged alphabetically by continents. Location is given in degrees and minutes of latitude and longitude; altitude in feet; temperature in degrees F; and precipitation in inches.

NORTH AMERICA

Station	Location	Alt.	Jan.	Feb.	Mar.	Apr.	May	June	July	Aug.	Sept.	Oct.	Nov.	Dec.	Year
Abitibi	48.43 N	850	0	1	15	32	46	56	64	61	52	40	25	8	33
	79.22 W		1.8	1.4	2.1	1.2	2.6	2.4	2.6	2.6	2.6	2.9	2.1	2.0	26.4
Acapulco	16.50 N	10	78	78	79	80	83	83	83	83	81	81	80	79	81
	99.56 W		0.6	0.0	0.0	0.0	1.7	1.7	6.0	6.3	14.7	5.9	1.9	0.7	39.5
Aklavik	68.14 N	25	−21	−11	−7	8	31	48	56	50	38	17	−6	−17	16
	134.50 W		0.6	0.6	0.4	0.5	0.6	0.7	1.3	1.0	1.0	1.0	0.7	0.5	8.9
Anchorage	61.13 N	132	12	20	24	35	45	54	57	55	48	36	23	14	35
	149.50 W		0.8	0.7	0.6	0.4	0.5	0.6	1.6	2.6	2.5	2.1	1.1	0.8	14.3
Angmagssalik	65.37 N	5	17	13	17	24	33	41	44	42	38	30	23	19	28
	37.33 W		3.5	1.7	2.2	2.4	2.8	2.1	2.1	2.5	4.0	6.2	3.4	2.7	35.6
Austin	39.30 N	6594	27	31	35	43	51	60	69	68	59	48	39	31	47
	117.00 W		1.2	1.3	1.5	1.5	1.6	0.6	0.4	0.6	0.5	0.6	0.7	1.2	11.7
Banff	51.25 N	4521	13	16	23	37	45	52	57	54	47	39	24	19	36
	115.30 W		1.4	0.8	1.3	1.2	2.4	3.0	2.5	2.4	1.7	1.2	1.8	1.1	20.7
Barbados	13.08 N	181	78	77	77	79	80	80	80	81	81	80	80	78	79
	59.36 W		3.4	2.2	2.0	1.6	2.6	4.3	5.1	7.4	7.8	7.6	6.4	4.2	54.8
Bermuda	32.18 N	150	63	63	63	66	70	76	80	81	79	74	69	65	71
	64.46 W		4.4	4.6	4.8	4.0	4.4	4.3	4.6	5.4	5.2	5.9	4.9	4.8	57.4
Boston	42.20 N	124	27	28	35	45	57	67	71	69	63	52	41	32	49
	71.09 W		3.4	3.1	3.7	3.7	2.9	3.2	3.3	3.3	3.5	2.8	2.9	3.4	38.9
Cap Haitien	19.46 N	49	72	74	74	76	78	80	79	79	80	78	76	75	77
	72.12 W		4.4	4.4	3.5	3.1	5.1	3.3	1.4	1.7	3.7	8.0	10.6	6.9	56.1
Chicago	41.30 N	610	25	27	37	48	59	69	75	73	66	54	39	28	50
	87.24 W		1.8	1.6	2.7	2.6	3.4	3.4	3.1	3.3	3.5	2.4	2.1	1.8	31.8
Chimax	15.29 N	4285	61	62	65	66	67	68	66	67	67	66	66	60	65
	90.16 W		5.8	4.5	4.4	4.9	7.9	12.3	9.9	10.4	10.5	13.2	8.8	6.5	99.1
Colon	9.22 N	10	80	79	80	80	80	80	80	79	80	79	79	80	80
	79.54 W		3.9	1.7	1.7	4.2	12.6	13.5	16.2	14.9	12.5	14.8	21.5	11.9	129.4
Columbus	39.56 N	918	29	30	40	51	62	71	75	73	67	55	42	32	52
	82.56 W		2.9	2.1	3.4	2.7	3.1	3.3	3.5	3.2	2.6	2.4	2.2	2.6	34.1
Corpus Christi	27.49 N	20	56	58	64	70	76	80	82	82	80	73	65	58	70
	97.25 W		1.1	1.2	1.6	1.8	3.1	2.7	1.8	1.5	4.5	2.5	1.8	2.0	25.5
Craig Harbor	76.12 N	12	−23	−23	−15	−4	17	33	41	38	29	12	5	−17	7
	79.35 W		0.4	0.2	0.5	0.5	0.5	0.8	0.9	1.8	0.5	1.4	0.8	0.2	8.5
Dawson	64.03 N	1052	−22	−12	4	28	46	57	59	54	42	26	1	−11	23
	139.25 W		0.8	0.7	0.5	0.6	0.9	1.2	1.5	1.4	1.5	1.1	1.2	1.0	12.5
Denver	39.32 N	5292	30	32	39	47	56	66	72	71	62	50	39	33	50
	105.00 W		0.3	0.6	1.3	2.1	2.0	1.3	1.6	1.5	1.2	1.0	0.6	0.7	14.0
Durango	24.03 N	5130	54	56	62	66	71	71	69	69	66	63	58	54	63
	104.39 W		0.3	0.0	0.2	0.2	0.2	2.6	4.1	3.2	4.1	1.1	0.6	1.1	17.7
El Paso	31.47 N	3762	45	49	56	63	72	80	81	79	74	63	52	45	63
	106.30 W		0.4	0.4	0.3	0.3	0.3	0.7	1.6	1.6	1.3	0.7	0.5	0.5	8.6

Station	Location	Alt.	Jan.	Feb.	Mar.	Apr.	May	June	July	Aug.	Sept.	Oct.	Nov.	Dec.	Year
Fort Chipewyan	58.46 N	714	−13	−7	5	27	42	53	59	56	44	32	14	−2	26
	111.14 W		0.7	0.5	0.7	0.7	0.8	1.4	2.3	1.6	1.2	0.9	0.9	0.8	12.6
Fort Stockton	30.54 N	3050	46	50	58	65	74	80	81	80	74	65	54	46	64
	102.51 W		0.5	0.6	0.5	1.0	1.7	1.5	1.9	1.8	2.4	1.4	0.8	0.7	14.7
Habana	23.08 N	79	72	72	74	77	79	82	82	82	81	79	76	72	77
	82.22 W		2.9	1.8	1.9	2.2	4.8	6.4	4.8	5.4	5.8	6.6	3.1	2.4	48
Halifax	44.39 N	88	23	23	30	39	49	58	65	64	58	49	39	28	44
	63.35 W		5.6	4.5	5.0	4.5	4.3	3.7	3.9	4.5	3.6	5.2	5.4	5.4	55.5
Helena	46.22 N	3893	20	23	32	44	52	60	68	57	56	45	33	25	44
	112.00 W		0.7	0.6	0.8	1.0	2.1	2.1	1.1	0.9	1.2	0.9	0.6	0.7	12.5
Honolulu	21.19 N	38	71	71	71	73	75	77	78	78	78	77	74	72	75
	157.52 W		3.8	4.2	3.8	2.2	1.8	1.1	1.3	1.4	1.5	1.9	4.2	4.2	31.6
Juneau	58.18 N	72	27	30	34	41	48	54	57	55	50	43	36	31	42
	134.24 W		7.2	5.6	5.4	5.5	5.2	4.0	5.0	7.3	10.2	11.2	9.1	7.6	83.2
Kamloops	50.41 N	1245	22	27	38	50	58	64	70	68	59	48	36	27	47
	120.29 W		0.9	0.7	0.3	0.4	0.9	1.3	1.1	1.1	1.0	0.6	0.9	1.0	10.4
La Paz	24.10 N	39	62	65	68	71	74	78	82	84	82	79	72	66	74
	110.20 W		0.2	0.1	0.0	0.0	0.0	0.2	0.4	1.2	1.4	0.6	0.5	1.1	5.7
Los Angeles	34.03 N	361	54	56	57	60	62	65	70	71	69	65	61	55	62
	118.15 W		3.2	3.3	2.6	1.0	0.4	0.1	0.0	0.0	0.2	0.5	1.1	2.4	14.8
Masset	53.58 N	30	35	37	39	43	49	53	58	58	53	47	41	38	46
	132.09 W		5.8	4.0	3.4	5.0	4.0	2.5	2.8	2.6	3.9	6.2	7.4	6.4	53.9
Mazatlán	23.12 N	13	68	68	70	72	76	81	83	83	83	80	75	70	76
	106.25 W		0.7	0.5	0.2	0.1	0.1	1.3	6.1	8.3	7.5	2.3	0.8	0.7	28.6
Medicine Hat	50.05 N	2144	11	14	26	45	55	62	68	66	56	46	29	21	42
	110.45 W		0.6	0.6	0.6	0.7	1.8	2.6	1.8	1.4	1.0	0.6	0.7	0.6	13.0
Mérida	20.58 N	72	72	74	78	80	83	82	81	81	81	78	76	73	78
	89.37 W		0.9	0.7	1.1	0.9	2.8	6.8	4.4	5.8	5.5	3.2	1.8	1.3	35.2
Miami	25.28 N	8	66	68	72	74	79	81	82	82	81	77	73	69	76
	80.07 W		2.4	2.0	2.4	3.4	7.1	7.4	5.3	6.4	8.9	9.0	3.3	1.7	59.2
Montgomery	32.30 N	240	48	51	58	65	73	80	82	81	76	66	56	49	66
	86.20 W		5.1	5.5	6.4	4.3	3.8	4.2	4.7	4.2	2.9	2.4	3.1	4.5	51.1
Montreal	45.30 N	187	13	14	26	41	55	65	70	67	59	47	33	20	42
	73.35 W		3.8	3.2	3.5	2.5	3.0	3.5	3.7	3.5	3.5	3.3	3.5	3.7	40.6
Morant Point	17.55 N	7	77	77	77	78	80	82	82	82	82	81	79	78	80
	76.12 W		4.2	3.0	2.6	3.2	6.2	5.2	2.9	4.7	6.3	10.1	9.3	4.7	62.4
New Orleans	30.00 N	9	54	57	63	70	76	82	84	83	80	73	63	57	70
	90.30 W		4.5	4.1	5.0	5.5	5.2	5.0	6.8	6.1	5.3	3.0	3.9	5.9	60.3
New York	40.25 N	10	33	33	41	50	61	70	75	73	67	57	46	36	54
	74.00 W		3.3	3.4	3.6	3.4	3.1	3.6	4.2	4.2	3.7	3.5	2.5	3.3	41.6
Omaha	41.16 N	1105	22	25	37	51	62	72	77	75	66	54	39	27	50
	95.56 W		0.8	0.9	1.2	2.0	3.0	4.0	3.1	3.2	3.4	1.9	1.3	0.9	25.5
Phoenix	33.16 N	1114	51	56	61	68	77	85	90	89	84	72	60	53	72
	112.02 W		0.8	0.9	0.7	0.4	0.1	0.1	1.0	0.9	0.7	0.4	0.7	0.9	7.6
Roseburg	43.07 N	505	41	45	49	53	58	64	69	68	63	56	47	43	55
	123.12 W		4.9	4.0	3.1	2.2	1.7	1.2	0.2	0.3	1.2	2.4	4.7	4.6	30.5
Roswell	33.21 N	3578	40	43	51	58	66	75	78	76	70	58	47	38	59
	104.22 W		0.4	0.5	0.6	1.0	1.1	1.6	2.2	2.0	1.9	1.2	1.0	0.6	14.1
St. Johns	47.34 N	125	23	22	28	35	43	51	59	59	54	45	37	29	40
	52.42 W		5.4	5.1	4.5	4.2	3.6	3.6	3.7	3.6	3.8	5.4	6.1	4.9	53.8

Station	Location	Alt.	Jan.	Feb.	Mar.	Apr.	May	June	July	Aug.	Sept.	Oct.	Nov.	Dec.	Year
Salt Lake City	40.27 N	4220	29	34	42	51	60	69	78	76	66	55	41	33	54
	111.31 W		1.3	1.6	2.0	1.8	1.9	0.7	0.6	0.9	1.0	1.4	1.3	1.3	15.8
San Diego	32.43 N	87	55	56	58	60	63	66	69	70	69	65	61	57	62
	117.10 W		1.7	2.3	1.5	0.8	0.3	0.1	0.0	0.1	0.2	0.6	0.8	2.6	10.9
San Francisco	37.27 N	8	49	51	53	54	56	57	57	58	60	59	56	51	55
	122.16 W		4.4	4.0	3.0	1.1	0.6	0.2	0.0	0.0	0.4	1.0	2.1	3.6	20.2
San José	9.58 N	3760	66	67	68	69	69	68	68	68	68	67	67	66	67
	84.02 W		0.6	0.2	0.8	1.8	9.0	9.5	8.3	9.5	12.0	11.8	5.8	1.6	70.9
San Juan	18.29 N	82	75	74	75	76	78	79	80	80	80	79	78	76	78
	66.07 W		4.1	2.8	3.0	4.4	5.1	5.3	5.7	5.9	6.1	5.5	7.0	5.7	60.5
Seattle	47.27 N	376	40	42	45	50	55	60	64	64	59	52	46	42	52
	122.15 W		4.9	3.5	2.8	2.0	1.7	1.4	0.6	0.7	1.6	2.7	4.8	5.2	31.8
Southwest Point	49.23 N	30	12	13	21	31	40	50	57	56	49	40	31	20	35
	63.43 W		2.3	2.5	2.3	1.8	1.8	2.5	3.0	3.1	3.5	2.7	3.6	2.9	31.9
Spokane	47.55 N	1943	27	31	40	48	56	62	70	68	59	48	38	31	48
	117.28 W		1.8	1.4	1.2	1.0	1.2	1.1	0.5	0.6	0.8	1.1	2.0	2.0	14.6
Springfield	37.17 N	1324	34	35	45	56	64	73	77	76	70	58	46	36	56
	93.22 W		2.4	1.8	3.3	3.8	4.8	5.1	3.5	4.2	3.4	3.2	2.6	2.0	40.2
Upernivik	72.47 N	6	−8	−9	−6	6	25	35	41	41	33	25	14	1	16
	56.07 W		0.4	0.5	0.7	0.6	0.6	0.5	0.9	1.1	1.1	1.1	1.1	0.5	9.1
Veracruz	19.10 N	49	71	73	75	79	81	82	82	82	80	76	75	71	77
	96.10 W		0.4	0.6	0.6	0.1	4.3	12.5	14.8	8.9	11.6	9.0	3.2	2.0	68.0
Victoria	48.24 N	228	39	40	44	48	53	57	60	60	56	50	45	41	49
	123.19 W		4.6	3.2	2.4	1.5	1.1	0.9	0.4	0.6	1.8	2.6	5.0	5.6	29.7
Washington	38.54 N	112	34	35	43	54	64	72	77	74	68	57	46	36	55
	77.03 W		3.2	3.0	3.5	3.3	3.6	3.9	4.4	4.0	3.1	3.0	2.5	3.0	40.5
Winnipeg	49.53 N	760	−3	1	16	39	52	63	67	63	54	42	23	7	35
	97.07 W		0.9	0.8	1.1	1.4	2.0	3.0	3.2	2.1	2.3	1.3	1.1	0.9	20.4
SOUTH AMERICA															
Andagoya	5.04 N	250	82	82	82	82	82	82	81	81	82	82	81	81	81
	76.55 W		25.0	21.4	19.5	26.1	25.5	25.8	23.3	25.3	24.6	22.7	22.4	19.5	281.1
Asunción	25.21 S	312	80	80	78	72	67	63	64	66	70	73	76	80	72
	57.35 W		5.7	5.2	4.8	5.5	4.7	2.8	2.3	1.6	3.3	5.6	5.8	6.1	53.4
Bahia	13.00 S	210	80	80	80	79	78	75	75	75	76	77	78	79	78
	38.31 W		3.8	2.9	6.5	12.0	10.9	8.1	7.0	3.2	3.3	2.1	3.7	1.9	65.4
Barra do Corda	5.35 S	266	78	77	77	77	77	76	76	78	80	80	80	79	78
	45.28 W		6.7	8.7	8.0	6.1	2.3	1.0	0.7	0.7	1.0	2.5	3.9	5.7	47.3
Belèm	1.27 S	33	78	78	78	79	79	79	79	79	80	80	80	80	79
	48.27 W		9.2	13.6	17.3	16.7	11.7	9.1	3.1	3.0	1.8	3.3	1.1	4.2	94.1
Bogotá	4.34 N	8678	56	58	59	59	59	58	57	57	57	58	58	58	58
	74.05 W		2.3	2.4	4.1	5.7	4.5	2.4	2.0	2.2	2.4	6.4	4.6	2.6	41.6
Buenos Aires	34.40 S	72	74	73	70	62	56	51	50	52	57	61	67	71	62
	58.30 W		3.0	2.5	4.6	3.0	2.8	2.7	2.2	2.4	3.0	3.6	2.8	3.9	36.5
Caracas	10.30 N	3050	69	69	69	73	74	73	72	73	73	71	71	69	71
	66.55 W		0.9	0.3	0.6	1.2	2.8	4.0	4.8	3.8	4.2	4.4	3.3	1.6	31.9
Cayenne	4.58 N	20	79	80	80	80	80	80	80	82	83	83	82	80	81
	52.18 W		14.4	12.3	15.8	18.9	21.9	15.5	6.9	2.8	1.2	1.3	4.6	10.7	126.3
Ciudad Bolivar	8.08 N	125	79	80	81	82	82	80	80	81	82	82	81	79	81
	63.33 W		0.6	0.4	0.3	1.2	5.1	5.5	7.1	6.7	4.7	3.1	3.8	1.6	40.1

Station	Location	Alt.	Jan.	Feb.	Mar.	Apr.	May	June	July	Aug.	Sept.	Oct.	Nov.	Dec.	Year
Copiapó	27.21 S	1236	71	71	68	63	59	54	54	57	60	63	66	68	63
	70.21 W		0.0	0.0	0.0	0.0	0.1	0.2	0.2	0.1	0.0	0.0	0.0	0.0	0.6
Cuyaba	15.36 S	541	80	80	80	79	77	74	75	77	81	81	81	80	79
	56.06 W		9.5	8.3	8.3	4.0	2.0	0.3	0.2	1.1	2.0	4.5	5.9	8.1	53.6
Cuzco	13.33 S	10581	54	54	54	52	52	50	48	51	53	54	55	54	53
	71.53 W		6.5	5.4	4.4	2.0	0.6	0.2	0.2	0.4	1.0	2.7	2.9	5.5	31.8
Evangelists Island	52.25 S	174	47	47	46	45	41	40	37	39	40	42	43	45	43
	75.12 W		12.8	8.9	12.3	11.6	8.8	8.5	8.9	8.6	7.7	9.0	10.3	9.6	117.0
Guayaquil	2.15 S	40	79	79	80	80	79	77	75	76	77	77	78	80	78
	79.52 W		9.7	10.5	7.4	5.3	2.1	0.8	0.4	0.0	0.2	0.4	0.3	1.9	39.0
Jujuy	24.11 S	4166	71	69	66	62	57	52	52	55	62	65	69	70	63
	65.22 W		6.5	5.5	5.4	1.3	0.5	0.2	0.2	0.1	0.4	1.5	2.6	5.2	29.4
La Paz	16.30 S	12001	51	51	51	50	48	45	44	46	50	51	53	52	49
	68.09 W		3.8	4.9	2.6	1.5	0.5	0.1	0.2	1.1	0.8	1.3	1.6	4.2	22.6
Lima	12.02 S	518	73	74	74	70	66	63	61	61	61	63	66	70	67
	77.02 W		0.0	0.0	0.0	0.0	0.1	0.2	0.4	0.4	0.4	0.2	0.1	0.0	1.8
Manáos	3.10 S	148	79	80	79	79	80	80	80	81	82	82	82	80	80
	60.00 W		9.8	9.2	9.8	8.6	7.0	3.8	2.3	1.4	2.0	4.1	6.1	7.8	71.9
Maracaibo	10.38 N	26	81	82	83	84	84	85	85	85	84	82	82	82	83
	71.37 W		0.1	0.0	0.3	0.5	2.5	2.5	2.2	2.4	3.0	4.8	3.3	0.6	22.2
Quito	0.10 S	9350	55	55	55	55	55	55	55	55	55	55	55	55	55
	78.35 W		4.2	4.3	5.2	7.4	5.0	1.5	0.9	1.5	3.0	3.7	3.8	3.8	44.3
Quixeramobim	5.16 S	679	83	82	81	81	80	80	80	80	82	83	83	83	86
	39.15 W		3.1	3.5	5.7	4.8	3.7	1.5	0.8	0.4	0.1	0.0	0.2	1.1	24.1
Rio Grande	32.02 S	6	73	74	71	67	62	55	55	57	59	62	67	71	64
	52.09 W		3.3	5.5	3.3	3.4	3.0	4.7	4.6	5.4	4.3	3.2	3.2	2.8	46.7
San Juan	18.30 S	2140	78	76	71	62	54	47	48	50	59	65	71	76	63
	66.12 W		0.7	0.3	0.3	0.1	0.1	0.0	0.0	0.0	0.1	0.3	0.2	0.4	2.5
Santa Cruz	49.50 S	85	61	57	54	48	41	34	33	38	43	48	55	55	47
	68.50 W		0.5	0.3	0.2	0.6	0.7	0.4	1.1	0.4	0.2	0.4	0.4	1.0	6.2
Santiago	33.25 S	1703	68	67	63	57	52	48	47	48	55	56	61	65	56
	70.45 W		0.0	0.1	0.2	0.6	2.3	3.2	3.4	2.4	1.2	0.6	0.2	0.2	14.4
Santos	23.56 S	26	77	78	77	73	69	67	66	66	67	70	73	76	72
	46.19 W		10.2	9.8	11.0	6.6	8.3	2.9	3.9	4.7	5.6	5.9	7.5	7.8	84.2
Ushuaia	54.50 S	8	50	49	47	41	37	33	34	35	39	43	44	49	47
	68.20 W		1.7	2.1	1.7	1.9	1.5	1.6	1.2	0.9	1.2	1.5	1.9	1.8	19.0
Valdivia	39.46 S	49	60	59	57	54	51	49	46	46	49	51	53	57	53
	73.12 W		2.9	3.2	6.4	9.3	15.3	17.5	15.4	13.5	7.3	5.0	4.4	4.8	105.0

EUROPE

Station	Location	Alt.	Jan.	Feb.	Mar.	Apr.	May	June	July	Aug.	Sept.	Oct.	Nov.	Dec.	Year
Aberdeen	57.10 N	79	39	39	40	43	48	53	56	56	53	48	42	39	46
	2.06 W		2.2	2.1	2.2	2.1	2.3	1.9	2.9	2.8	2.6	3.1	3.1	3.1	30.4
Amsterdam	52.23 N	5	38	38	42	47	56	59	63	63	59	51	42	40	50
	4.55 E		2.0	1.5	1.9	1.6	1.9	2.0	2.9	3.2	2.5	3.4	2.3	2.7	27.9
Archangel	64.28 N	20	8	10	17	30	41	53	60	55	46	34	21	12	32
	40.31 E		0.9	0.8	0.9	0.8	1.4	2.0	2.4	2.5	2.2	1.8	1.3	1.1	18.1
Astrakhan	46.21 N	−82	19	23	33	48	64	73	77	74	63	50	36	27	49
	48.02 E		0.5	0.3	0.4	0.5	0.6	0.9	0.5	0.5	0.6	0.5	0.6	0.6	6.5

Station	Location	Alt.	Jan.	Feb.	Mar.	Apr.	May	June	July	Aug.	Sept.	Oct.	Nov.	Dec.	Year
Barcelona	41.23 N	131	46	48	51	56	62	68	74	73	69	61	54	48	59
	2.08 E		1.4	1.3	1.7	2.0	1.5	1.3	1.1	1.4	3.0	3.0	1.8	1.7	14.5
Batum	41.40 N	20	43	44	47	52	60	68	73	74	68	62	53	48	58
	41.35 E		10.2	6.0	6.2	5.0	2.8	5.9	6.0	8.2	11.9	8.8	12.2	10.0	93.3
Bergen	60.24 N	144	34	34	36	42	49	55	58	57	52	45	39	36	45
	5.19 E		8.8	7.1	6.1	4.4	4.7	4.2	5.6	7.7	9.3	9.2	8.7	8.7	84.6
Berlin	52.32 N	105	32	34	39	48	57	63	66	65	58	49	39	34	49
	13.07 E		1.7	1.4	1.6	1.5	1.9	2.4	3.1	2.3	1.7	1.8	1.7	1.9	23.0
Berufjord	64.41 N	5	30	30	30	34	39	45	48	47	44	38	33	31	37
	14.22 W		5.0	3.9	3.2	3.2	2.8	2.6	2.6	3.1	4.8	5.0	4.3	5.4	45.9
Bologna	44.32 N	203	35	40	47	55	63	71	77	75	68	58	46	38	56
	11.17 E		1.7	1.8	2.2	2.5	2.5	2.1	1.8	1.5	2.1	3.5	3.1	2.2	27.0
Bordeaux	44.50 N	154	41	44	48	54	60	65	70	70	66	57	48	42	55
	0.34 E		2.5	2.0	2.3	2.5	2.8	2.8	1.9	2.0	2.6	3.6	3.2	2.7	30.9
Brest	48.24 N	89	45	45	47	51	56	60	64	65	62	56	50	47	54
	4.30 W		3.3	2.6	2.1	2.1	1.8	2.0	2.0	2.2	2.8	3.9	3.6	3.4	31.8
Bucharest	44.29 N	302	27	33	42	52	62	69	73	72	64	54	41	34	52
	26.08 E		1.3	1.1	1.7	1.7	2.5	3.5	2.7	2.0	1.6	1.7	1.9	1.6	23.3
Budapest	47.31 N	426	31	34	43	52	62	68	71	69	61	52	41	35	52
	19.01 E		1.5	1.2	1.8	2.3	2.9	2.9	2.1	2.0	2.0	2.6	2.1	1.9	25.3
Chkalov	51.50 N	360	3	6	16	38	58	66	71	67	55	39	24	11	38
	55.10 E		1.1	0.8	1.0	0.9	1.4	2.0	1.7	1.3	1.3	1.2	1.2	1.2	15.2
Copenhagen	55.40 N	16	32	32	34	42	51	59	62	61	55	47	40	34	46
	12.34 E		1.3	1.3	1.5	1.3	1.5	1.8	2.3	2.6	1.8	2.1	1.7	1.8	20.7
Corinth	37.54 N	16	50	50	53	59	66	75	81	80	74	68	59	52	64
	22.53 E		2.6	2.1	1.5	1.2	0.9	0.5	0.2	0.2	0.9	1.9	2.2	2.9	17.1
Coruna	42.23 N	82	49	50	51	53	57	61	64	65	63	58	53	51	56
	8.23 W		3.2	3.1	3.2	2.5	2.2	1.4	0.9	1.2	2.2	3.5	4.2	4.4	32.0
Gdańsk	54.23 N	49	29	30	35	42	51	59	64	62	56	47	37	31	45
	18.36 E		1.2	1.1	1.3	1.4	1.8	2.2	2.7	2.7	2.2	1.7	1.9	1.5	21.7
Gibraltar	36.06 N	90	55	56	57	60	65	69	73	75	72	66	60	56	64
	5.21 W		4.6	4.5	4.7	2.7	1.6	0.5	0.0	0.1	1.3	3.3	6.4	5.4	35.1
Graz	47.04 N	1211	29	32	40	48	58	62	67	65	59	50	41	31	49
	15.26 E		1.0	1.1	1.6	2.7	3.3	4.9	4.8	4.6	3.7	3.4	2.1	1.5	34.8
Härnösand	62.35 N	49	20	20	25	34	43	54	59	57	50	39	30	22	38
	17.49 E		1.5	1.1	1.5	1.0	1.7	1.7	2.4	3.1	2.7	2.7	2.1	1.7	23.2
Helsinki	60.10 N	39	21	20	25	34	46	57	62	60	51	42	32	25	40
	24.57 E		1.8	1.5	1.4	1.4	1.7	1.8	2.2	2.9	2.5	2.6	2.5	2.0	24.3
Istanbul	40.58 N	59	43	40	47	53	62	70	74	75	68	63	54	47	58
	28.50 E		3.3	2.8	2.2	1.6	1.2	1.3	1.0	1.4	3.7	2.1	2.6	4.6	27.8
Kazan	55.47 N	292	8	11	21	38	55	63	68	63	52	38	24	13	38
	49.08 E		0.9	0.7	0.8	0.9	1.4	2.4	2.2	2.1	1.6	1.5	1.3	1.0	16.9
Lisbon	38.43 N	312	51	52	55	58	62	67	71	72	69	62	57	51	60
	9.09 W		3.0	3.1	3.0	2.4	2.4	0.7	0.2	0.2	1.3	2.9	3.9	3.4	26.5
London	51.28 N	18	41	41	43	47	55	60	63	62	58	51	44	41	51
	0.19 W		1.9	1.6	1.6	1.6	1.8	2.0	2.2	2.2	2.0	2.6	2.3	2.3	23.9
Madrid	40.24 N	2188	40	44	48	54	61	69	77	76	67	57	47	41	57
	3.43 W		1.2	1.4	1.6	1.6	1.7	1.4	0.4	0.5	1.4	1.8	2.0	1.6	16.6
Milan	45.28 N	482	32	38	46	55	63	70	75	73	66	56	44	36	54
	9.10 E		2.4	2.3	2.7	3.4	4.1	3.3	2.8	3.2	3.5	4.7	4.3	3.0	39.8

Station	Location	Alt.	Jan.	Feb.	Mar.	Apr.	May	June	July	Aug.	Sept.	Oct.	Nov.	Dec.	Year
Moscow	55.45 N	480	12	15	23	38	53	62	66	63	52	40	28	17	39
	37.35 E		1.1	0.9	1.2	1.5	1.9	2.0	2.8	2.9	2.2	1.4	1.6	1.5	21.0
Munich	48.11 N	1670	28	31	37	46	54	60	63	62	56	46	36	30	46
	11.33 E		1.7	1.4	1.9	2.7	3.7	4.6	4.7	4.2	3.2	2.2	1.9	1.9	34.2
Naples	40.52 N	489	47	49	52	57	64	71	76	76	71	63	55	49	61
	14.15 E		4.2	3.7	3.4	3.4	2.4	1.7	0.5	1.1	3.2	5.4	5.6	5.3	39.9
Odessa	46.18 N	210	25	28	35	48	59	68	73	71	62	52	41	31	49
	30.27 E		0.9	0.7	1.1	1.1	1.3	2.3	2.1	1.2	1.4	1.1	1.6	1.3	16.1
Oslo	59.55 N	82	25	26	29	40	51	60	63	60	52	42	33	26	42
	10.43 E		1.3	1.1	1.2	1.3	1.5	1.9	2.7	3.2	2.4	2.4	1.7	1.6	22.1
Paris	48.48 N	164	38	39	44	50	56	62	66	65	60	52	43	40	51
	2.30 E		1.5	1.2	1.6	1.7	2.1	2.3	2.2	2.2	2.0	2.3	1.8	1.7	22.6
Ponta Delgada	37.45 N	72	58	57	58	59	62	66	70	71	69	66	62	60	63
	25.41 W		2.7	2.6	2.5	2.1	2.2	1.2	0.8	1.6	2.3	2.9	3.5	3.1	27.3
Prague	50.06 N	1217	30	32	39	48	57	63	67	65	59	49	39	32	48
	14.17 E		0.9	0.8	1.1	1.5	2.4	2.8	2.6	2.2	1.7	1.2	1.2	0.9	19.1
Reykjavik	64.09 N	52	32	33	35	39	45	49	52	51	46	39	35	33	41
	21.57 W		3.9	3.3	2.7	2.4	1.9	1.9	1.9	2.0	3.5	3.4	3.7	3.5	34.1
Sarajevo	43.52 N	2091	31	32	42	49	56	62	66	65	59	50	42	34	49
	18.26 E		2.4	2.3	2.9	2.9	3.4	3.9	2.5	2.4	3.1	3.7	3.2	2.9	35.5
Thorshavn	62.20 N	85	38	38	37	41	44	49	51	51	48	44	41	38	43
	7.15 W		6.7	5.3	4.9	3.7	3.3	2.6	3.2	3.6	4.7	6.1	6.5	6.6	57.2
Tromsö	69.35 N	20	26	24	27	32	39	47	52	51	44	36	30	27	36
	19.00 E		4.3	4.4	3.1	2.3	1.9	2.2	2.2	2.8	4.8	4.6	4.4	3.8	40.8
Vaigach	70.24 N	43	1	1	0	10	22	35	42	42	38	20	18	6	20
	58.47 E		0.3	0.2	0.2	0.2	0.3	0.9	1.1	1.3	1.3	0.8	0.5	0.4	7.5
Valencia	51.50 N	30	44	44	45	48	52	57	59	59	57	52	48	45	51
	10.20 W		5.5	5.2	4.5	3.7	3.2	3.2	3.8	4.8	4.1	5.6	5.5	6.6	55.6
Zurich	47.23 N	1617	32	34	41	47	56	60	64	63	57	48	39	34	48
	8.33 E		2.3	1.9	2.9	3.4	4.0	4.9	5.0	4.6	3.3	3.2	2.5	2.9	40.5
ASIA (excluding Indonesia and the Philippine Islands)															
Aden	12.46 N	98	76	77	79	83	87	90	88	86	88	84	80	77	83
	45.03 E		0.3	0.2	0.4	0.2	0.1	0.1	0.3	0.1	0.1	0.1	0.1	0.1	1.8
Akmolinsk	51.12 N	1148	0	3	12	33	56	66	70	65	53	36	19	8	35
	71.23 E		0.6	0.5	0.5	0.6	1.0	1.8	1.4	1.5	1.0	1.0	0.7	0.6	11.2
Alexandrovsk	50.54 N	52	0	4	15	32	42	52	60	62	54	40	23	8	33
	142.10 E		1.7	0.9	1.5	1.4	1.4	1.5	2.5	3.1	3.6	2.7	2.0	2.2	24.5
Allahabad	25.30 N	309	60	65	78	88	92	91	84	83	83	78	68	60	77
	81.58 E		0.7	0.5	0.3	0.1	0.3	4.5	11.4	11.1	6.0	2.2	0.2	0.2	37.5
Ankara	39.57 N	2976	31	31	41	52	62	68	72	74	65	57	47	36	53
	32.53 E		0.9	1.1	1.2	1.3	2.0	1.0	0.5	0.3	0.5	0.6	0.9	1.5	11.8
Baghdad	33.15 N	220	47	53	60	70	81	90	94	94	87	76	61	50	72
	44.25 E		1.1	1.1	1.2	0.8	0.2	0.0	0.0	0.0	0.0	0.1	0.8	1.2	6.6
Bahrein	26.14 N	18	61	62	68	76	84	88	91	92	88	82	74	65	78
	50.35 E		0.4	0.7	0.5	0.2	0.1	0.0	0.0	0.0	0.0	0.0	0.2	0.8	2.9
Bangalore	12.59 N	3021	68	72	77	80	78	74	72	72	72	72	70	68	73
	77.28 E		0.2	0.3	0.6	1.2	4.5	3.0	4.1	5.8	7.4	6.2	2.4	0.4	36.0

Station	Location	Alt.	Jan.	Feb.	Mar.	Apr.	May	June	July	Aug.	Sept.	Oct.	Nov.	Dec.	Year
Bangkok	13.44 N	7	77	80	84	85	85	84	83	83	82	81	79	77	82
	100.30 E		0.3	0.8	1.4	2.3	7.8	6.3	6.3	6.9	12.0	8.1	2.6	0.2	55.0
Barnaul	53.20 N	535	−1	2	15	34	52	64	68	63	52	36	17	5	34
	83.47 E		0.8	0.5	0.6	0.6	1.3	1.7	2.1	1.8	1.1	1.3	1.1	1.0	13.9
Bombay	19.00 N	37	74	75	78	82	85	82	80	79	79	81	79	76	79
	72.48 E		0.1	0.0	0.1	0.0	0.7	20.6	27.3	16.0	11.8	2.4	0.4	0.0	79.4
Bulun	70.45 N	40	−40	−31	−16	5	25	47	53	49	34	11	−20	−32	7
	127.47 E		0.3	0.2	0.3	0.4	0.5	1.0	1.5	1.6	1.5	0.9	0.4	0.3	8.9
Cape Chelyuskin	77.43 N	20	−14	−13	−19	−5	15	30	35	33	27	13	−7	−15	7
	104.17 E		0.2	0.1	0.0	0.1	0.1	0.5	1.0	0.9	0.3	0.3	0.2	0.1	3.8
Changchun	43.55 N	704	1	9	24	44	57	68	75	72	59	44	24	7	40
	125.18 E		0.3	0.2	0.7	0.8	2.2	4.7	7.6	5.2	2.8	1.3	0.8	0.3	26.9
Cherrapunji	25.15 N	4309	53	55	61	64	66	68	69	69	69	66	61	55	63
	91.44 E		0.5	2.7	9.4	28.2	46.3	95.9	98.5	79.8	38.0	21.3	3.2	0.3	424.1
Colombo	6.54 N	23	79	80	82	83	83	82	81	81	81	80	80	80	81
	79.52 E		3.5	2.2	4.4	9.8	12.4	7.7	4.7	3.7	5.5	13.7	12.2	5.3	84.9
Darjeeling	27.04 N	7376	40	42	50	56	58	60	62	61	59	55	48	42	53
	88.25 E		0.6	1.1	1.8	3.8	8.7	24.9	32.3	26.1	18.4	4.5	0.8	0.2	122.7
Delhi	28.39 N	718	59	63	74	85	92	93	88	86	85	80	69	61	78
	77.15 E		1.0	0.8	0.5	0.3	0.6	3.2	7.7	7.5	4.6	0.4	0.1	0.5	27.2
Haifa	32.49 N	33	54	58	60	66	71	76	80	82	80	75	65	58	69
	35.00 E		6.1	3.5	2.1	1.0	0.3	0.0	0.0	0.0	0.1	0.8	3.6	6.4	23.9
Hankow	30.35 N	118	40	43	50	62	71	80	85	85	77	67	55	45	63
	114.17 E		1.8	1.9	3.6	5.8	7.0	9.0	7.1	4.1	3.0	3.1	1.9	1.2	49.5
Hong Kong	22.18 N	109	60	59	63	70	77	81	82	82	81	76	69	63	72
	114.10 E		1.3	1.6	2.9	5.4	11.5	15.5	15.0	14.2	10.1	4.6	1.7	1.1	84.9
Hué	16.30 N	23	69	68	74	80	83	85	84	85	81	78	73	70	78
	107.35 E		4.0	4.8	1.8	2.4	3.6	2.8	3.4	4.0	16.2	26.3	22.4	10.2	101.9
Igarka	67.32 N	40	−17	−14	−4	13	24	45	60	53	43	23	−1	−22	13
	86.50 E		1.0	0.6	0.7	0.8	0.9	1.7	2.1	2.0	2.0	1.8	1.5	0.6	15.7
Irkutsk	52.30 N	1610	−5	1	16	35	48	59	65	60	48	33	13	1	31
	104.20 E		0.6	0.5	0.4	0.6	1.2	2.3	2.9	2.4	1.6	0.7	0.6	0.8	14.5
Kaifeng	34.54 N	377	30	37	47	58	70	79	83	79	70	60	47	36	58
	114.25 E		1.5	0.4	0.6	2.4	2.1	2.2	6.0	5.7	2.7	0.7	0.8	0.4	25.5
Kars	40.40 N	5725	8	14	24	38	49	57	62	63	55	45	30	18	39
	43.10 E		0.7	0.7	1.1	1.7	2.7	1.9	1.7	1.3	1.1	1.3	1.2	0.8	16.3
Kashgar	39.24 N	4255	19	32	46	61	70	75	80	78	69	57	39	26	54
	76.07 E		0.2	0.0	0.2	0.3	0.3	0.3	0.3	0.1	0.1	0.2	0.1	0.1	2.2
Leh	34.12 N	11503	17	19	31	43	50	58	63	61	54	43	32	22	41
	77.35 E		0.4	0.3	0.3	0.2	0.2	0.2	0.5	0.5	0.3	0.2	0.0	0.2	3.2
Madras	13.05 N	22	75	77	80	84	89	88	86	84	84	81	78	76	82
	80.15 E		1.1	0.3	0.3	0.6	1.8	2.0	3.8	4.5	4.9	11.2	13.6	5.4	49.6
Mandalay	21.59 N	250	69	74	82	89	88	85	85	85	84	82	76	70	81
	95.59 E		0.1	0.1	0.2	1.1	5.8	5.5	3.3	4.6	5.7	4.7	1.6	0.4	33.2
Miyako	39.38 N	98	31	32	37	47	54	61	68	72	65	55	45	36	50
	141.59 E		2.7	2.6	3.5	3.8	4.7	5.0	5.3	7.0	8.5	6.7	3.2	2.5	55.5
Osaka	34.39 N	10	40	40	46	56	64	72	79	82	75	63	53	44	59
	135.26 E		2.0	2.3	4.0	5.2	5.2	8.2	6.4	4.2	7.5	5.0	2.7	1.9	54.6
Peshawar	33.59 N	1113	50	53	63	74	84	91	90	88	82	71	59	51	71
	71.43 E		1.5	1.2	2.0	1.7	0.7	0.3	1.2	2.1	0.8	0.2	0.4	0.6	12.8

Station	Location	Alt.	Jan.	Feb.	Mar.	Apr.	May	June	July	Aug.	Sept.	Oct.	Nov.	Dec.	Year
Quetta	30.15 N	5500	40	41	51	60	67	74	78	75	67	56	47	42	58
	67.05 E		2.1	2.1	1.8	1.1	0.3	0.2	0.5	0.6	0.1	0.1	0.3	0.8	10.0
Rangoon	16.47 N	18	77	79	84	87	84	81	80	80	81	82	80	77	81
	96.13 E		0.2	0.2	0.3	1.6	12.0	18.0	21.4	19.9	15.3	6.9	2.8	0.4	99.0
Saigon	10.47 N	36	79	81	84	86	84	82	81	82	81	81	81	79	82
	106.40 E		0.6	0.1	0.5	1.7	8.5	13.0	12.2	10.6	13.2	10.5	4.5	2.2	77.6
Sapporo	43.04 N	56	21	23	29	41	51	59	66	69	61	49	37	26	44
	141.21 E		3.5	2.5	2.4	2.2	2.7	2.8	3.3	3.7	5.0	4.6	4.4	4.0	41.1
Seistan	31.03 N	2001	46	51	60	71	80	88	91	88	80	69	57	48	69
	61.30 E		0.3	0.4	0.5	0.1	0.0	0.0	0.0	0.0	0.0	0.1	0.1	0.4	1.9
Shanghai	31.15 N	23	38	39	46	56	66	73	80	80	73	63	52	42	59
	121.30 E		2.8	2.0	3.9	4.4	3.3	6.6	7.4	4.7	3.9	3.7	1.7	1.3	45.8
Singapore	1.18 N	10	78	79	80	81	82	81	81	81	80	80	79	79	80
	103.55 E		8.5	6.1	6.5	6.9	7.2	6.7	6.8	8.5	7.1	8.2	10.0	10.4	92.9
Suchow	39.50 N	5577	15	26	36	49	60	69	73	71	60	50	30	16	46
	99.07 E		0.0	0.1	0.0	0.1	0.2	0.3	0.5	1.7	0.4	0.0	0.1	0.1	3.5
Taihoku	25.02 N	30	60	59	63	69	75	80	83	82	79	73	68	62	71
	121.31 E		3.4	5.3	6.7	6.4	8.9	11.1	8.3	11.8	10.2	5.3	2.7	3.0	82.7
Tashkent	41.25 N	1610	30	34	47	58	70	77	81	77	67	54	43	36	56
	69.10 E		1.8	1.4	2.6	2.6	1.1	0.5	0.1	0.1	0.2	1.1	1.4	1.7	14.6
Tehran	35.50 N	3800	34	42	48	61	71	80	85	83	77	66	51	42	62
	51.35 E		1.6	1.0	1.9	1.4	0.5	0.1	0.2	0.0	0.1	0.3	1.0	1.3	9.3
Tomsk	56.30 N	390	−3	1	14	30	45	59	66	60	48	32	11	1	30
	85.10 E		1.1	0.8	0.8	0.7	1.5	2.7	3.0	2.3	1.4	2.4	1.4	1.9	19.9
Tortkul	41.28 N	341	23	30	42	58	72	80	84	79	68	52	38	29	55
	61.05 E		0.4	0.3	0.7	0.6	0.3	0.1	0.0	0.0	0.0	0.1	0.2	0.3	3.0
Uelen	61.10 N	126	−5	−8	−3	8	22	35	42	40	35	24	10	1	17
	169.50 W		0.5	0.5	0.4	0.4	0.4	0.5	1.5	1.8	2.1	1.8	0.5	0.8	11.2
Ulan-Bator-Khoto	47.55 N	4347	−11	−3	12	33	46	58	63	60	47	31	8	−6	28
	106.50 E		0.1	0.0	0.1	0.2	0.3	1.0	2.9	1.9	0.8	0.2	0.2	0.1	7.8
Yakutsk	62.00 N	330	−46	−35	−10	16	41	59	66	60	42	16	−21	−40	12
	129.50 E		0.9	0.2	0.4	0.6	1.1	2.1	1.7	2.6	1.2	1.4	0.6	0.9	13.7

AFRICA AND ASSOCIATED ISLANDS

Station	Location	Alt.	Jan.	Feb.	Mar.	Apr.	May	June	July	Aug.	Sept.	Oct.	Nov.	Dec.	Year
Akasa	4.15 N	20	78	79	80	80	79	77	76	76	76	77	78	79	78
	6.15 E		2.6	6.5	10.0	8.6	17.0	18.6	10.1	9.3	19.3	24.7	10.6	6.5	143.8
Alexandria	31.12 N	105	58	59	63	67	72	76	79	81	79	76	69	62	70
	29.53 E		2.0	0.9	0.4	0.1	0.0	0.0	0.0	0.0	0.0	0.2	1.3	2.3	7.2
Algiers	36.24 N	72	53	55	58	61	66	71	77	78	75	69	62	56	65
	3.30 E		4.2	3.5	3.5	2.3	1.3	0.6	0.1	0.3	1.1	3.1	4.6	5.4	30.0
Antsirabe	19.50 S	4921	67	66	65	64	59	55	54	56	60	64	66	66	62
	47.00 E		13.0	10.3	7.8	3.1	1.2	0.4	0.4	0.7	0.9	3.2	6.3	11.8	59.1
Azizia	32.32 N	518	53	56	63	68	76	82	87	86	84	75	65	54	71
	13.01 E		1.6	0.8	0.9	0.3	0.3	0.0	0.0	0.0	0.1	0.6	1.1	2.7	8.4
Bahrdar	11.36 N	6037	65	64	71	70	68	68	65	65	65	65	65	62	66
	37.25 E		0.0	0.0	0.4	1.1	3.1	4.6	16.6	11.3	9.7	3.9	1.0	0.1	52.2
Beira	19.50 S	23	82	82	80	78	74	70	69	70	74	77	79	80	76
	34.51 E		11.6	8.2	10.0	4.1	2.3	1.4	1.0	1.0	0.8	1.5	5.2	10.1	57.2

Station	Location	Alt.	Jan.	Feb.	Mar.	Apr.	May	June	July	Aug.	Sept.	Oct.	Nov.	Dec.	Year
Bloemfontein	29.07 S	4583	74	71	68	61	53	46	46	51	58	65	68	72	61
	26.12 E		3.6	3.1	3.5	1.8	1.0	0.4	0.3	0.4	0.8	1.5	2.1	2.3	21.2
Brazzaville	4.17 S	951	79	79	80	80	79	74	72	74	78	79	79	78	78
	15.16 E		4.8	5.1	5.9	8.3	5.2	0.5	0.0	0.2	1.2	5.5	7.2	6.0	49.9
Bulawayo	20.09 S	4440	71	70	69	66	61	57	57	61	67	72	72	72	66
	28.36 E		5.6	3.9	3.3	1.7	0.4	0.0	0.1	0.0	0.2	0.7	3.2	5.1	24.2
Capetown	33.56 S	40	70	71	68	64	59	56	55	56	58	61	65	68	62
	18.20 E		0.7	0.6	0.9	1.9	3.7	4.4	3.6	3.3	2.3	1.6	1.1	0.8	24.8
Casablanca	33.37 N	164	53	54	57	59	64	68	72	73	71	67	60	56	63
	7.35 W		2.0	2.0	2.4	1.2	0.8	0.2	0.0	0.0	0.1	1.2	2.8	2.6	15.3
Dakar	14.40 N	98	71	72	72	73	76	81	83	82	83	83	79	74	77
	17.26 W		0.0	0.1	0.0	0.0	0.1	0.7	3.4	9.7	5.0	1.4	0.1	0.2	21.0
Dar es Salaam	6.50 S	30	83	82	82	80	80	76	75	75	75	77	80	82	79
	39.17 E		2.8	2.4	4.7	11.0	7.3	1.5	1.2	1.0	1.1	1.5	3.1	3.7	41.3
Douala	4.04 N	43	80	80	79	80	80	78	76	76	77	77	79	79	78
	19.41 E		1.7	3.6	8.4	8.6	11.6	21.4	29.7	27.2	20.5	16.7	6.3	2.5	158.2
Durban	29.52 S	50	76	76	75	72	68	64	63	65	67	69	72	74	70
	31.03 E		4.6	5.3	6.0	3.6	2.6	1.8	1.7	1.8	2.7	5.1	4.7	5.2	45.1
El Fāsher	13.38 N	2395	69	72	78	83	86	87	83	81	82	82	75	69	79
	25.21 E		0.0	0.0	0.0	0.1	0.4	0.8	4.0	5.0	1.2	0.2	0.0	0.0	11.7
Elizabethville	11.40 S	4055	72	72	72	69	66	61	61	65	71	75	74	72	69
	27.34 E		9.8	9.8	8.5	1.8	0.2	0.0	0.0	0.0	0.1	1.3	4.8	11.2	47.5
Entebbe	0.05 N	3863	73	72	72	70	71	71	70	70	72	72	72	72	72
	32.25 E		2.6	3.6	5.8	9.8	8.5	5.1	3.0	3.0	3.1	3.5	4.9	5.1	58.0
Fort Lapperrine	23.01 N	3280	55	58	65	74	79	84	84	83	80	73	65	58	71
	5.10 E		0.2	0.1	0.1	0.2	0.6	0.2	0.1	0.4	0.1	0.1	0.0	0.2	2.3
Kano	12.02 N	1539	72	77	85	89	88	83	80	78	80	81	78	73	80
	8.32 E		0.0	0.1	0.1	0.5	2.5	4.8	8.2	12.1	5.5	0.5	0.0	0.0	34.3
Kumasi	6.41 N	840	78	81	81	81	81	78	76	76	77	79	80	79	79
	1.37 W		0.6	2.3	5.6	5.7	7.0	9.1	4.8	2.9	7.2	8.0	3.9	1.2	58.2
Las Palmas	28.07 N	23	64	64	64	66	68	70	72	75	74	73	69	67	69
	15.26 W		0.9	1.3	0.8	0.3	0.1	0.0	0.1	0.0	0.1	0.8	0.7	1.0	6.1
Mbabane	26.19 S	3816	69	66	66	62	59	54	54	57	61	65	66	67	62
	31.09 E		9.9	7.6	7.9	2.6	1.3	0.5	0.9	1.1	2.2	5.0	6.8	8.4	54.2
Mongalla	5.11 N	1469	83	84	86	83	82	80	79	79	80	81	82	82	82
	31.47 E		0.1	0.6	1.5	4.0	5.2	4.4	5.6	5.2	4.7	4.0	1.5	0.4	37.2
Mongu	15.17 S	3488	77	77	76	75	71	67	66	71	79	82	79	77	75
	23.05 E		8.7	8.5	6.0	1.1	0.2	0.0	0.0	0.0	0.1	1.4	4.1	7.7	37.8
Monrovia	6.18 N	230	79	81	81	82	81	79	78	78	78	80	79	81	80
	10.45 W		0.1	0.1	4.4	11.7	13.4	36.1	24.8	18.6	29.9	25.2	8.2	2.9	175.4
Nairobi	1.16 S	5490	66	67	68	67	65	63	61	62	64	67	66	65	65
	36.50 E		1.6	2.0	5.1	7.7	5.5	1.7	0.6	1.1	1.0	2.4	4.2	2.5	35.4
New Antwerp	1.20 N	1230	79	80	79	78	79	78	76	76	77	77	78	78	78
	18.40 E		4.1	3.5	4.1	5.6	6.2	6.1	6.3	6.3	6.3	6.6	7.6	9.3	72.0
Nova Lisboa	12.46 S	5627	68	67	68	67	65	63	65	66	70	71	68	68	67
	15.44 E		8.7	8.9	8.4	6.1	1.1	0.0	0.0	0.1	0.5	5.1	8.9	9.5	57.6
O'Okiep	29.36 S	3035	72	73	71	66	58	53	53	54	58	63	66	70	63
	17.52 E		0.2	0.3	0.4	0.6	1.0	1.1	0.8	0.9	0.6	0.4	0.2	0.2	6.7
Pôrto Amelia	13.06 S	197	81	81	80	79	77	75	74	74	76	78	80	82	78
	40.32 E		6.3	7.5	7.7	5.4	0.7	1.3	0.4	0.1	0.1	0.6	0.8	4.3	34.9

Station	Location	Alt.	Jan.	Feb.	Mar.	Apr.	May	June	July	Aug.	Sept.	Oct.	Nov.	Dec.	Year
Pretoria	25.45 S	4350	73	72	69	64	58	53	52	57	64	70	7)	72	64
	28.12 E		5.4	4.2	3.6	1.2	0.6	0.1	0.3	0.5	0.6	2.5	4.7	4.4	28.1
Salisbury	17.50 S	4865	70	69	69	66	61	57	56	60	66	70	71	70	66
	31.01 E		7.4	6.4	4.5	1.1	0.5	0.1	0.0	0.1	0.3	1.1	3.7	5.9	31.1
Stanleyville	0.30 N	1400	78	78	78	78	79	77	76	76	76	77	77	77	77
	25.11 E		2.2	3.5	6.0	6.4	5.9	4.5	4.8	5.5	7.7	8.0	7.7	4.0	66.2
Tamatave	18.10 S	30	80	80	80	77	74	71	69	70	72	74	77	78	75
	49.23 E		13.7	13.3	15.9	13.1	10.2	10.3	12.2	6.1	5.2	4.0	4.2	10.4	118.6
Timbuktu	16.45 N	886	71	74	82	89	94	94	91	88	90	88	80	71	85
	2.55 W		0.0	0.0	0.0	0.1	0.2	0.8	2.7	2.6	1.3	1.1	0.0	0.0	8.8
Walvis Bay	22.58 S	10	67	67	66	65	63	61	58	57	58	59	63	65	62
	14.30 E		0.0	0.0	0.1	0.2	0.0	0.0	0.0	0.0	0.1	0.1	0.0	0.0	0.5

OCEANIA (including Indonesia, the Philippines, Australia, New Zealand, and certain islands of the Pacific)

Station	Location	Alt.	Jan.	Feb.	Mar.	Apr.	May	June	July	Aug.	Sept.	Oct.	Nov.	Dec.	Year
Adelaide	34.56 S	140	74	74	70	64	58	53	52	54	57	62	67	71	63
	138.36 E		0.8	0.6	1.1	1.8	2.8	3.0	2.6	2.4	1.8	1.8	1.0	0.8	20.6
Alice Springs	23.38 S	1900	83	82	77	68	60	54	53	58	65	74	79	82	70
	133.35 E		1.7	1.5	1.3	0.7	0.6	0.6	0.4	0.4	0.4	0.7	1.0	1.5	10.8
Auckland	36.52 S	260	67	67	66	61	57	54	52	52	54	57	60	64	59
	174.46 E		2.6	3.0	3.1	3.3	4.4	4.8	5.0	4.2	3.6	3.6	3.3	2.9	43.8
Balikpapan	1.13 S	3	79	79	79	79	79	79	78	79	79	80	79	79	79
	116.58 E		8.1	7.1	9.3	7.3	8.5	7.4	6.8	6.3	5.2	5.5	6.4	7.8	85.7
Bourke	30.10 S	456	84	83	78	68	58	54	51	56	63	70	76	82	68
	146.00 E		2.0	1.9	1.6	1.4	1.1	1.0	0.9	0.9	1.0	1.1	1.3	1.1	15.2
Brisbane	27.28 S	125	77	77	74	70	65	60	59	61	65	70	73	76	69
	153.02 E		6.5	6.4	5.7	3.9	2.8	2.7	2.2	2.0	2.0	2.5	3.8	4.9	45.4
Broome	17.57 S	63	86	85	84	83	76	71	70	73	77	81	85	86	80
	122.15 E		6.2	6.1	3.8	1.4	0.6	0.1	0.2	0.2	0.1	0.1	0.1	3.7	22.6
Cape York	11.00 S	69	80	81	80	80	80	78	77	76	77	80	81	82	79
	142.30 E		22.9	18.6	16.9	8.1	3.7	0.5	0.6	0.4	0.1	0.1	2.0	8.2	82.0
Daru	9.04 S	26	82	81	81	81	80	79	77	77	79	80	81	82	80
	143.12 E		11.9	11.4	12.7	13.1	9.6	3.9	4.0	2.0	1.6	2.0	4.1	7.6	83.9
Darwin	12.28 S	98	84	83	84	84	82	79	77	79	83	85	86	85	83
	130.51 E		15.8	12.9	10.0	4.2	0.7	0.2	0.1	0.1	0.5	2.2	4.9	10.0	61.7
Dunedin	45.47 S	236	58	58	55	52	47	44	42	44	48	51	53	56	51
	170.44 E		3.3	2.8	3.0	2.8	3.3	3.2	3.0	3.1	2.8	3.1	3.3	3.5	37.2
Fanning Island	3.54 N	17	82	82	82	82	82	82	82	83	82	83	84	82	82
	159.23 W		10.3	10.3	10.8	13.5	12.5	10.1	8.4	4.6	3.2	3.7	3.1	8.5	99.0
Guam	13.24 N	67	81	81	81	82	82	82	81	81	81	81	82	81	81
	144.38 E		3.5	3.1	3.0	2.3	4.5	6.2	15.7	15.3	13.5	12.7	8.2	4.9	92.9
Hokitika	42.41 S	9	60	61	58	55	50	46	45	46	50	53	55	58	53
	170.49 E		9.8	7.3	9.7	9.2	9.8	9.7	9.0	9.4	9.2	11.8	10.6	10.6	116.0
Jakarta	6.11 S	23	79	79	80	80	81	80	79	80	80	81	80	79	80
	126.50 E		12.0	12.6	8.4	5.5	4.3	3.6	2.6	1.7	2.7	4.7	5.7	7.8	71.9
Jolo	6.03 N	43	79	79	79	80	80	80	80	81	80	80	80	79	80
	121.00 E		4.8	4.1	4.2	5.4	7.5	8.5	6.3	6.8	7.1	8.9	7.9	6.7	78.2
Kajoemas	7.56 S	3051	68	68	69	69	69	69	68	69	71	72	71	69	69
	114.09 E		18.8	17.6	18.1	8.6	5.8	2.5	1.6	0.6	0.5	1.9	8.0	15.6	99.6

Station	Location	Alt.	Jan.	Feb.	Mar.	Apr.	May	June	July	Aug.	Sept.	Oct.	Nov.	Dec.	Year
Kiandra	35.52 S	4640	56	57	52	46	38	35	32	34	40	44	52	56	45
	148.32 E		4.1	3.2	4.0	4.4	5.3	8.7	6.6	5.9	6.9	6.6	4.9	3.9	64.5
Manggar	2.52 S	16	79	80	80	81	81	81	82	82	82	81	80	80	81
	108.16 E		12.1	8.2	10.4	9.1	10.0	7.9	6.7	5.1	4.1	6.3	9.8	13.9	103.6
Manila	14.24 N	46	77	78	80	83	83	82	81	81	80	80	78	77	80
	121.00 E		0.8	0.4	0.8	1.3	4.4	9.2	17.3	16.0	14.3	6.7	5.2	3.0	79.4
Manokwari	0.53 S	98	80	80	80	80	80	80	80	80	80	81	81	81	80
	133.58 E		12.1	9.3	13.1	11.1	7.8	7.4	5.5	5.6	5.0	4.8	6.5	10.3	98.5
Medan	3.43 N	98	78	79	80	81	81	81	80	80	79	79	79	78	79
	98.40 E		5.4	3.6	4.1	5.2	6.9	5.2	5.3	7.0	8.3	10.2	9.7	9.0	79.9
Melbourne	37.39 S	115	67	67	64	59	54	50	49	51	54	58	61	65	58
	144.58 E		1.9	1.8	2.2	2.3	2.2	2.1	1.9	1.8	2.4	2.7	2.2	2.3	25.8
Menado	1.30 N	10	77	77	78	78	79	79	79	80	79	79	78	78	79
	124.50 E		18.3	14.3	11.1	7.9	6.5	6.5	4.8	3.8	3.4	4.8	8.6	14.3	104.3
Nauru	0.30 S	26	82	82	82	83	83	82	82	83	83	83	83	82	83
	167.00 E		8.6	11.1	6.5	5.5	5.7	5.0	6.7	6.1	6.5	5.3	6.8	10.5	84.3
Nouméa	22.07 S	30	79	79	79	76	73	70	69	69	70	72	76	78	74
	160.50 E		3.8	4.8	5.8	5.1	4.5	3.6	3.7	2.7	2.5	2.1	2.4	2.6	43.6
Pago Pago	14.19 S	10	83	82	83	81	81	80	80	80	80	80	82	82	81
	170.41 E		24.5	20.5	19.2	16.5	15.4	12.3	10.0	8.2	13.1	14.9	19.2	19.8	193.6
Perth	31.57 S	177	74	74	71	67	61	57	55	56	58	61	66	71	64
	115.50 E		0.3	0.3	0.7	1.7	4.9	6.6	6.4	5.6	3.3	2.1	0.8	0.6	33.3
Rabaul	4.13 S	52	82	82	82	82	82	82	81	80	82	82	82	82	81
	152.15 E		14.8	10.4	10.2	10.0	5.2	3.3	5.4	4.7	3.5	5.1	7.1	10.1	89.8
Sydney	33.51 S	138	72	71	69	65	59	55	53	55	59	63	67	70	63
	151.13 E		3.6	4.4	4.9	5.4	5.1	4.8	5.0	3.0	2.9	2.9	2.8	2.8	47.6
Wellington	41.17 S	142	62	62	61	57	53	50	48	49	51	54	57	61	55
	174.47 E		3.3	3.2	3.3	3.8	4.7	4.9	5.6	4.4	4.0	4.1	3.5	3.2	48.1

Index

WITHDRAWN FROM
OHIO NORTHERN
UNIVERSITY LIBRARY